Mathematical Analysis and Applications II

Mathematical Analysis and Applications II

Special Issue Editor

Hari M. Srivastava

MDPI • Basel • Beijing • Wuhan • Barcelona • Belgrade • Manchester • Tokyo • Cluj • Tianjin

Special Issue Editor
Hari M. Srivastava
University of Victoria
Canada

Editorial Office
MDPI
St. Alban-Anlage 66
4052 Basel, Switzerland

This is a reprint of articles from the Special Issue published online in the open access journal *Axioms* (ISSN 2075-1680) (available at: https://www.mdpi.com/journal/axioms/special_issues/mathematical_analysis_II).

For citation purposes, cite each article independently as indicated on the article page online and as indicated below:

LastName, A.A.; LastName, B.B.; LastName, C.C. Article Title. *Journal Name* **Year**, *Article Number*, Page Range.

ISBN 978-3-03928-384-2 (Pbk)
ISBN 978-3-03928-385-9 (PDF)

© 2020 by the authors. Articles in this book are Open Access and distributed under the Creative Commons Attribution (CC BY) license, which allows users to download, copy and build upon published articles, as long as the author and publisher are properly credited, which ensures maximum dissemination and a wider impact of our publications.

The book as a whole is distributed by MDPI under the terms and conditions of the Creative Commons license CC BY-NC-ND.

Contents

About the Special Issue Editor vii

Hari M. Srivastava
Mathematical Analysis and Applications II
Reprinted from: *Axioms* **2020**, *9*, 16, doi:10.3390/axioms9010016 1

Giuseppe Dattoli, Silvia Licciardi, Rosa Maria Pidatella, Elio Sabia
Repeated Derivatives of Hyperbolic Trigonometric Functions and Associated Polynomials
Reprinted from: *Axioms* **2019**, *8*, 138, doi:10.3390/axioms8040138 5

Hari Mohan Srivastava, Gürhan İçöz and Bayram Çekim
Approximation Properties of an Extended Family of the Szász–Mirakjan Beta-Type Operators
Reprinted from: *Axioms* **2019**, *8*, 111, doi:10.3390/axioms8040111 15

Andriy Bandura and Oleh Skaskiv
Slice Holomorphic Functions in Several Variables with Bounded L-Index in Direction
Reprinted from: *Axioms* **2019**, *8*, 88, doi:10.3390/axioms8030088 29

Wolfram Koepf, Insuk Kim and Arjun K. Rathie
On a New Class of Laplace-Type Integrals Involving Generalized Hypergeometric Functions
Reprinted from: *Axioms* **2019**, *8*, 87, doi:10.3390/axioms8030087 41

Yang-Hi Lee and Gwang Hui Kim
Generalized Hyers–Ulam Stability of the Additive Functional Equation
Reprinted from: *Axioms* **2019**, *8*, 76, doi:10.3390/axioms8020076 63

Nabil Mlaiki, Katarina Kukić, Milanka Gardašević-Filipović and Hassen Aydi
On Almost b-Metric Spaces and Related Fixed Point Results
Reprinted from: *Axioms* **2019**, *8*, 70, doi:10.3390/axioms8020070 75

Julalak Prabseang, Jessada Tariboon and Kamsing Nonlaopon
(p,q)-Hermite–Hadamard Inequalities for Double Integral and (p,q)-Differentiable Convex Functions
Reprinted from: *Axioms* **2019**, *8*, 68, doi:10.3390/axioms8020068 87

Rekha Srivastava, Humera Naaz, Sabeena Kazi and Asifa Tassaddiq
Some New Results Involving the Generalized Bose–Einstein and Fermi–Dirac Functions
Reprinted from: *Axioms* **2019**, *8*, 63, doi:10.3390/axioms8020063 97

Maxie D. Schmidt
A Short Note on Integral Transformations and Conversion Formulas forSequence Generating Functions
Reprinted from: *Axioms* **2019**, *8*, 62, doi:10.3390/axioms8020062 109

Tariq Qawasmeh, Abdalla Tallafha and Wasfi Shatanawi
Fixed Point Theorems through Modifiedω-Distance and Application to Nontrivial Equations
Reprinted from: *Axioms* **2019**, *8*, 57, doi:10.3390/axioms8020057 119

Alfonsina Tartaglione
A Note on the Displacement Problem of Elastostatics with Singular Boundary Values
Reprinted from: *Axioms* **2019**, *8*, 46, doi:10.3390/axioms8020046 131

Mohammad Asim, A. Rauf Khan and Mohammad Imdad
Fixed Point Results in Partial Symmetric Spaces with an Application
Reprinted from: *Axioms* **2019**, *8*, 13, doi:10.3390/axioms8010013 . . . . **137**

Mohammad Masjed-Jamei and Wolfram Koepf
A New Identity for Generalized Hypergeometric Functions and Applications
Reprinted from: *Axioms* **2019**, *8*, 12, doi:10.3390/axioms8010012 . . . . **153**

Aiman Mukheimer
Extended Partial S_b-Metric Spaces
Reprinted from: *Axioms* **2018**, *7*, 87, doi:10.3390/axioms7040087 . . . . **163**

Nabil Mlaiki, Nihal Taş and Nihal Yılmaz Özgür
On the Fixed-Circle Problem and Khan Type Contractions
Reprinted from: *Axioms* **2018**, *7*, 80, doi:10.3390/axioms7040080 . . . . **173**

Haitham Qawaqneh, Mohd Noorani, Wasfi Shatanawi and Habes Alsamir
Common Fixed Point Theorems for Generalized Geraghty (α, ψ, ϕ)-Quasi Contraction Type Mapping in Partially Ordered Metric-Like Spaces
Reprinted from: *Axioms* **2018**, *7*, 74, doi:10.3390/axioms7040074 . . . . **183**

Paolo Emilio Ricci
Differential Equations for Classical and Non-Classical Polynomial Sets: A Survey [†]
Reprinted from: *Axioms* **2019**, *8*, 50, doi:10.3390/axioms8020050 . . . . **201**

About the Special Issue Editor

Hari M. Srivastava has held the position of Professor Emeritus in the Department of Mathematics and Statistics at the University of Victoria in Canada since 2006, having joined the faculty there in 1969, first as an Associate Professor (1969–1974) and then as a Full Professor (1974–2006). He began his university-level teaching career right after having received his M.Sc. degree in 1959 at the age of 19 years from the University of Allahabad in India. He earned his Ph.D. degree in 1965 while he was a full-time member of the teaching faculty at the Jai Narain Vyas University of Jodhpur in India. He holds numerous visiting research and honorary chair positions at many universities and research institutes in different parts of the world. Having received several D.Sc. (honoris causa) degrees as well as honorary memberships and honorary fellowships of many scientific academies and learned societies around the world, he is also actively associated editorially with numerous international scientific research journals. His current research interests include several areas of Pure and Applied Mathematical Sciences, such as Real and Complex Analysis, Fractional Calculus and Its Applications, Integral Equations and Transforms, Higher Transcendental Functions and Their Applications, q-Series and q-Polynomials, Analytic Number Theory, Analytic and Geometric Inequalities, Probability and Statistics, and Inventory Modelling and Optimization. He has published 33 books, monographs, and edited volumes, 33 book (and encyclopedia) chapters, 48 papers in international conference proceedings, and more than 1250 scientific research articles in peer-reviewed international journals, as well as forewords, editorials and prefaces to many books and journals, and so on. He is a Clarivate Analytics [Thomson-Reuters] (Web of Science) Highly Cited Researcher. For further details about his other professional achievements and scholarly accomplishments, as well as honors, awards, and distinctions, including the lists of his most recent publications such as journal articles, books, monographs and edited volumes, book chapters, encyclopedia chapters, papers in conference proceedings, forewords to books and journals, et cetera), the interested reader should look into the following regularly updated website: http://www.math.uvic.ca/~harimsri/.

Editorial

Mathematical Analysis and Applications II

Hari M. Srivastava [1,2,3]

1. Department of Mathematics and Statistics, University of Victoria, Victoria, BC V8W 3R4, Canada; harimsri@math.uvic.ca
2. Department of Medical Research, China Medical University Hospital, China Medical University, Taichung 40402, Taiwan
3. Department of Mathematics and Informatics, Azerbaijan University, 71 Jeyhun Hajibeyli Street, Baku AZ1007, Azerbaijan

Received: 13 January 2020; Accepted: 3 February 2020; Published: 6 February 2020

Web Site: http://www.math.uvic.ca/faculty/harimsri/

The present volume contains the invited, accepted and published submissions (see [1–17]) to a Special Issue of the MDPI's journal, *Axioms*, on the subject-area of "Mathematical Analysis and Applications II". A successful predecessor of this volume happens to be the Special Issue of the MDPI's journal, *Axioms*, on the subject-area of "Mathematical Analysis and Applications" (see, for details, [18]). In fact, encouraged by the noteworthy success of these two Special Issues, *Axioms* has already started the publication of a Topical Collection, titled "Mathematical Analysis and Applications" (Collection Editor: H. M. Srivastava), with an open submission deadline.

In recent years, investigations involving the theory and applications of mathematical analytic tools and techniques are remarkably wide-spread in many diverse areas of the mathematical, physical, chemical, engineering and statistical sciences. In this Special Issue, we chose to invite and welcome review, expository and original research articles dealing with the recent advances in mathematical analysis and its multidisciplinary applications.

The suggested topics of interest for the call of papers for this Special Issue included, but by no means limited to, the following keywords:

- Mathematical (or Higher Transcendental) Functions and Their Applications.
- Fractional Calculus and Its Applications.
- q-Series and q-Polynomials.
- Analytic Number Theory.
- Special Functions of Mathematical Physics and Applied Mathematics.
- Geometric Function Theory of Complex Analysis.

Here, in this Editorial, we choose first to briefly describe the status of the Special Issue as follows:

1. Publications: 17.
2. Rejections: 22.
3. Article Type: Research Article (16); Review (1).

Authors' geographical distribution:

- Saudi Arabia (5).
- Italy (3).
- Taiwan (3).
- Germany (2).
- India (2).
- Turkey (2).
- Jordan (2).
- Korea (2).

- Thailand (2).
- Ukraine (1).
- Azerbaijan (1).
- Malaysia (1).
- USA (1).
- Iran (1).
- Thailand (1).
- Serbia (1).
- Tunisia (1).

Papers included in this volume deal extensively with various theoretical as well as applied topics of mathematical analysis of current research interests. Some of the notable contributions in this volume happen to have successfully addressed such topics of mathematical analysis and applications as (for example) Hyperbolic Trigonometric Functions and Associated Polynomials, Szász-Mirakjan Beta-Type Approximation Operators, Holomorphic Functions in One and More Variables, Hypergeometric Functions and Their Generalizations, Hyers-Ulam-Type Stability Problems, Fixed Point Results, Bose–Einstein and Fermi–Dirac Functions, Hermite–Hadamard-Type Inequalities, Elastostatics with Singular Boundary Values, Sequence Generating Functions and Classical and Non-Classical Polynomial Sets.

I take this opportunity to thank all of the participating authors, and the referees and the peer-reviewers, for their *invaluable* contributions toward the remarkable success of each of the above-mentioned Special Issues. I do also greatly appreciate the editorial and managerial help and assistance provided efficiently and generously by Ms. Luna Shen and many of her colleagues and associates in the Editorial Office of *Axioms*.

Funding: This research received no external funding.

Conflicts of Interest: The author declares no conflicts of interest.

References

1. Dattoli, G.; Licciardi, S.; Pidatella, R.M.; Sabia, E. Repeated Derivatives of Hyperbolic Trigonometric Functions and Associated Polynomials. *Axioms* **2019**, *8*, 138.
2. Srivastava, H.M.; İçöz, G.; Çekim, B. Approximation Properties of an Extended Family of the Szász-Mirakjan Beta-Type Operators. *Axioms* **2019**, *8*, 111.
3. Bandura, A.; Skaskiv, O. Slice Holomorphic Functions in Several Variables with Bounded L-Index in Direction. *Axioms* **2019**, *8*, 88.
4. Koepf, W.; Kim, I.; Rathie, A.K. On a New Class of Laplace-Type Integrals Involving Generalized Hypergeometric Functions. *Axioms* **2019**, *8*, 87.
5. Lee, Y.-H.; Kim, G.H. Generalized Hyers-Ulam Stability of the Additive Functional Equation. *Axioms* **2019**, *8*, 76.
6. Mlaiki, N.; Kukić, K.; Gardaşević-Filipović, M.; Aydi, H. On Almost b-Metric Spaces and Related Fixed Point Results. *Axioms* **2019**, *8*, 70.
7. Prabseang, J.; Nonlaopon, K.; Tariboon, J. (p,q)-Hermite-Hadamard Inequalities for Double Integral and (p,q)-Differentiable Convex Functions. *Axioms* **2019**, *8*, 68.
8. Srivastava, R.; Naaz, H.; Kazi, S.; Tassaddiq, A. Some New Results Involving the Generalized Bose-Einstein and Fermi-Dirac Functions. *Axioms* **2019**, *8*, 63.
9. Schmidt, M.D. A Short Note on Integral Transformations and Conversion Formulas for Sequence Generating Functions. *Axioms* **2019**, *8*, 62.
10. Qawasmeh, T.; Tallafha, A.; Shatanawi, W. Fixed Point Theorems through Modified ω-Distance and Application to Nontrivial Equations. *Axioms* **2019**, *8*, 57.
11. Tartaglione, A. A Note on the Displacement Problem of Elastostatics with Singular Boundary Values. *Axioms* **2019**, *8*, 46.
12. Asim, M.; Khan, A.R.; Imdad, M. Fixed Point Results in Partial Symmetric Spaces with an Application. *Axioms* **2019**, *8*, 13.

13. Masjed-Jamei, M.; Koepf, W. A New Identity for Generalized Hypergeometric Functions and Applications. *Axioms* **2019**, *8*, 12.
14. Mukheimer, A. Extended Partial S_b-Metric Spaces. *Axioms* **2018**, *7*, 87.
15. Mlaiki, N.; Taş, N.; Özgür, N.Y. On the Fixed-Circle Problem and Khan Type Contractions. *Axioms* **2018**, *7*, 80.
16. Qawaqneh, H.; Noorani, M.; Shatanawi, W.; Alsamir, H. Common Fixed Point Theorems for Generalized Geraghty (α, Ψ, φ)-Quasi Contraction Type Mapping in Partially Ordered Metric-Like Spaces. *Axioms* **2018**, *7*, 74.
17. Ricci, P.E. Differential Equations for Classical and Non-Classical Polynomial Sets: A Survey. *Axioms* **2019**, *8*, 50.
18. Srivastava, H.M. (Ed.) *Mathematical Analysis and Applications*; Printed Edition of the Special Issue, Published in Axioms; MDPI Publishers; Basel, Switzerland, 2019; 217p; ISBN 978-3-03897-400-0; ISBN 978-3-03897-401-7.

© 2020 by the authors. Licensee MDPI, Basel, Switzerland. This article is an open access article distributed under the terms and conditions of the Creative Commons Attribution (CC BY) license (http://creativecommons.org/licenses/by/4.0/).

Article

Repeated Derivatives of Hyperbolic Trigonometric Functions and Associated Polynomials

Giuseppe Dattoli [1], Silvia Licciardi [1,*], Rosa Maria Pidatella [2] and Elio Sabia [1]

1 ENEA—Frascati Research Center, Via Enrico Fermi 45, 00044 Frascati (Rome), Italy; giuseppe.dattoli@enea.it (G.D.); elio.sabia@enea.it (E.S.)
2 Department of Mathematics and Computer Sciences, University of Catania, Viale A. Doria 6, 95125 Catania, Italy; rosa@dmi.unict.it
* Correspondence: silviakant@gmail.com; Tel.: +39-392-509-6741

Received: 8 November 2019; Accepted: 5 December 2019; Published: 6 December 2019

Abstract: Elementary problems as the evaluation of repeated derivatives of ordinary transcendent functions can usefully be treated with the use of special polynomials and of a formalism borrowed from combinatorial analysis. Motivated by previous researches in this field, we review the results obtained by other authors and develop a complementary point of view for the repeated derivatives of sec(.), tan(.) and for their hyperbolic counterparts.

Keywords: operators theory 44A99, 47B99, 47A62; special functions 33C52, 33C65, 33C99, 33B10, 33B15; Stirling numbers and Touchard polynomials 11B73

1. Introduction

The problem of finding closed forms for the repeated derivatives of trigonometric functions as tangent and secant, even though being an apparently elementary issue, has been solved in relatively recent times in ref. [1]. Inspired by this work, a significant amount of research has been subsequently developed. In ref. [2] the proof of the results of [1] was reformulated in terms of a procedure exploiting the "Zeons" Algebra [3]. In [4,5] the authors addressed this study by employing a class of polynomials (the derivative polynomials (DP) introduced in refs. [6,7]) to reformulate the derivation and eventually get a set of fairly simple formulae, providing the successive derivatives of tan, cot, sec, csc, . . . along with those of their hyperbolic counterparts.

In this note we develop a point of view not dissimilar from that of ref. [4,5]. We provide straightforward results in terms of a single family of Legendre like polynomials and comment on the two forms of DP introduced in [3–5].

The repeated derivatives of composite functions $F(x) = f(g(x))$ is a well-established topic in calculus. The formulation of a procedure allowing the derivation of a formula comprising all the possible cases was established in the XIX century [8] and opened important avenue of research in combinatorics [9] and umbral calculus [10,11] as well.

The problem is particularly interesting, encompasses different topics in analysis, including special polynomials like those belonging to the Touchard family [12] and special numbers like the generalized Stirling forms [13,14] of crucial importance in combinatorial analysis. The repeated derivatives of the Gaussian function are those of a composite function in which $f(.)$ is an exponential and $g(.)$ a quadratic function. The relevant expression leads to the Hermite polynomials as auxiliary tool, to get a synthetic expression for any order of the derivative [15]. Within the same context, Bell polynomials emerge whenever one is interested in the derivatives of $F(x) = e^{g(x)}$ [9,16,17]. The generalization to the case of $f(.)$, provided by a generic infinitely differentiable function, and $g(.)$, a quadratic form, has been discussed in refs. [18,19] where the problem has been solved by the use of generalized nested forms of Hermite polynomials.

The hyperbolic secant is a composite function too in which $f(.) = (.)^{-1}$, $g(.) = \cosh(.)$. In refs. [2,4,5,20] interesting speculations have been presented on the study of the relevant repeated derivatives and of the associated auxiliary polynomials. In this paper we address the same problem, within a different context. Before entering the specific elements of the discussion we review the formalism we are going to exploit in this paper.

A pivotal role within the context of repeated derivatives is played by the Stirling number of second kind [13,14]. They will be introduced using the Touchard polynomials which are defined through the Rodriguez type formula [15,21] where, for $n \geq 0$, we set

$$T_n(x) = e^{-x}(x\hat{D}_x)^n e^x \tag{1}$$

where $\hat{D}_x$ is the derivative operator $\frac{d}{dx}$. They can be written in explicit form by the use of the following expression [13]

$$(x\hat{D}_x)^n = \sum_{r=0}^{n} S_2(n,r) x^r \hat{C}_x^r, \tag{2}$$

with $S_2(n,r)$ being the Stirling number of second kind (we introduce the notation $S_2(l,m)$ instead of the usually symbol $\left\{ \begin{matrix} l \\ m \end{matrix} \right\}$; to indicate functions domains, we will use conventional symbols as $\mathbb{N}, \mathbb{Z}, \mathbb{R}$ for Natural, Integer or Real numbers respectively including the zero). Ref. [9,12] defined as

$$S_2(l,m) = \frac{1}{m!} \sum_{j=0}^{m} (-1)^{m-j} \binom{m}{j} j^l. \tag{3}$$

According to Equations (1) and (2), the Touchard polynomials are explicitly given by

$$T_n(x) = \sum_{r=0}^{n} S_2(n,r) x^r \tag{4}$$

and the numbers $S_2(n,k)$ are therefore the coefficients of the polynomials.

A fairly straightforward application of the previous computational tool is provided by the evaluation of a closed expression for higher order derivatives.

Theorem 1. *We have that, $\forall m \in \mathbb{N}$,*

$$\hat{D}_x^m f(e^x) = \sum_{r=0}^{m} S_2(m,r) e^{xr} f^{(r)}(e^x) \tag{5}$$

where $f^{(r)}(\xi)$ denotes the r-order derivative.

Proof. Let

$$I_m(x) = \hat{D}_x^m f(e^x), \tag{6}$$

by setting $e^x = \xi$, we find

$$\hat{D}_x^m f(e^x) = (\xi \hat{D}_\xi)^m f(\xi) = \sum_{r=0}^{m} S_2(m,r) \xi^r f^{(r)}(\xi) \tag{7}$$

and in conclusion we obtain

$$\hat{D}_x^m f(e^x) = \sum_{r=0}^{m} S_2(m,r) e^{xr} f^{(r)}(e^x)$$

in which it is understood that $f^{(r)}(e^x) = \hat{D}_\xi^r f(\xi) |_{\xi=e^x}$. □

We next define a family of auxiliary polynomials $P_n(x,y)$ and show how they can be very useful for the computation of the repeated derivatives of $\tan^{-1}(x)$ (we use the notation $\tan^{-1}(x)$ for $\arctan(x)$).

Theorem 2. *For any* $x, y \in \mathbb{R}$, $n \in \mathbb{N}$,

$$P_n(x,y) = n! \sum_{r=0}^{\lfloor \frac{n}{2} \rfloor} \frac{x^{n-2r} y^r (n-r)!}{(n-2r)! r!}, \tag{8}$$

a family of two variable polynomials loosely ascribed to the Legendre [22,23] *family, then*

$$\hat{D}_x^n \left(\tan^{-1}(x) \right) = \frac{1}{1+x^2} P_{n-1} \left(-\frac{2x}{1+x^2}, -\frac{1}{1+x^2} \right), \quad \forall n \in \mathbb{N}_0. \tag{9}$$

Proof. We start from the computation of the quantity

$$K_n(x) = \hat{D}_x^n \left(\frac{1}{1+x^2} \right), \quad \forall x \in \mathbb{R}, \forall n \in \mathbb{N} \tag{10}$$

which upon the use of the Laplace transform method can be written as

$$K_n(x) = \hat{D}_x^n \int_0^\infty e^{-s(1+x^2)} ds = \int_0^\infty e^{-s} \hat{D}_x^n e^{-sx^2} ds. \tag{11}$$

The n-th order derivative inside the integral sign can be explicetely worked in terms of two variable Hermite polynomials [15]

$$H_n(x,y) = n! \sum_{r=0}^{\lfloor \frac{n}{2} \rfloor} \frac{x^{n-2r} y^r}{(n-2r)! r!}, \tag{12}$$

namely the auxiliary polynomials for the repeated derivatives of Gaussian functions according to the identity

$$\hat{D}_x^n e^{-ax^2} = H_n(-2ax, -a) e^{-ax^2}. \tag{13}$$

Accordingly we find

$$K_n(x) = \int_0^\infty e^{-s} H_n(-2xs, -s) e^{-sx^2} ds = \int_0^\infty e^{-\sigma} H_n \left(-2\frac{x\sigma}{1+x^2}, -\frac{\sigma}{1+x^2} \right) \frac{1}{1+x^2} d\sigma =$$

$$= \frac{n!}{1+x^2} \sum_{r=0}^{\lfloor \frac{n}{2} \rfloor} \frac{\left(\frac{-2x}{1+x^2} \right)^{n-2r} \left(\frac{-1}{1+x^2} \right)^r}{(n-2r)! r!} \int_0^\infty e^{-\sigma} \sigma^{n-r} d\sigma = \frac{1}{1+x^2} P_n \left(-\frac{2x}{1+x^2}, -\frac{1}{1+x^2} \right). \tag{14}$$

The proof of the identity (9) is readily achieved by noting that

$$\hat{D}_x \left(\tan^{-1}(x) \right) = \left(\frac{1}{1+x^2} \right), \tag{15}$$

which eventually provides

$$\hat{D}_x^n \left(\tan^{-1}(x) \right) = K_{n-1}(x) = \frac{1}{1+x^2} P_{n-1} \left(-\frac{2x}{1+x^2}, -\frac{1}{1+x^2} \right), \quad \forall n \in \mathbb{N}_0.$$

□

Corollary 1. *If we introduce the further notation*

$$K_n^\nu(x) := \hat{D}_x^n \left(\frac{1}{(1+x^2)^\nu} \right), \tag{16}$$

an extension of the procedure we have just envisaged allows the further results

$$K_n^\nu(x) = \frac{1}{(1+x^2)^\nu} P_n^\nu\left(-\frac{2x}{1+x^2}, -\frac{1}{1+x^2}\right),$$

$$P_n^\nu(w,z) = \frac{n!}{\Gamma(\nu)} \sum_{r=0}^{\lfloor \frac{n}{2} \rfloor} \frac{w^{n-2r} z^r \, \Gamma(\nu+n-r)}{(n-2r)! r!}, \qquad \forall \nu \in \mathbb{R}.$$ (17)

Proof. $\forall \nu \in \mathbb{R}$

$$K_n^\nu(x) = \hat{D}_x^n \left(\frac{1}{(1+x^2)^\nu}\right) = \hat{D}_x^n \frac{1}{\Gamma(\nu)} \int_0^\infty s^{\nu-1} e^{-s(1+x^2)} ds = \frac{1}{\Gamma(\nu)} \int_0^\infty s^{\nu-1} e^{-s} \hat{D}_x^n e^{-sx^2} ds =$$

$$= \frac{1}{\Gamma(\nu)} \int_0^\infty s^{\nu-1} e^{-s} H_n(-2xs, -s) e^{-sx^2} ds =$$

$$= \frac{1}{\Gamma(\nu)} \int_0^\infty \left(\frac{\sigma}{1+x^2}\right)^{\nu-1} e^{-\sigma} H_n\left(-2\frac{x\sigma}{1+x^2}, -\frac{\sigma}{1+x^2}\right) \frac{1}{1+x^2} d\sigma =$$

$$= \frac{n!}{\Gamma(\nu)(1+x^2)^\nu} \sum_{r=0}^{\lfloor \frac{n}{2} \rfloor} \frac{\left(\frac{-2x}{1+x^2}\right)^{n-2r} \left(\frac{-1}{1+x^2}\right)^r}{(n-2r)! r!} \int_0^\infty e^{-\sigma} \sigma^{\nu+n-r-1} d\sigma =$$

$$= \frac{1}{(1+x^2)^\nu} P_n^\nu\left(-\frac{2x}{1+x^2}, -\frac{1}{1+x^2}\right).$$

□

Remark 1. *Corollary 1 can be exploited to write the repeated derivatives of $\cos^{-1}(x)$ in the form*

$$\hat{D}_x^n \cos^{-1}(x) = -\hat{D}_x^{n-1}\left(\frac{1}{\sqrt{1-x^2}}\right) = -\frac{1}{\sqrt{1-x^2}} P_{n-1}^{\frac{1}{2}}\left(\frac{2x}{1-x^2}, \frac{1}{1-x^2}\right), \qquad \forall n \in \mathbb{N}_0.$$ (18)

The previous examples have shown the interplay between special polynomials of the $P_n(x,y)$ and the derivatives of the inverse of trigonometric functions.

In the following we will see that the same polynomials are of central importance for the evaluation of the repetead derivatives of trigonometric functions. They play the role of auxiliary polynomials for the lorentzian type functions and inverse trigonometric functions as well. The relevant elements of contact with the *DP* will be discussed in the final section of the paper.

2. Higher Order Derivatives of Trigonometric Functions

The starting point of the discussion of this section is the derivation of a closed form for the quantity $\hat{D}_x^m(\text{sech}(x))$, $\forall m \in \mathbb{N}_0$.

Proposition 1. $\forall m \in \mathbb{N}_0, \forall x \in [0, 2\pi]$,

$$\hat{D}_x^m(\text{sech}(x)) = \text{sech}(x) \sum_{k=0}^m S_2(m,k) \, e^{(k-1)x} \left(e^x P_k\left(-\text{sech}(x), -\frac{1}{2e^x}\text{sech}(x)\right) + k \, P_{k-1}\left(-\text{sech}(x), -\frac{1}{2e^x}\text{sech}(x)\right)\right).$$ (19)

Proof. Let

$$\text{sech}(x) = \frac{2}{e^x + e^{-x}}$$ (20)

then, after setting $e^x = \xi$, we obtain $\forall m \in \mathbb{N}_0$

$$\hat{D}_x^m (\text{sech}(x)) = \hat{D}_{\log \xi}^m \left(\frac{2\xi}{1+\xi^2} \right) = 2 \left(\xi \hat{D}_\xi \right)^m \left(\frac{\xi}{1+\xi^2} \right). \tag{21}$$

The straightforward applications of the proof of the Theorem 1 yields (we omit the argument of the P_n polynomials for simplicity)

$$2 \left(\xi \hat{D}_\xi \right)^m \left(\frac{\xi}{1+\xi^2} \right) = \frac{2}{1+\xi^2} \sum_{k=0}^{m} S_2(m,k) \, \xi^k \left(\xi P_k(.,.) + k P_{k-1}(.,.) \right) \tag{22}$$

which, eventually, leads to

$$\hat{D}_x^m (\text{sech}(x)) = \text{sech}(x) \sum_{k=0}^{m} S_2(m,k) \, e^{(k-1)x} \left(e^x \, P_k \left(-\text{sech}(x), -\frac{1}{2e^x} \text{sech}(x) \right) \right.$$
$$\left. + k \, P_{k-1} \left(-\text{sech}(x), -\frac{1}{2e^x} \text{sech}(x) \right) \right).$$

□

We can go even further and obtain a closed form for $\hat{D}_x^m (\text{sech}(x))^\nu$, $\forall m \in \mathbb{N}, \forall \nu \in \mathbb{R}$.

Corollary 2. $\forall m \in \mathbb{N}, \forall \nu \in \mathbb{R}$

$$\hat{D}_x^m (\text{sech}^\nu(x)) = \text{sech}^\nu(x) \sum_{r=0}^{m} S_2(m,r) \, e^{xr} \sum_{s=0}^{r} \binom{r}{s} \frac{n!}{(n-r+s)!} e^{-x(r-s)} \, P_s^\nu \left(-\text{sech}(x), -\frac{1}{2e^x} \text{sech}(x) \right) \tag{23}$$

where $P_m^k(.,.)$ are the polynomials in Equation (17).

By using the same techniques of the Proposition 1, Corollary 2 and the opportune identity $\sec(x) = \frac{2}{e^{ix}+e^{-ix}}$, we can move on and provide the higher order derivatives for other trigonometric circular functions.

Example 1. $\forall m \in \mathbb{N}_0, \forall x \in [0, 2\pi]$

$$\hat{D}_x^m (\sec(x)) = i^m \sec(x) \sum_{r=0}^{m} S_2(m,r) \, e^{ix(r-1)} \left(e^{ix} P_r \left(-\sec(x), -\frac{1}{2e^{ix}} \sec(x) \right) \right.$$
$$\left. + r \, P_{r-1} \left(-\sec(x), -\frac{1}{2e^{ix}} \sec(x) \right) \right). \tag{24}$$

Example 2. $\forall m \in \mathbb{N}, \forall x \in [0, 2\pi] : x \neq \frac{\pi}{2} + k\pi, k \in \mathbb{Z}$

$$\hat{D}_x^m (\tan(x)) = \sum_{s=0}^{m} \binom{m}{s} \sin\left(x + (m-s)\frac{\pi}{2}\right) \hat{D}_x^s (\sec(x)). \tag{25}$$

Example 3. $\forall m \in \mathbb{N}, \forall x \in [0, 2\pi] : x \neq k\pi, k \in \mathbb{Z}$

$$\hat{D}_x^m (\cot(x)) = \sum_{s=0}^{m} \binom{m}{s} \cos\left(x + (m-s)\frac{\pi}{2}\right) \hat{D}_x^s \left(\sec\left(x - \frac{\pi}{2}\right) \right). \tag{26}$$

Remark 2. *The use of the Leibniz rule in Equations (25) or (26) can be avoided by setting*

$$\tan(x) = i \frac{1-\xi^2}{1+\xi^2} = i \left(\frac{2}{1+\xi^2} - 1 \right), \quad \xi = e^{ix} \tag{27}$$

and eventually ending up with

$$\hat{D}_x^m(\tan(x)) = i^{m+1}\left(\sec(x)\sum_{r=0}^{m} S_2(m,r)\, e^{ix(r-1)} P_r\left(-\sec(x), -\frac{1}{2e^{ix}}\sec(x)\right) - \delta_{m,0}\right), \qquad (28)$$

where $\delta_{m,0}$ is Kronecker's delta, indeed

$$\hat{D}_x^m(\tan(x)) = i^m\left(\zeta\hat{D}_\zeta\right)^m\left(i\left(\frac{2}{1+\zeta^2}-1\right)\right) = 2\, i^{m+1}\left(\zeta\hat{D}_\zeta\right)^m\left(\frac{1}{1+\zeta^2}-\frac{1}{2}\right) =$$

$$= 2\, i^{m+1}\sum_{r=0}^{m} S_2(m,r)\,\zeta^r\left(\frac{1}{1+\zeta^2}-\frac{1}{2}\right)^{(r)} =$$

$$= 2\, i^{m+1}\sum_{r=0}^{m} S_2(m,r)\,\zeta^r\left(\left(\frac{1}{1+\zeta^2}\right)^{(r)}+\left(-\frac{1}{2}\right)^{(r)}\right) =$$

$$= 2\, i^{m+1}\left(\left(\sum_{r=0}^{m} S_2(m,r)\,\zeta^r\,\frac{1}{1+\zeta^2}P_r(.,.)\right)+\left(\sum_{r=0}^{m} S_2(m,r)\,\zeta^r\left(-\frac{1}{2}\right)^{(r)}\right)\right) =$$

$$= i^{m+1}\left(\sec(x)\sum_{r=0}^{m} S_2(m,r)\, e^{ix(r-1)} P_r\left(-\sec(x), -\frac{1}{2e^{ix}}\sec(x)\right) - \delta_{m,0}\right).$$

In the forthcoming section we will discuss the comparison with previous papers and present possible developments along the lines we have indicated.

3. Final Comments

In the previous two sections we have dealt with a general procedure useful to derive closed formulae for the repeated derivatives of circular and hyperbolic functions and of their inverse. We have underscored the importance of the polynomials $P_n(x,y)$ which play a role analogous to the DP introduced in refs. [6,7] and used in [4,5] for analogous occurrences.

The $P_n(x,y)$ are two variable polynomials defined in terms of the Laplace transform of the Hermite Kampé dé Fériét family [24]. They have been loosely defined Legendre-like and can be reduced to more familiar forms, by noting that

$$P_n(x,y) = y^{\frac{n}{2}} P_n\left(\frac{x}{\sqrt{-y}}\right),$$

$$P_n(z) = n!\sum_{r=0}^{\lfloor\frac{n}{2}\rfloor}\frac{(-1)^r z^{n-2r}(n-r)!}{(n-2r)!r!}$$
(29)

with $P_n(z)$ satisfying the generating function

$$\sum_{n=0}^{\infty}\frac{t^n}{n!}P_n(z) = \frac{1}{1-tz+t^2}, \qquad \forall t, z \in \mathbb{R}:|\,t^2-t\,z\,|<1 \qquad (30)$$

and

$$\frac{1}{n!}P_n(2x,-1) = \sum_{k=0}^{\lfloor\frac{n}{2}\rfloor}\frac{(-1)^k(n-k)!(2x)^{n-2k}}{(n-2k)!k!} = U_n(x). \qquad (31)$$

The link with previous papers addressing the same problem treated here, in particular with that developed in refs. References [4,5] can be obtained by using the same steps suggested by Cvijović or Boyadzhiev. We assume that an identity of the type

$$\hat{D}_x^n(\tan(x)) = \Pi_n(\tan(x)) \tag{32}$$

with $\Pi_n(.)$ being not yet specified polynomials, (in refs. [4,5] they are denoted by $P_n(x)$, we have used the capital Greek letter to avoid confusion with the $P_n(x,y)$ polynomials defined in this paper) does holds. The polynomials $\Pi_n(.)$ can be determined by the change of variable $\tan(x) = \xi$ which allows to transform Equation (32) into

$$\left[\left(1+\xi^2\right)\hat{D}_\xi\right]^n(\xi) = \Pi_n(\xi) \tag{33}$$

which is a kind of Rodrigues type relation [25] defining the polynomials $\Pi_n(\xi)$.

The relevant generating function can be obtained by multiplying both sides of Equation (33) by $\dfrac{t^n}{n!}$, by summing up over the index n and ending up with

$$e^{t\left[(1+\xi^2)\hat{D}_\xi\right]}\xi = \sum_{n=0}^{\infty} \frac{t^n}{n!} \Pi_n(\xi). \tag{34}$$

The use of the Lie derivative identity [26]

$$e^{t\left[(1+\xi^2)\hat{D}_\xi\right]}f(\xi) = f\left(\frac{\xi\cos(t) + \sin(t)}{\cos(t) - \xi\sin(t)}\right) \tag{35}$$

finally yields

$$\sum_{n=0}^{\infty} \frac{t^n}{n!} \Pi_n(\xi) = \frac{\xi + \tan(t)}{1 - \xi \tan(t)} \tag{36}$$

which is the generating function given in [4,5].

The second family of DP can be defined by the use of the same procedure. According to refs. [4,5] they are implicitly defined by the condition

$$\hat{D}_x^n(\sec(x)) = \sec(x) Q_n(\tan(x)) \tag{37}$$

which, by the use of the same change of variable leading to Equation (33), yields

$$Q_n(\xi) = \frac{1}{\sqrt{1+\xi^2}} \left((1+\xi^2)\hat{D}_\xi\right)^n \sqrt{1+\xi^2} \tag{38}$$

which can be straightforwardly exploited to derive the relevant properties. The use of the identity (35) eventually yields the associated generating functions

$$\sum_{n=0}^{\infty} \frac{t^n}{n!} Q_n(\xi) = \frac{\sec(t)}{1 - \xi \tan(t)} \tag{39}$$

which is the same reported in refs. [4,5].

By recalling that the Hoppe formula writes

$$\begin{aligned}\hat{D}_t^m f(g(t)) &= \sum_{k=0}^m \frac{1}{k!} f^{(k)}(\sigma)\Big|_{\sigma = g(t)} A_{m,k}, \\ A_{m,k} &= \sum_{j=0}^k \binom{k}{j}(-1)^{k-j} g(t)^{k-j} \hat{D}_t^m (g(t)^j), \quad \forall m,k \in \mathbb{N},\end{aligned} \tag{40}$$

for the case of $\sec(x)$, we find

$$\hat{D}_x^m \sec(x) = \sum_{k=0}^m \sec^{k+1}(x) \sum_{j=0}^k \binom{k}{j}(-1)^j \cos^{k-j}(x)\, \hat{D}_x^m(\cos^j(x)) \tag{41}$$

which appears impractical since it needs the repeated derivatives of an integer power of the cosine function.

Before closing the paper we consider worth to underscore the possible use of DP to evaluate the derivatives of the type

$$\lambda_{m,j}(x) = \hat{D}_x^m(\cos^j(x)). \tag{42}$$

Making the ansatz that

$$\hat{D}_x^m(\cos^j(x)) = (-1)^j \Lambda_{m,j}(\tan(x)) \tag{43}$$

and, by using the same procedure leading to Equation (33), we can identify the function

$$\Lambda_{m,j}(\xi) = \left[\left(1+\xi^2\right)\hat{D}_\xi\right]^m \left(\frac{1}{\left(\sqrt{1+\xi^2}\right)^j}\right) \tag{44}$$

which is specified by the generating function

$$\sum_{m=0}^{\infty} \frac{t^m}{m!} \Lambda_{m,j}(\xi) = \frac{[1 - \xi \tan(t)]^j}{\left[(1+\xi^2)\left(1+\tan^2(t)\right)\right]^{\frac{j}{2}}}. \tag{45}$$

It is worth to consider the alternative assumption

$$\hat{D}_x^m(\cos^j(x)) = (-1)^m \Delta_{m,j}(\cos(x)) \tag{46}$$

which, after the change of variable $\xi = \cos(x)$, yields for the function $\Delta_{m,j}(.)$ the operational definition

$$\Delta_{m,j}(\xi) = \left(-\sqrt{1-\xi^2}\,\hat{D}_\xi\right)^m \xi^j \tag{47}$$

The use of the Lie derivative identity [26]

$$e^{-t\left(\sqrt{1-\xi^2}\,\hat{D}_\xi\right)} f(\xi) = f\left(\xi \cos(t) - \sqrt{1-\xi^2} \sin(t)\right) \tag{48}$$

yields, for $\Delta_{m,j}(\xi)$, the generating function

$$\sum_{m=0}^{\infty} \frac{t^m}{m!} \Delta_{m,j}(\xi) = \left(\xi - \sqrt{1-\xi^2}\,\tan(t)\right)^j \cos^j(t). \tag{49}$$

The two methods we have discussed in this paper, namely the procedure based on the DP of refs. [4,5] or on the support polynomials $P_n(x,y)$, are complementary. There are no prevailing reasons to prefer one or the other method. The formulae associated with the first are more synthetic but $\Pi_n(\xi)$, $Q_n(\xi)$ are not given explicitly and should be evaluated recursively (which becomes cumbersome for larger order of the derivative). On the other side, the use of the other procedure leads to less appealing formulae in terms of polynomials which are, however, explicitly given.

Author Contributions: Conceptualization: G.D.; methodology: G.D., S.L.; data curation: S.L.; validation: G.D., S.L., R.M.P.; formal analysis: G.D., S.L., R.M.P.; writing—original draft preparation: G.D., E.S.; writing—review and editing: S.L.

Funding: This research received no external funding.

Acknowledgments: The work of S. Licciardi was supported by an Enea Research Center individual fellowship. R.M. Pidatella wants to thank the fund of University of Catania "Metodi gruppali e umbrali per modelli di diffusione e trasporto" included in "Piano della Ricerca 2016/2018 Linea di intervento 2" for partial support of this work.

Conflicts of Interest: The authors declare no conflict of interest.

References

1. Adamchik, V.S. On the Hurwitz function for rational arguments. *Appl. Math. Comput.* **2007**, *187*, 3–12. [CrossRef]
2. Neto, A.F. Higher Order Derivatives of Trigonometric Functions, Stirling Numbers of the Second Kind and Zeon Algebra. *J. Integer Seq.* **2014**, *17*, 14.9.3.
3. Feinsilver, P. Zeon algebra, Fock space, and Markov chains. *Commun. Stoch. Anal.* **2008**, *2*, 6. [CrossRef]
4. Cvijović, D. Derivative Polynomials and Closed-Form Higher Derivative Formulae. *Appl. Math. Comput.* **2009**, *215*, 3002–3006. [CrossRef]
5. Boyadzhiev, K.N. Derivative polynomials for tanh, tan, sech and sec in explicit form. *Fibonacci Q.* **2007**, *45*, 291–303.
6. Hoffman, M.E. Derivative polynomials for tangent and secant. *Am. Math. Mon.* **1995**, *102*, 23–30. [CrossRef]
7. Hoffman, M.E. Derivative polynomials, Euler polynomials and associated integer sequences. *Electron. J. Comb.* **1999**, *6*, R21.
8. Johnson, W.P. The Curious History of Faa di Bruno's Formula. *Am. Math. Mon.* **2002**, *109*, 217–234.
9. Comtet, L. *Advanced Combinatorics: The Art of Finite and Infinite Expansions*; Springer: Berlin/Heidelberg, Germany, 1974.
10. Roman, S. *The Umbral Calculus*; Dover Publications: New York, NY, USA, 2005.
11. Licciardi, S. Umbral Calculus, a Different Mathematical Language. Ph.D. Thesis, Department of Mathematics and Computer Sciences, XXIX Cycle, University of Catania, Catania, Italy, 2018.
12. Dattoli, G.; Germano, B.; Martinelli, M.R.; Ricci, P.E. Touchard like polynomials and generalized Stirling numbers. *Appl. Math. Comput.* **2012**, *218*, 6661–6665. [CrossRef]
13. Graham,R.L.; Knuth, D.E.; Patashnik, O. *Concrete Mathematics: A Foundation for Computer Science*, 2nd ed.; Addison-Wesley Longman Publishing Co., Inc.: Boston, MA, USA, 1994.
14. Mansour, T.; Schork, M. *Commutation Relations, Normal Ordering, and Stirling Numbers*; Chapman and Hall/CRC: Boca Raton, FL, USA, 2015; p. 504, ISBN 9781466579880.
15. Dattoli, G. Generalized polynomials, operational identities and their applications. *J. Comp. Appl. Math.* **2000**, *118*. 111–123. [CrossRef]
16. Riordan, J. *An Introduction to Combinatorial Analysis*; Wiley: New York, NY, USA, 1980.
17. Bell, E.T. Exponential Polynomials. *Ann. Math.* **1934**, *35*, 258–277. [CrossRef]
18. Dattoli, G.; Srivastava, H.M.; Sacchetti, D. Mixed and higher-order generating functions: An extension to new families. *Int. J. Math. Stat. Sci.* **1999**, *8*, 79–87.
19. Babusci, D.; Dattoli, G.; Gorska, K.; Penson, K. Repeated derivatives of composite functions and generalizations of the Leibniz rule. *Appl. Math. Comput.* **2014**, *241*, 193. [CrossRef]
20. Wintucky, E.G. *Formulas for nth Order Derivatives of Hyperbolic and Trigonometric Functions*; NASA Lewis Research Center: Cleveland, OH, USA, 1971.
21. Andrews, L.C. *Special Functions For Engeneers and Applied Mathematicians*; Mc Millan: New York, NY, USA, 1985.
22. Dattoli, G.; Germano, B.; Martinelli, M.R.; Ricci, P.E. A novel theory of Legendre polynomials. *Math. Comput. Modell.* **2011**, *54*, 80–87. [CrossRef]
23. Dattoli, G.; Germano, B.; Martinelli, M.R.; Ricci, P.E. Sheffer and Non-Sheffer Polynomial Families. *Int. J. Math. Math. Sci.* **2012**, *2012*, 323725. [CrossRef]
24. Appél, P.; Kampé de Fériét, J. *Fonctions Hypergeometriques and Hyperspheriques. Polynomes d'Hermite*; Gauthiers-Villars: Paris, France, 1926.
25. Rainville, E.D. *Special Functions*; Macmillan: New York, NY, USA, 1960.
26. Dattoli, G.; Ottaviani, P.L.; Torre, A.; Vazquez, L. Evolution operator equations-integration with algebraic and finite-difference methods-applications to physical problems in classical and quantum mechanics and quantum field theory. *La Rivista del Nuovo Cimento* **1997**, *20*, 3–133. [CrossRef]

 © 2019 by the authors. Licensee MDPI, Basel, Switzerland. This article is an open access article distributed under the terms and conditions of the Creative Commons Attribution (CC BY) license (http://creativecommons.org/licenses/by/4.0/).

Article

Approximation Properties of an Extended Family of the Szász–Mirakjan Beta-Type Operators

Hari Mohan Srivastava [1,2,3,*], Gürhan İçöz [4] and Bayram Çekim [4]

[1] Department of Mathematics and Statistics, University of Victoria, Victoria, BC V8W 3R4, Canada
[2] Department of Medical Research, China Medical University Hospital, China Medical University, Taichung 40402, Taiwan
[3] Department of Mathematics and Informatics, Azerbaijan University, 71 Jeyhun Hajibeyli Street, AZ1007 Baku, Azerbaijan
[4] Department of Mathematics, Gazi University, Ankara TR-06500, Turkey; gurhanicoz@gazi.edu.tr (G.İ.); bayramcekim@gazi.edu.tr (B.Ç.)
* Correspondence: harimsri@math.uvic.ca

Received: 14 September 2019; Accepted: 30 September 2019; Published: 10 October 2019

Abstract: Approximation and some other basic properties of various linear and nonlinear operators are potentially useful in many different areas of researches in the mathematical, physical, and engineering sciences. Motivated essentially by this aspect of approximation theory, our present study systematically investigates the approximation and other associated properties of a class of the Szász-Mirakjan-type operators, which are introduced here by using an extension of the familiar Beta function. We propose to establish moments of these extended Szász-Mirakjan Beta-type operators and estimate various convergence results with the help of the second modulus of smoothness and the classical modulus of continuity. We also investigate convergence via functions which belong to the Lipschitz class. Finally, we prove a Voronovskaja-type approximation theorem for the extended Szász-Mirakjan Beta-type operators.

Keywords: gamma and beta functions; Szász-Mirakjan operators; Szász-Mirakjan Beta type operators; extended Gamma and Beta functions; confluent hypergeometric function; Modulus of smoothness; modulus of continuity; Lipschitz class; local approximation; Voronovskaja type approximation theorem

2010 Mathematics Subject Classification: Primary 33B15; 33C05; 41A25; 41A35; Secondary 33C20

---

1. Introduction, Definitions and Preliminaries

In approximation theory and its related fields, approximation and other basic properties of various linear and nonlinear operators are investigated, because mainly of the potential for their usefulness in many areas of researches in the mathematical, physical, and engineering sciences. Our study in this article is motivated essentially by the demonstrated applications of such results as those associated with various approximation operators. With this objective in view, we begin by providing the following definitions and other (chiefly historical) background material related to our presentation here.

For a given continuous function, $f \in C[0,\infty)$, and for $x \in [0,\infty)$, Otto Szász [1] defined a family of operators in the year 1950, which we recall here as follows,

$$S_n(f;x) = e^{-nx} \sum_{k=0}^{\infty} \frac{(nx)^k}{k!} f\left(\frac{k}{n}\right) \qquad (n \in \mathbb{N} := \{1,2,3,\cdots\}), \tag{1}$$

This family of operators was considered earlier in 1941 by G. M. Mirakjan (see [2]). There are several integral and other modifications, variations, and basic (or q-) extensions of the Szász-Mirakjan-type operators. These include the Bézier, Kantorovich, Durrmeyer, and other types of modifications and extensions of the Szász-Mirakjan operators (see, for details, [3–15]). In particular, Gupta and Noor [6] introduced an integral modification of the Szász-Mirakjan operators in Equation (1) by considering a weight function in terms of the Beta basis functions as given below.

$$T_n(f;x) = \sum_{k=1}^{\infty} s_{n,k}(x) \int_0^{\infty} b_{n,k}(t) \; f(t) \; dt + s_{n,0}(x) \, f(0), \qquad (2)$$

where

$$s_{n,k}(x) = e^{-nx} \frac{(nx)^k}{k!},$$

$$b_{n,k}(t) = \frac{1}{B(k, n+1)} \frac{t^{k-1}}{(1+t)^{n+k+1}} = \frac{(n+1)_k}{(k-1)!} \frac{t^{k-1}}{(1+t)^{n+k+1}},$$

$$B(\alpha, \beta) = \frac{\Gamma(\alpha)\Gamma(\beta)}{\Gamma(\alpha+\beta)} = B(\beta, \alpha),$$

and $(\lambda)_\ell$ ($\ell \in \mathbb{N}_0 := \mathbb{N} \cup \{0\}$) represents the Pochhammer symbol given by

$$(\lambda)_\ell := \frac{\Gamma(\lambda+\ell)}{\Gamma(\lambda)} = \begin{cases} 1 & (\ell = 0) \\ \lambda(\lambda+1)(\lambda+2)\cdots(\lambda+\ell-1) & (\ell \in \mathbb{N}) \end{cases}$$

in terms of the classical (Euler's) Gamma function $\Gamma(z)$ and the classical Beta function $B(\alpha, \beta)$.

Gupta and Noor [6] observed that the operators in Equation (2) reproduce not only the constant function, but linear functions as well. Owing to this valuable property of the operators in Equation (2), many authors investigated the different approximation properties of the summation-integral operators in Equation (2) (see, for example, [16–18]). Gupta and Noor [6] also derived some direct results for the operators T_n, a pointwise rate of convergence, a Voronovskaja-type asymptotic formula, and an error estimate in simultaneous approximation.

In recent years, some extensions of such well-known special functions as, for example, the classical Gamma and Beta functions, have been considered by several authors. For example, in 1994, Chaudhry and Zubair [19] introduced the following extension of the Gamma function,

$$\Gamma_p(x) := \int_0^{\infty} t^{x-1} \exp\left(-t - \frac{p}{t}\right) dt \qquad (\Re(p) > 0). \qquad (3)$$

Subsequently, in 1997, Chaudhry et al. [20] presented the following extension of Euler's Beta function,

$$B_p(x, y) := \int_0^1 t^{x-1} (1-t)^{y-1} \exp\left(-\frac{p}{t(1-t)}\right) dt \qquad (4)$$

$$(\Re(p) > 0; \; \Re(x) > 0; \; \Re(y) > 0).$$

Obviously, each of the definitions in Equations (3) and (4) also remains valid when $p = 0$, in which case we have the following relationships,

$$\Gamma_0(x) = \Gamma(x) \qquad \text{and} \qquad B_0(x, y) = B(x, y).$$

Özergin et al. [21] considered the following generalizations of the Gamma and Beta functions,

$$\Gamma_p^{(\alpha,\beta)}(x) := \int_0^\infty t^{x-1} {}_1F_1\left(\alpha;\beta;-t-\frac{p}{t}\right) dt$$

$$(\Re(\alpha) > 0;\ \Re(\beta) > 0;\ \Re(p) > 0;\ \Re(x) > 0)$$

and

$$B_p^{(\alpha,\beta)}(x,y) := \int_0^1 t^{x-1}(1-t)^{y-1} {}_1F_1\left(\alpha;\beta;-\frac{p}{t(1-t)}\right) dt$$

$$(\Re(\alpha) > 0;\ \Re(\beta) > 0;\ \Re(p) > 0;\ \Re(x) > 0;\ \Re(y) > 0),$$

respectively. Here, as usual, ${}_1F_1$ denotes the (Kummer's) confluent hypergeometric function. It is obvious that

$$\Gamma_p^{(\alpha,\alpha)}(x) = \Gamma_p(x) \quad \text{and} \quad \Gamma_0^{(\alpha,\alpha)}(x) = \Gamma(x)$$

and that

$$B_p^{(\alpha,\alpha)}(x,y) = B_p(x,y) \quad \text{and} \quad B_0^{(\alpha,\alpha)}(x,y) = B(x,y).$$

Finding different integral representations of the generalized Beta function is important and useful for later use. It is also useful to discuss the relationships between the classical Gamma and Beta functions and their generalizations. In fact, by definition, it is easily seen that

$$\int_0^\infty \Gamma_p(x)\, dp = \Gamma(x+1) \quad (\Re(x) > -1).$$

and that (see, for example, [21])

$$\int_0^\infty B_p(x,y)\, dp = B(x+1,y+1) \tag{5}$$

$$(\Re(x) > -1;\ \Re(y) > -1).$$

Note that various further extensions and generalizations of the classical Gamma and Beta functions, as well as their corresponding hypergeometric and related functions, were introduced and studied by, among others, Lin et al. [22] and Srivastava et al. [23].

We now introduce the following generalization of Szász-Mirakjan Beta-type operators via the above extension of the Beta function as follows,

$$S_n^*(f,x) = \sum_{k=0}^\infty e^{-nx} \frac{(nx)^k}{k!} \frac{1}{B(k+2,n+1)} \cdot \int_0^\infty \int_0^\infty \frac{t^k}{(1+t)^{n+k+1}} \exp\left(-\frac{p(1+t)^2}{t}\right) f(t)\, dt\, dp \tag{6}$$

for $x \in [0,\infty)$, and for a function $f \in C_\nu[0,\infty)$, provided that the double integral in Equation (6) is convergent when $n > \nu$. Here, and in what follows, we have

$$C_\nu[0,\infty) := \{f : f \in C[0,\infty) \text{ and } |f(t)| \leq M(1+t)^\nu \ (M > 0;\ \nu > 0)\}.$$

17

We note that, by setting $t = \frac{u}{1+u}$ in Equation (4), we get

$$B_p(x,y) = \int_0^\infty \frac{u^{x-1}}{(1+u)^{x+y}} \exp\left(-\frac{p(1+u)^2}{u}\right) du. \qquad (7)$$

So, if we take in consideration Equations (5) and (7) in the definition Equation (6), then we can say that the operators, S_n^*, are a generalization of the operators, T_n, given by Equation (2).

In this article, we investigate the moments of the general Szász-Mirakjan Beta-type operators S_n^* and find the rate of convergence with the help of the classical and second moduli of continuity. We also derive a Voronovskaja-type approximation theorem associated with these general operators, S_n^*.

2. A Set of Auxiliary Results

In this section, we give the moments of Szász-Mirakjan Beta-type operators, S_n^*, defined by Equation (6). We first recall for the S_n that

$$S_n(1,x) = 1, \quad S_n(t,x) = x \quad \text{and} \quad S_n(t^2,x) = x^2 + \frac{x}{n}, \qquad (8)$$

just as in [1].

Lemma 1. *The moments of the Szász-Mirakjan Beta-type operators, S_n^*, defined by (6) are given by*

$$S_n^*(1,x) = 1, \qquad (9)$$

$$S_n^*(t,x) = x + \frac{2}{n} \qquad (10)$$

and

$$S_n^*(t^2,x) = \frac{n}{n-1}x^2 + \frac{6}{n-1}x + \frac{6}{n(n-1)}. \qquad (11)$$

Proof. By using the known formulas in Equation (8), we find from the definition (6) that

$$S_n^*(1,x) = \sum_{k=0}^\infty e^{-nx} \frac{(nx)^k}{k!} \frac{1}{B(k+2,n+1)}$$

$$\cdot \int_0^\infty \int_0^\infty \frac{t^k}{(1+t)^{n+k+1}} \exp\left(-\frac{p(1+t)^2}{t}\right) dt\, dp$$

$$= \sum_{k=0}^\infty e^{-nx} \frac{(nx)^k}{k!} \frac{1}{B(k+2,n+1)} \int_0^\infty B_p(k+1,n)\, dp$$

$$= \sum_{k=0}^\infty e^{-nx} \frac{(nx)^k}{k!} \frac{B(k+2,n+1)}{B(k+2,n+1)}$$

$$= S_n(1,x) = 1.$$

For $n > 1$, we have

$$S_n^*(t, x) = \sum_{k=0}^{\infty} e^{-nx} \frac{(nx)^k}{k!} \frac{1}{B(k+2, n+1)}$$

$$\cdot \int_0^{\infty} \int_0^{\infty} \frac{t^{k+1}}{(1+t)^{n+k+1}} \exp\left(-\frac{p(1+t)^2}{t}\right) dt\, dp$$

$$= \sum_{k=0}^{\infty} e^{-nx} \frac{(nx)^k}{k!} \frac{1}{B(k+2, n+1)}$$

$$\cdot \int_0^{\infty} B_p(k+2, n-1)\, dp$$

$$= \sum_{k=0}^{\infty} e^{-nx} \frac{(nx)^k}{k!} \frac{B(k+3, n)}{B(k+2, n+1)}$$

$$= \sum_{k=0}^{\infty} e^{-nx} \frac{(nx)^k}{k!} \frac{\Gamma(k+3)\Gamma(n)}{\Gamma(n+k+3)} \frac{\Gamma(n+k+3)}{\Gamma(k+2)\Gamma(n+1)}$$

$$= \sum_{k=0}^{\infty} e^{-nx} \frac{(nx)^k}{k!} \frac{k+2}{n}$$

$$= S_n(t, x) + \frac{2}{n} S_n(1, x) = x + \frac{2}{n}$$

and, for $n > 2$, we find that

$$S_n^*(t^2, x) = \sum_{k=0}^{\infty} e^{-nx} \frac{(nx)^k}{k!} \frac{1}{B(k+2, n+1)}$$

$$\cdot \int_0^{\infty} \int_0^{\infty} \frac{t^{k+2}}{(1+t)^{n+k+1}} \exp\left(-\frac{p(1+t)^2}{t}\right) dt\, dp$$

$$= \sum_{k=0}^{\infty} e^{-nx} \frac{(nx)^k}{k!} \frac{1}{B(k+2, n+1)}$$

$$\cdot \int_{p=0}^{\infty} B_p(k+3, n-2)\, dp$$

$$= \sum_{k=0}^{\infty} e^{-nx} \frac{(nx)^k}{k!} \frac{B(k+4, n-1)}{B(k+2, n+1)}$$

$$= \sum_{k=0}^{\infty} e^{-nx} \frac{(nx)^k}{k!} \frac{\Gamma(k+4)\Gamma(n-1)}{\Gamma(n+k+3)} \frac{\Gamma(n+k+3)}{\Gamma(k+2)\Gamma(n+1)}$$

$$= \sum_{k=0}^{\infty} e^{-nx} \frac{(nx)^k}{k!} \frac{(k+3)(k+2)}{n(n-1)}$$

$$= \frac{n}{n-1} S_n(t^2, x) + \frac{5}{n-1} S_n(t, x) + \frac{6}{n(n-1)} S_n(1, x)$$

$$= \frac{n}{n-1} x^2 + \frac{6}{n-1} x + \frac{6}{n(n-1)}.$$

The proof of Lemma 1 is thus completed. □

Lemma 2. *The central moments of the Szász-Mirakjan Beta-type operators, S_n^*, defined by Equation (6) are given by*

$$S_n^*(t - x, x) = \frac{2}{n} \qquad (12)$$

and

$$S_n^*\left((t-x)^2,x\right) = \frac{1}{n-1}x^2 + \frac{2(n+2)}{n(n-1)}x + \frac{6}{n(n-1)} =: \varepsilon_n(x). \quad (13)$$

Proof. The assertions (12) and (13) of Lemma 2 follow easily from those of Lemma 1, so we omit the details involved. □

3. Local Approximation

Let $C_B[0,\infty)$ be the set of all real-valued continuous and bounded functions f on $[0,\infty)$, which is endowed with the norm given by

$$\|f\| = \sup_{x\in[0,\infty)} |f(x)|.$$

Then Peetre's K-functional is defined by

$$K_2(f;\delta) = \inf\left\{\|f-g\| + \delta\|g''\| : g \in C_B^2[0,\infty)\right\},$$

where

$$C_B^2[0,\infty) := \{g : g \in C_B[0,\infty) \text{ and } g',g'' \in C_B[0,\infty)\}.$$

There exists a positive constant $C > 0$ such that (see, for example, [24])

$$K_2(f,\delta) \leq C\omega_2\left(f,\sqrt{\delta}\right), \quad (14)$$

where $\delta > 0$ and ω_2 denotes the second-order modulus of smoothness for $f \in C_B[0,\infty)$, which is defined by

$$\omega_2\left(f;\sqrt{\delta}\right) = \sup_{0<h\leq\delta} \sup_{x\in[0,\infty)} |f(x+2h) - 2f(x+h) + f(x)|.$$

The usual modulus of continuity for $f \in C_B[0,\infty)$ is given by

$$\omega(f;\delta) = \sup_{0<h\leq\delta} \sup_{x\in[0,\infty)} |f(x+h) - f(x)|.$$

Lemma 3 below provides an auxiliary inequality which is useful in proving our next theorem (see Theorem 1).

Lemma 3. *For all $g \in C_B^2[0,\infty)$, it is asserted that*

$$\left|\widetilde{S}_n^*(g,x) - g(x)\right| \leq \delta_n(x)\|g''(x)\|, \quad (15)$$

where

$$\delta_n(x) = \frac{1}{n-1}x^2 + \frac{2(n+2)}{n(n-1)}x + \frac{2(5n-2)}{n^2(n-1)} \quad (16)$$

and

$$\widetilde{S}_n^*(f,x) = S_n^*(f,x) + f(x) - f\left(x + \frac{2}{n}\right) \quad (17)$$

for $f \in C_B[0,\infty)$.

Proof. First of all, we find from (17) that

$$\widetilde{S}_n^*(t-x,x) = S_n^*(t-x,x) - \frac{2}{n} = \frac{2}{n} - \frac{2}{n} = 0. \quad (18)$$

Now, by using the Taylor's formula, we have

$$g(t) - g(x) = (t-x)g'(x) + \int_x^t (t-u)g''(u)\,du,$$

which, in view of Equation (18), yields

$$\widetilde{S}_n^*(g,x) - g(x) = \widetilde{S}_n^*(t-x,x)g'(x) + \widetilde{S}_n^*\left(\int_x^t (t-u)g''(u)\,du, x\right)$$

$$= S_n^*\left(\int_x^t (t-u)g''(u)\,du, x\right) - \int_x^{x+\frac{2}{n}}\left(x+\frac{2}{n}-u\right)g''(u)\,du.$$

On the other hand, as

$$\left|\int_x^t (t-u)g''(u)\,du\right| \leq \int_x^t |t-u|\cdot|g''(u)|\,du$$

$$\leq \|g''\|\int_x^t |t-u|\,du$$

$$\leq (t-x)^2 \|g''\|$$

and

$$\left|\int_x^{x+\frac{2}{n}}\left(x+\frac{2}{n}-u\right)g''(u)\,du\right| \leq \left(\frac{2}{n}\right)^2 \|g''\|,$$

we conclude that

$$\left|\widetilde{S}_n^*(g,x) - g(x)\right| \leq S_n^*((t-x)^2, x)\|g''\| + \frac{4}{n^2}\|g''\|$$

$$= \left(\frac{1}{n-1}x^2 + \frac{2(n+2)}{n(n-1)}x + \frac{6}{n(n-1)} + \frac{4}{n^2}\right)\|g''\|$$

$$= \left(\frac{1}{n-1}x^2 + \frac{2(n+2)}{n(n-1)}x + \frac{2(5n-2)}{n^2(n-1)}\right)\|g''\| = \delta_n(x)\|g''\|.$$

This is the result asserted by Lemma 3. □

We now state and prove our main results in this section.

Theorem 1. *Let $f \in C_B[0,\infty)$. Then, for every $x \in [0,\infty)$, there exists a constant $L > 0$, such that*

$$|S_n^*(f,x) - f(x)| \leq L\omega_2\left(f; \sqrt{\delta_n(x)}\right) + \omega\left(f; \frac{2}{n}\right),$$

where $\omega_2(f;\delta)$ is the second-order modulus of smoothness, $\omega(f;\delta)$ is the usual modulus of continuity, and $\delta_n(x)$ is given by Equation (16).

Proof. We observe from Equation (17) that

$$|S_n^*(f,x) - f(x)| \leq \left|\widetilde{S}_n^*(f,x) - f(x)\right| + \left|f(x) - f\left(x+\frac{2}{n}\right)\right|$$

$$\leq \left|\widetilde{S}_n^*(f-g,x) - (f-g)(x)\right| + \left|f(x) - f\left(x+\frac{2}{n}\right)\right|$$

$$+ \left|\widetilde{S}_n^*(g,x) - g(x)\right|$$

for $g \in C_B^2[0,\infty)$. Thus, by applying Lemma 3 for $g \in C_B^2[0,\infty)$, we get

$$|S_n^*(f,x) - f(x)| \leq 4\|f-g\| + \delta_n(x)\|g''\| + \omega\left(f;\frac{2}{n}\right),$$

which, by taking the infimum on the right-hand side over all $g \in C_B^2[0,\infty)$ and using (14), yields

$$|S_n^*(f,x) - f(x)| \leq 4K_2(f;\delta_n(x)) + \omega\left(f;\frac{2}{n}\right)$$
$$\leq L\omega_2\left(f;\sqrt{\delta_n(x)}\right) + \omega\left(f;\frac{2}{n}\right).$$

where $L = 4M > 0$. This evidently completes the demonstration of Theorem 1. □

Theorem 2. *Let E be any bounded subset of the interval $[0,\infty)$, and suppose that $0 < \alpha \leq 1$. If $f \in C_B[0,\infty)$ is locally $\text{Lip}_M(\alpha)$, that is, if the following inequality holds true,*

$$|f(y) - f(x)| \leq M|y-x|^\alpha \quad (y \in E; \, x \in [0,\infty)),$$

then, for each $x \in [0,\infty)$,

$$|S_n^*(f,x) - f(x)| \leq M\big([\varepsilon_n(x)]^{\frac{\alpha}{2}} + 2[d(x,E)]^\alpha\big), \tag{19}$$

$\varepsilon_n(x)$ *is given by Equation* (13), *M is a constant depending on α and f, and $d(x,E)$ is the distance between x and E defined as follows:*

$$d(x,E) = \inf\{|y-x| : y \in E\}.$$

Proof. Let $\overline{E}$ denote the closure of E in $[0,\infty)$. Then there exists a point $x_0 \in \overline{E}$ such that

$$|x - x_0| = d(x,E).$$

By the above-mentioned definition of $\text{Lip}_M(\alpha)$, we get

$$|S_n^*(f,x) - f(x)| \leq S_n^*(|f(y) - f(x)|, x)$$
$$\leq S_n^*(|f(y) - f(x_0)|, x) + S_n^*(|f(x) - f(x_0)|, x)$$
$$\leq M\{S_n^*(|y-x_0|^\alpha, x) + |x-x_0|^\alpha\}$$
$$\leq M\{S_n^*(|y-x|^\alpha + |x-x_0|^\alpha, x) + |x-x_0|^\alpha\}$$
$$\leq M\{S_n^*(|y-x_0|^\alpha, x) + 2|x-x_0|^\alpha\}.$$

Now, if we use the Hölder inequality with

$$p = \frac{2}{\alpha} \quad \text{and} \quad q = \frac{2}{2-\alpha},$$

we find that

$$|S_n^*(f,x) - f(x)| \leq M\left([S_n^*((y-x_0)^2, x)]^{\frac{\alpha}{2}} + 2[d(x,E)]^\alpha\right)$$
$$= M\left([\varepsilon_n(x)]^{\frac{\alpha}{2}} + 2[d(x,E)]^\alpha\right).$$

We have thus completed our demonstration of the result asserted by Theorem 2. □

4. A Voronovskaja-Type Approximation Theorem

By applying Equations (5) to (7), as well as Lemma 1, we first prove the following result.

Lemma 4. *It is asserted that*

$$S_n^*(t^3, x) = \frac{n^2}{(n-1)(n-2)} x^3 + \frac{12n}{(n-1)(n-2)} x^2$$
$$+ \frac{36}{(n-1)(n-2)} x + \frac{24}{n(n-1)(n-2)} \quad (20)$$

and

$$S_n^*(t^4, x) = \frac{n^3}{(n-1)(n-2)(n-3)} x^4 + \frac{20n^2}{(n-1)(n-2)(n-3)} x^3$$
$$+ \frac{120n}{(n-1)(n-2)(n-3)} x^2 + \frac{240}{(n-1)(n-2)(n-3)} x \quad (21)$$
$$+ \frac{120}{n(n-1)(n-2)(n-3)}.$$

Furthermore, the following result holds true,

$$S_n^*((t-x)^4, x) = \frac{3(n+6)}{(n-1)(n-2)(n-3)} x^4 + \frac{4(3n^2 + 32n + 12)}{n(n-1)(n-2)(n-3)} x^3$$
$$+ \frac{12(n^2 + 21n + 18)}{n(n-1)(n-2)(n-3)} x^2 + \frac{144(n+2)}{n(n-1)(n-2)(n-3)} x \quad (22)$$
$$+ \frac{120}{n(n-1)(n-2)(n-3)}.$$

Proof. We begin by recalling the following moments of the Szász-Mirakjan operators,

$$S_n(t^3, x) = x^3 + \frac{3x^2}{n} + \frac{x}{n^2} \quad (23)$$

and

$$S_n(t^4, x) = x^4 + \frac{6x^3}{n} + \frac{7x^2}{n^2} + \frac{x}{n^3}. \quad (24)$$

Using Equations (8) and the above formulas (23) and (24), we thus find for $n > 3$ that

$$S_n^*(t^3, x) = \sum_{k=0}^{\infty} e^{-nx} \frac{(nx)^k}{k!} \frac{1}{B(k+2, n+1)}$$
$$\cdot \int_0^{\infty} \int_0^{\infty} \frac{t^{k+3}}{(1+t)^{n+k+1}} \exp\left(-\frac{p(1+t)^2}{t}\right) dt\, dp$$
$$= \sum_{k=0}^{\infty} e^{-nx} \frac{(nx)^k}{k!} \frac{1}{B(k+2, n+1)} \int_0^{\infty} B_p(k+4, n-3)\, dp$$
$$= \sum_{k=0}^{\infty} e^{-nx} \frac{(nx)^k}{k!} \frac{B(k+5, n-2)}{B(k+2, n+1)}$$
$$= \sum_{k=0}^{\infty} e^{-nx} \frac{(nx)^k}{k!} \frac{\Gamma(k+5)\Gamma(n-2)}{\Gamma(n+k+3)} \frac{\Gamma(n+k+3)}{\Gamma(k+2)\Gamma(n+1)}$$
$$= \sum_{k=0}^{\infty} e^{-nx} \frac{(nx)^k}{k!} \frac{(k+4)(k+3)(k+2)}{n(n-1)(n-2)}$$
$$= \frac{n^2}{(n-1)(n-2)} S_n(t^3, x) + \frac{9n}{(n-1)(n-2)} S_n(t^2, x)$$
$$+ \frac{26}{(n-1)(n-2)} S_n(t, x) + \frac{24}{n(n-1)(n-2)} S_n(1, x)$$
$$= \frac{n^2}{(n-1)(n-2)} \left(x^3 + \frac{3x^2}{n} + \frac{x}{n^2}\right) + \frac{9n}{(n-1)(n-2)} \left(x^2 + \frac{x}{n}\right)$$
$$+ \frac{26}{(n-1)(n-2)} x + \frac{24}{n(n-1)(n-2)}$$
$$= \frac{n^2}{(n-1)(n-2)} x^3 + \frac{12n}{(n-1)(n-2)} x^2 + \frac{36}{(n-1)(n-2)} x$$
$$+ \frac{24}{n(n-1)(n-2)}.$$

On the other hand, for $n > 4$, we find that

$$S_n^*(t^4, x) = \sum_{k=0}^{\infty} e^{-nx} \frac{(nx)^k}{k!} \frac{1}{B(k+2, n+1)}$$
$$\cdot \int_0^{\infty} \int_0^{\infty} \frac{t^{k+4}}{(1+t)^{n+k+1}} \exp\left(-\frac{p(1+t)^2}{t}\right) dt\, dp$$
$$= \sum_{k=0}^{\infty} e^{-nx} \frac{(nx)^k}{k!} \frac{1}{B(k+2, n+1)} \int_0^{\infty} B_p(k+5, n-4)\, dp$$
$$= \sum_{k=0}^{\infty} e^{-nx} \frac{(nx)^k}{k!} \frac{B(k+6, n-3)}{B(k+2, n+1)}$$
$$= \sum_{k=0}^{\infty} e^{-nx} \frac{(nx)^k}{k!} \frac{\Gamma(k+6)\Gamma(n-3)}{\Gamma(n+k+3)} \frac{\Gamma(n+k+3)}{\Gamma(k+2)\Gamma(n+1)}$$
$$= \sum_{k=0}^{\infty} e^{-nx} \frac{(nx)^k}{k!} \frac{(k+5)(k+4)(k+3)(k+2)}{n(n-1)(n-2)(n-3)},$$

that is, that

$$S_n^*(t^4,x) = \frac{n^3}{(n-1)(n-2)(n-3)} S_n(t^4,x) + \frac{14n^2}{(n-1)(n-2)(n-3)} S_n(t^3,x)$$
$$+ \frac{71n}{(n-1)(n-2)(n-3)} S_n(t^2,x) + \frac{154}{(n-1)(n-2)(n-3)} S_n(t,x)$$
$$+ \frac{120}{n(n-1)(n-2)(n-3)} S_n(1,x)$$
$$= \frac{n^3}{(n-1)(n-2)(n-3)} x^4 + \frac{20n^2}{(n-1)(n-2)(n-3)} x^3$$
$$+ \frac{120n}{(n-1)(n-2)(n-3)} x^2 + \frac{240}{(n-1)(n-2)(n-3)} x$$
$$+ \frac{120}{n(n-1)(n-2)(n-3)},$$

which, together, complete the proof of Lemma 4. □

Theorem 3. *Let* $f, f', f'' \in C_v[0,\infty)$ *for* $v \geq 4$. *Then, the following Voronovskaja-type approximation result holds true,*

$$\lim_{n \to \infty} \{n [S_n^*(f,x) - f(x)]\} = 2f'(x) + \left(\frac{x^2}{2} + x\right) f''(x). \tag{25}$$

Proof. By Taylor's expansion of $f(t)$ at the point $t = x$, we have

$$f(t) = f(x) + f'(x)(t-x) + \frac{1}{2} f''(x)(t-x)^2 + \Psi(t,x)(t-x)^2, \tag{26}$$

where $\Psi(t,x)$ is remainder term, $\Psi(\cdot,x) \in C_v[0,\infty)$ and $\Psi(t,x) \to 0$ as $t \to x$.

Applying the Szász-Mirakjan Beta-type operators S_n^* to Equation (26) and, using Lemma 2, we obtain

$$S_n^*(f,x) - f(x) = f'(x) S_n^*(t-x,x) + \frac{1}{2} f''(x) S_n^*((t-x)^2,x)$$
$$+ S_n^*(\Psi(t,x)(t-x)^2,x) \tag{27}$$
$$= \frac{2}{n} f'(x) + \frac{1}{2} \left(\frac{1}{n-1} x^2 + \frac{2(n+2)}{n(n-1)} x + \frac{6}{n(n-1)}\right) f''(x)$$
$$+ S_n^*(\Psi(t,x)(t-x)^2,x).$$

We now apply the Cauchy-Schwarz inequality to the third term on the right-hand side of Equation (27). We thus find that

$$n \left| S_n^*(\Psi(t,x)(t-x)^2,x) \right| \leq \sqrt{n^2 S_n^*((t-x)^4,x)} \cdot \sqrt{S_n^*([\Psi(t,x)]^2,x)}.$$

Let

$$\eta(t,x) := [\Psi(t,x)]^2.$$

In this case, we observe that $\eta(x,x) = 0$ and also that $\eta(\cdot,x) \in C_v[0,\infty)$. Then, it follows that

$$\lim_{n \to \infty} \{S_n^*([\Psi(t,x)]^2,x)\} = \lim_{n \to \infty} \{S_n^*(\eta(t,x),x)\} = \eta(x,x) = 0$$

uniformly with respect to $x \in [0, b]$ $(b > 0)$ and the following limit,

$$\lim_{n \to \infty} \left\{ n^2 S_n^*((t-x)^4, x) \right\}$$

is finite. Consequently, we have

$$\lim_{n \to \infty} \left\{ n S_n^*(\Psi(t,x) (t-x)^2, x) \right\} = 0.$$

Thus, in the limit when $n \to \infty$ in Equation (27), we obtain

$$\lim_{n \to \infty} \left\{ n \left[S_n^*(f, x) - f(x) \right] \right\} = 2 f'(x) + \left(\frac{x^2}{2} + x \right) f''(x).$$

The proof of Theorem 3 is thus completed. □

5. Concluding Remarks and Observations

We find it worthwhile to reiterate the fact that, in approximation theory and related fields, the approximation and some other basic properties of various linear and nonlinear operators are investigated because mainly of the potential for their usefulness in many areas of researches in the mathematical, physical, and engineering sciences. This article has been motivated essentially by the demonstrated applications of such results as those associated with various approximation operators.

In our present investigation, we have systematically studied a number of approximation properties of a class of the Szász-Mirakjan Beta-type operators, which we have introduced here by using an extension of the familiar Beta function $B(\alpha, \beta)$. We have established the moments of these extended Szász-Mirakjan Beta-type operators and estimated several convergence results with the help of the second modulus of smoothness and the classical modulus of continuity. We have also investigated convergence via functions belonging to the Lipschitz class. Finally, we have proved a Voronovskaja-type approximation theorem for the general Szász-Mirakjan Beta-type operators.

Using the other substantially more general forms of the classical Beta function $B(\alpha, \beta)$, which we have indicated in Section 1 of this article (see, for example, [22,23]), one can analogously develop further extensions and generalizations of the various results which we have presented here. In many of these suggested areas of further researches on the subject of this article, some other, possibly deeper, mathematical analytic tools and techniques will have to be called for.

Author Contributions: All three authors contributed equally to this investigation.

Funding: This research received no external funding.

Conflicts of Interest: The authors declare no conflict of interest.

References

1. Szász, O. Generalizations of S. Bernstein's polynomials to the infinite interval. *J. Res. Nat. Bur. Stand.* **1950**, *45*, 239–245. [CrossRef]
2. Mirakjan, G.M. Approximation des fonctions continues au moyen de polynômes de la forme $e^{-nx} \sum_{k=0}^{m_n} C_{k,n} x^k$. *C. R. (Doklady) Acad. Sci. URSS (New Ser.)* **1941**, *31*, 201–205.
3. Duman, O.; Özarslan, M.A. Szász-Mirakjan-type operators providing a better error estimation. *Appl. Math. Lett.* **2007**, *20*, 1184–1188. [CrossRef]
4. Gal, S.G. Approximation with an arbitrary order by generalized Szász-Mirakjan operators. *Stud. Univ. Babeş-Bolyai Math.* **2014**, *59*, 77–81.
5. Gupta, V.; Mahmudov, N. Approximation properties of the q-Szász-Mirakjan-Beta operators. *Indian J. Ind. Appl. Math.* **2012**, *3*, 41–53.
6. Gupta, V.; Noor, M.A. Convergence of derivatives for certain mixed Szász-Beta operators. *J. Math. Anal. Appl.* **2006**, *321*, 1–9. [CrossRef]

7. İçöz, G.; Mohapatra, R.N. Approximation properties by q-Durrmeyer-Stancu operators. *Anal. Theory Appl.* **2013**, *29*, 373–383.
8. Gupta, V.; Srivastava, H.M. A general family of the Srivastava-Gupta operators preserving linear functions. *Eur. J. Pure Appl. Math.* **2018**, *11*, 575–579. [CrossRef]
9. Gupta, V.; Srivastava, G.S.; Sahai, A. On simultaneous approximation by Szász-beta operators. *Soochow J. Math.* **1995**, *21*, 1–11.
10. Păltănea, R. Modified Szász-Mirakjan operators of integral form. *Carpathian J. Math.* **2008**, *24*, 378–385.
11. Srivastava, H.M.; Mursaleen, M.; Alotaibi, A.M.; Nasiruzzaman, M.; Al-Abied, A.A.H. Some approximation results involving the q-Szász-Mirakjan-Kantorovich type operators via Dunkl's generalization. *Math. Methods Appl. Sci.* **2017**, *40*, 5437–5452. [CrossRef]
12. Srivastava, H.M.; Özger, F.; Mohiuddine, S.A. Construction of Stancu-type Bernstein operators based on Bézier bases with shape parameter λ. *Symmetry* **2019**, *11*, 316. [CrossRef]
13. Srivastava, H.M.; Zeng, X.-M. Approximation by means of the Szász-Bézier integral operators. *Int. J. Pure Appl. Math.* **2004**, *14*, 283–294.
14. Xie, L.-S.; Xie, T.-F. Approximation theorems for localized Szász-Mirakjan operators. *J. Approx. Theory* **2008**, *152*, 125–134. [CrossRef]
15. Zeng, X.-M. Approximation of absolutely continuous functions by Stancu Beta operators. *Ukrainian Math. J.* **2012**, *63*, 1787–1794. [CrossRef]
16. Duman, O.; Özarslan, M.A.; Aktuğlu, H. Better error estimates for Szász-Mirakjan-Beta operators. *J. Comput. Anal. Appl.* **2008**, *10*, 53–59.
17. Özarslan, M.A.; Aktuğlu, H. A-Statistical approximation of generalized Szász-Mirakjan-Beta operators. *Appl. Math. Lett.* **2011**, *24*, 1785–1790. [CrossRef]
18. Qi, Q.-L.; Zhang, Y.-P. Pointwise approximation for certain mixed Szász-Beta operators. In *Further Progress in Analysis, Proceedings of the Sixth International Conference (ISAAC 2002) on Clifford Algebras and Their Applications in Mathematical Physics, Tennessee Technological University, Cookeville, TN, USA, 20–25 May 2002*; World Scientific Publishing Company: Singapore, 2009; pp. 152–164.
19. Chaudhry, M.A.; Zubair, S.M. Generalized incomplete gamma functions with applications. *J. Comput. Appl. Math.* **1994**, *55*, 99–124. [CrossRef]
20. Chaudhry, M.A.; Qadir, A.; Rafique, M.; Zubair, S.M. Extension of Euler's beta function. *J. Comput. Appl. Math.* **1997**, *78*, 19–32. [CrossRef]
21. Özergin, E.; Özarslan, M.A.; Altın, A. Extension of gamma, beta and hypergeometric functions. *J. Comput. Appl. Math.* **2011**, *235*, 4601–4610. [CrossRef]
22. Lin, S.-D.; Srivastava, H.M.; Yao, J.-C. Some classes of generating relations associated with a family of the generalized Gauss type hypergeometric functions. *Appl. Math. Inform. Sci.* **2015**, *9*, 1731–1738.
23. Srivastava, H.M.; Parmar, R.K.; Chopra, P. A class of extended fractional derivative operators and associated generating relations involving hypergeometric functions. *Axioms* **2012**, *1*, 238–258. [CrossRef]
24. DeVore, R.A.; Lorentz, G.G. *Constructive Approximation*; Springer: Berlin/Heidelberg, Germany; New York, NY, USA, 1993.

 © 2019 by the authors. Licensee MDPI, Basel, Switzerland. This article is an open access article distributed under the terms and conditions of the Creative Commons Attribution (CC BY) license (http://creativecommons.org/licenses/by/4.0/).

Article

Slice Holomorphic Functions in Several Variables with Bounded L-Index in Direction

Andriy Bandura [1],* and Oleh Skaskiv [2]

[1] Department of Advanced Mathematics, Ivano-Frankivsk National Technical University of Oil and Gas, 76019 Ivano-Frankivsk, Ukraine
[2] Department of Mechanics and Mathematics, Ivan Franko National University of Lviv, 79000 Lviv, Ukraine
* Correspondence: andriykopanytsia@gmail.com

Received: 25 June 2019; Accepted: 25 July 2019; Published: 26 July 2019

Abstract: In this paper, for a given direction $\mathbf{b} \in \mathbb{C}^n \setminus \{\mathbf{0}\}$ we investigate slice entire functions of several complex variables, i.e., we consider functions which are entire on a complex line $\{z^0 + t\mathbf{b} : t \in \mathbb{C}\}$ for any $z^0 \in \mathbb{C}^n$. Unlike to quaternionic analysis, we fix the direction $\mathbf{b}$. The usage of the term slice entire function is wider than in quaternionic analysis. It does not imply joint holomorphy. For example, it allows consideration of functions which are holomorphic in variable z_1 and continuous in variable z_2. For this class of functions there is introduced a concept of boundedness of L-index in the direction $\mathbf{b}$ where $L: \mathbb{C}^n \to \mathbb{R}_+$ is a positive continuous function. We present necessary and sufficient conditions of boundedness of L-index in the direction. In this paper, there are considered local behavior of directional derivatives and maximum modulus on a circle for functions from this class. Also, we show that every slice holomorphic and joint continuous function has bounded L-index in direction in any bounded domain and for any continuous function $L: \mathbb{C}^n \to \mathbb{R}_+$.

Keywords: bounded index; bounded L-index in direction; slice function; entire function; bounded l-index

MSC: 32A10; 32A17; 32A37; 30H99; 30A05

---

1. Introduction

In recent years, analytic functions of several variables with bounded index have been intensively investigated. The main objects of investigations are such function classes: entire functions of several variables [1–3], functions analytic in a polydisc [4], in a ball [5] or in the Cartesian product of the complex plane and the unit disc [6].

For entire functions and analytic function in a ball there were proposed two approaches to introduce a concept of index boundedness in a multidimensional complex space. They generate so-called functions of bounded L-index in a direction, and functions of bounded $\mathbf{L}$-index in joint variables.

Let us introduce some notations and definitions.

Let $\mathbb{R}_+ = (0, +\infty)$, $\mathbb{R}_+^* = [0, +\infty)$, $\mathbf{0} = (0, \ldots, 0)$, $\mathbf{b} = (b_1, \ldots, b_n) \in \mathbb{C}^n \setminus \{\mathbf{0}\}$ be a given direction, $L: \mathbb{C}^n \to \mathbb{R}_+$ be a continuous function, $F: \mathbb{C}^n \to \mathbb{C}$ an entire function. The slice functions on a line $\{z^0 + t\mathbf{b} : t \in \mathbb{C}\}$ for fixed $z^0 \in \mathbb{C}^n$ we will denote as $g_{z^0}(t) = F(z^0 + t\mathbf{b})$ and $l_{z^0}(t) = L(z^0 + t\mathbf{b})$.

Definition 1 ([7]). *An entire function $F: \mathbb{C}^n \to \mathbb{C}$ is called a function of bounded L-index in a direction $\mathbf{b}$, if there exists $m_0 \in \mathbb{Z}_+$ such that for every $m \in \mathbb{Z}_+$ and for all $z \in \mathbb{C}^n$ one has*

$$\frac{|\partial_{\mathbf{b}}^m F(z)|}{m! L^m(z)} \leq \max_{0 \leq k \leq m_0} \frac{|\partial_{\mathbf{b}}^k F(z)|}{k! L^k(z)}, \tag{1}$$

where $\partial_{\mathbf{b}}^0 F(z) = F(z), \partial_{\mathbf{b}} F(z) = \sum_{j=1}^{n} \frac{\partial F(z)}{\partial z_j} b_j, \partial_{\mathbf{b}}^k F(z) = \partial_{\mathbf{b}} \left(\partial_{\mathbf{b}}^{k-1} F(z) \right), k \geq 2.$

The least such integer number m_0, obeying (1), is called the L-index in the direction **b** of the function F and is denoted by $N_{\mathbf{b}}(F, L)$. If such m_0 does not exist, then we put $N_{\mathbf{b}}(F, L) = \infty$, and the function F is said to be of unbounded L-index in the direction **b** in this case. If $L(z) \equiv 1$, then the function F is said to be of bounded index in the direction **b** and $N_{\mathbf{b}}(F) = N_{\mathbf{b}}(F, 1)$ is called the index in the direction **b**. Let $l : \mathbb{C} \to \mathbb{R}_+$ be a continuous function. For $n = 1$, $\mathbf{b} = 1$, $L(z) \equiv l(z)$, $z \in \mathbb{C}$ inequality (1) defines a function of bounded l-index with the l-index $N(F, l) \equiv N_1(F, l)$ [8,9], and if in addition $l(z) \equiv 1$, then we obtain a definition of index boundedness with index $N(F) \equiv N_1(F, 1)$ [10,11]. It is also worth to mention paper [12], which introduces the concept of generalized index. It is quite close to the bounded l-index. Let $N_{\mathbf{b}}(F, L, z^0)$ stands for the L-index in the direction **b** of the function F at the point z^0, i.e., it is the least integer m_0, for which inequality (1) is satisfied at this point $z = z^0$. By analogy, the notation $N(f, l, z^0)$ is defined if $n = 1$, i.e., in the case of functions of one variable.

The concept of L-index boundedness in direction requires to consider a slice $\{z^0 + t\mathbf{b} : t \in \mathbb{C}\}$. We fixed $z^0 \in \mathbb{C}^n$ and used considerations from one-dimensional case. Then we construct uniform estimates above all z^0. This is a nutshell of the method.

In view of this, Prof S. Yu. Favorov (2015) posed the following **problem** in a conversation with one of the authors.

Problem 1 ([13]). *Let $\mathbf{b} \in \mathbb{C}^n \setminus \{0\}$ be a given direction, $L : \mathbb{C}^n \to \mathbb{R}_+$ be a continuous function. Is it possible to replace the condition "F is holomorphic in $\mathbb{C}^n$" by the condition "F is holomorphic on all slices $z^0 + t\mathbf{b}$" and to deduce all known properties of entire functions of bounded L-index in direction for this function class?*

There is a negative answer to Favorov's question [13]. This relaxation of restrictions by the function F does not allow the proving of some theorems. Here by $\overline{D}$ we denote a closure of domain D. There was proved the following proposition.

Proposition 1 ([13], Theorem 5). *For every direction $\mathbf{b} \in \mathbb{C}^n \setminus \{0\}$ there exists a function $F(z)$ and a bounded domain $D \subset \mathbb{C}^n$ with following properties:*

(1) *F is holomorphic function of bounded index on every slice $\{z^0 + t\mathbf{b} : t \in \mathbb{C}\}$ for each fixed $z^0 \in \mathbb{C}^n$;*
(2) *F is not entire function in $\mathbb{C}^n$;*
(3) *F does not satisfy (1) in $\overline{D}$, i.e., for any $p \in \mathbb{Z}_+$ there exists $m \in \mathbb{Z}_+$ and $z_p \in \overline{D}$*

$$\frac{|\partial_{\mathbf{b}}^m F(z_p)|}{m!} > \max \left\{ \frac{|\partial_{\mathbf{b}}^k F(z_p)|}{k!} : 0 \leq k \leq p \right\}.$$

Let D be a bounded domain in $\mathbb{C}^n$. If inequality (1) holds for all $z \in D$ instead $\mathbb{C}^n$, then F is called function of bounded L-index in the direction **b** in the domain D. The least such integer m_0 is called the L-index in the direction $\mathbf{b} \in \mathbb{C}^n \setminus \{0\}$ in the domain D and is denoted by $N_{\mathbf{b}}(F, L, D) = m_0$.

Proposition 2 ([13], Theorem 2). *Let D be a bounded domain in $\mathbb{C}^n$, $\mathbf{b} \in \mathbb{C}^n \setminus \{0\}$ be arbitrary direction. If $L : \mathbb{C}^n \to \mathbb{R}_+$ is continuous function and $F(z)$ is an entire function such that $(\forall z^0 \in \overline{D}) : F(z^0 + t\mathbf{b}) \not\equiv 0$, then $N_{\mathbf{b}}(F, L, D) < \infty$.*

Hence, if we replace holomorphy in $\mathbb{C}^n$ by holomorphy on the slices $\{z^0 + t\mathbf{b} : t \in \mathbb{C}\}$, then conclusion of Proposition 2 is not valid. Thus, Proposition 1 shows impossibility to replace joint holomorphy by slice holomorphy without additional hypothesis. The proof of Proposition 2 uses continuity in joint variables (see [13], Equation (6)). It leads to the following question (see [14], where it is also formulated. There was considered a case $L(z) \equiv 1$).

Problem 2. *What are additional conditions providing validity of Proposition 2 for slice holomorphic functions?*

A main goal of this investigation is to deduce an analog of Proposition 2 for slice holomorphic functions. Please note that the positivity and continuity of the function L are weak restrictions to deduce constructive results. Thus, we assume additional restrictions by the function L.

Let us denote

$$\lambda_{\mathbf{b}}(\eta) = \sup_{z \in \mathbb{C}^n} \sup_{t_1, t_2 \in \mathbb{C}} \left\{ \frac{L(z + t_1 \mathbf{b})}{L(z + t_2 \mathbf{b})} : |t_1 - t_2| \leq \frac{\eta}{\min\{L(z + t_1 \mathbf{b}), L(z + t_2 \mathbf{b})\}} \right\}.$$

By $Q_{\mathbf{b}}^n$ we denote a class of positive continuous function $L : \mathbb{C}^n \to \mathbb{R}_+$, satisfying the condition

$$(\forall \eta \geq 0) : \lambda_{\mathbf{b}}(\eta) < +\infty, \tag{2}$$

Moreover, it is sufficient to require validity of (2) for one value $\eta > 0$.

For a positive continuous function $l(t)$, $t \in \mathbb{C}$, and $\eta > 0$ we define $\lambda(\eta) \equiv \lambda_1^{\mathbf{b}}(\eta)$ in the cases when $\mathbf{b} = 1$, $n = 1$, $L \equiv l$. As in [15], let $Q \equiv Q_1^1$ be a class of positive continuous functions $l(t)$, $t \in \mathbb{C}$, obeying the condition $0 < \lambda(\eta) < +\infty$ for all $\eta > 0$.

Besides, we denote by $\langle a, c \rangle = \sum\limits_{j=1}^{n} a_j \overline{c_j}$ the scalar product in $\mathbb{C}^n$, where $a, c \in \mathbb{C}^n$.

Let $\widetilde{\mathcal{H}}_{\mathbf{b}}^n$ be a class of functions which are holomorphic on every slices $\{z^0 + t\mathbf{b} : t \in \mathbb{C}\}$ for each $z^0 \in \mathbb{C}^n$ and let $\mathcal{H}_{\mathbf{b}}^n$ be a class of functions from $\widetilde{\mathcal{H}}_{\mathbf{b}}^n$ which are joint continuous. The notation $\partial_{\mathbf{b}} F(z)$ stands for the derivative of the function $g_z(t)$ at the point 0, i.e., for every $p \in \mathbb{N}$ $\partial_{\mathbf{b}}^p F(z) = g_z^{(p)}(0)$, where $g_z(t) = F(z + t\mathbf{b})$ is entire function of complex variable $t \in \mathbb{C}$ for given $z \in \mathbb{C}^n$. In this research, we will often call this derivative as directional derivative because if F is entire function in $\mathbb{C}^n$ then the derivatives of the function $g_z(t)$ matches with directional derivatives of the function F.

Please note that if $F \in \mathcal{H}_{\mathbf{b}}^n$ then for every $p \in \mathbb{N}$ $\partial_{\mathbf{b}} F \in \mathcal{H}_{\mathbf{b}}^n$. It can be proved by using of Cauchy's formula.

Together the hypothesis on joint continuity and the hypothesis on holomorphy in one direction do not imply holomorphy in whole n-dimensional complex space. We give some examples to demonstrate it. For $n = 2$ let $f : \mathbb{C} \to \mathbb{C}$ be an entire function, $g : \mathbb{C} \to \mathbb{C}$ be a continuous function. Then $f(z_1)g(z_2)$, $f(z_1) \pm g(z_2)$, $f(z_1 \cdot g(z_2))$ are functions which are holomorphic in the direction $(1, 0)$ and are joint continuous in $\mathbb{C}^2$. Moreover, if we have performed an affine transformation

$$\begin{cases} z_1 = b_2 z_1' + b_1 z_2', \\ z_2 = b_2 z_1' - b_1 z_2' \end{cases}$$

then the appropriate new functions are also holomorphic in the direction (b_1, b_2) and are joint continuous in $\mathbb{C}^2$, where $b_1 \neq 0$, $b_2 \neq 0$.

A function $F \in \widetilde{\mathcal{H}}_{\mathbf{b}}^n$ is said to be of *bounded L-index in the direction* $\mathbf{b}$, if there exists $m_0 \in \mathbb{Z}_+$ such that for all $m \in \mathbb{Z}_+$ and each $z \in \mathbb{C}^n$ inequality (1) is true. All notations, introduced above for entire functions of bounded L-index in direction, keep for functions from $\widetilde{\mathcal{H}}_{\mathbf{b}}^n$.

2. Sufficient Sets

Now we prove several assertions that establish a connection between functions of bounded L-index in direction and functions of bounded l-index of one variable. The similar results for entire functions of several variables were obtained in [7,16]. The next proofs use ideas from the mentioned papers. The proofs of Propositions 3, 4 and Theorems 1, 2 literally repeat arguments from proofs of corresponding propositions for entire functions [7,16]. Therefore, we omit these proofs.

Proposition 3. *If a function $F \in \widetilde{\mathcal{H}}_\mathbf{b}^n$ has bounded L-index in the direction $\mathbf{b}$ then for every $z^0 \in \mathbb{C}^n$ the entire function $g_{z^0}(t)$ is of bounded l_{z^0}-index and $N(g_{z^0}, l_{z^0}) \leq N_\mathbf{b}(F, L)$.*

Proposition 4. *If a function $F \in \widetilde{\mathcal{H}}_\mathbf{b}^n$ has bounded L-index in the direction $\mathbf{b}$ then*

$$N_\mathbf{b}(F, L) = \max\left\{N(g_{z^0}, l_{z^0}) : z^0 \in \mathbb{C}^n\right\}.$$

Theorem 1. *A function $F \in \widetilde{\mathcal{H}}_\mathbf{b}^n$ has bounded L-index in the direction $\mathbf{b}$ if and only if there exists a number $M > 0$ such that for all $z^0 \in \mathbb{C}^n$ the function $g_{z^0}(t)$ is of bounded l_{z^0}-index with $N(g_{z^0}, l_{z^0}) \leq M < +\infty$, as a function of variable $t \in \mathbb{C}$. Thus, $N_\mathbf{b}(F, L) = \max\{N(g_{z^0}, l_{z^0}) : z^0 \in \mathbb{C}^n\}$.*

Theorem 2. *Let $\mathbf{b} \in \mathbb{C}^n \setminus \{\mathbf{0}\}$ be a given direction, $A_0 \subset \mathbb{C}^n$ such that $\{z + t\mathbf{b} : t \in \mathbb{C}, z \in A_0\} = \mathbb{C}^n$. A function $F \in \widetilde{\mathcal{H}}_\mathbf{b}^n$ has bounded L-index in the direction $\mathbf{b}$ if and only if there exists a number $M > 0$ such that for all $z^0 \in A_0$ the function $g_{z^0}(t)$ is of bounded l_{z^0}-index with $N(g_{z^0}, l_{z^0}) \leq M < +\infty$, as a function of one variable $t \in \mathbb{C}$ and $N_\mathbf{b}(F, L) = \max\{N(g_{z^0}, l_{z^0}) : z^0 \in A_0\}$.*

Remark 1. *An arbitrary hyperplane $A_0 = \{\tilde{z} \in \mathbb{C}^n : \langle \tilde{z}, c \rangle = 1\}$, where $\langle c, \mathbf{b} \rangle \neq 0$, satisfies conditions of Theorem 2.*

Corollary 1. *If $F \in \widetilde{\mathcal{H}}_\mathbf{b}^n$ is of bounded L-index in the direction $\mathbf{b}$ and j_0 is chosen such that $b_{j_0} \neq 0$, then $N_\mathbf{b}(F, L) = \max\{N(g_{z^0}, l_{z^0}) : z^0 \in \mathbb{C}^n, z_{j_0}^0 = 0\}$, and if $\sum_{j=0}^n b_j \neq 0$, then $N_\mathbf{b}(F, L) = \max\left\{N(g_{z^0}, l_{z^0}) : z^0 \in \mathbb{C}^n, \sum_{j=0}^n z_j^0 = 0\right\}$.*

We note that for a given $z \in \mathbb{C}^n$ the choice of $z^0 \in \mathbb{C}^n$ and $t \in \mathbb{C}$ such that $\sum_{j=1}^n z_j^0 = 0$ and $z = z^0 + t\mathbf{b}$, is unique.

Theorem 3 requires replacement of the space $\widetilde{\mathcal{H}}_\mathbf{b}^n$ by the space $\mathcal{H}_\mathbf{b}^n$. In other words, we use joint continuity in its proof.

Theorem 3. *Let $\overline{A} = \mathbb{C}^n$, i.e., A be an everywhere dense set in $\mathbb{C}^n$ and let a function $F \in \mathcal{H}_\mathbf{b}^n$. The function F is of bounded L-index in the direction $\mathbf{b}$ if and only if there exists $M > 0$ such that for all $z^0 \in A$ a function $g_{z^0}(t)$ is of bounded l_{z^0}-index $N(g_{z^0}, l_{z^0}) \leq M < +\infty$ and $N_\mathbf{b}(F, L) = \max\{N(g_{z^0}, l_{z^0}) : z^0 \in A\}$.*

Proof. The necessity follows from Theorem 1.

Sufficiency. Since $\overline{A} = \mathbb{C}^n$, then for every $z^0 \in \mathbb{C}^n$ there exists a sequence $z^{(m)}$, that $z^{(m)} \to z^0$ as $m \to +\infty$ and $z^{(m)} \in A$ for all $m \in \mathbb{N}$. However, $F(z + t\mathbf{b})$ is of bounded l_z-index for all $z \in \overline{A}$ as a function of variable t. That is why in view the definition of bounded l_z-index there exists $M > 0$ that for all $z \in A, t \in \mathbb{C}, p \in \mathbb{Z}_+$ $\frac{|g_z^{(p)}(t)|}{p! l^p(t)} \leq \max\left\{\frac{|g_z^{(k)}(t)|}{k! l_z^k(t)} : 0 \leq k \leq M\right\}.$

Substituting instead of z a sequence $z^{(m)} \in A$, $z^{(m)} \to z^0$, we obtain that for every $m \in \mathbb{N}$

$$\frac{|\partial_\mathbf{b}^p F(z^{(m)} + t\mathbf{b})|}{p! L^p(z^{(m)} + t\mathbf{b})} \leq \max\left\{\frac{|\partial_\mathbf{b}^k F(z^{(m)} + t\mathbf{b})|}{k! L^k(z^{(m)} + t\mathbf{b})} : 0 \leq k \leq M\right\}.$$

However, F and $\partial_{\mathbf{b}}^p F$ are continuous in $\mathbb{C}^n$ for all $p \in \mathbb{N}$ and L is a positive continuous function. Thus, in the obtained expression the limiting transition is possible as $m \to +\infty$ ($z^{(m)} \to z^0$). Evaluating the limit as $m \to +\infty$ we obtain that for all $z^0 \in \mathbb{C}^n, t \in \mathbb{C}, m \in \mathbb{Z}_+$

$$\frac{|\partial_{\mathbf{b}}^p F(z^0 + t\mathbf{b})|}{p! L^p(z^0 + t\mathbf{b})} \leq \max\left\{ \frac{|\partial_{\mathbf{b}}^k F(z^0 + t\mathbf{b})|}{k! L^k(z^0 + t\mathbf{b})} : 0 \leq k \leq M \right\}.$$

This inequality implies that $F(z + t\mathbf{b})$ is of bounded $L(z + t\mathbf{b})$-index as a function of variable t for every given $z \in \mathbb{C}^n$. Applying Theorem 1 we obtain the desired conclusion. Theorem 3 is proved. □

Remark 1 and Theorem 3 imply the following corollary.

Corollary 2. *Let $\mathbf{b} \in \mathbb{C}^n \setminus \{0\}$ be a given direction, $A_0 \subset \mathbb{C}^n$ such that its closure $\overline{A_0} = \{z \in \mathbb{C}^n : \langle z, c \rangle = 1\}$, where $\langle c, \mathbf{b} \rangle \neq 0$. And let a function $F \in \mathcal{H}_{\mathbf{b}}^n$ and its derivatives $\partial_{\mathbf{b}}^p F \in \mathcal{H}_{\mathbf{b}}^n$ for all $p \in \mathbb{N}$. The function $F(z)$ is of bounded L-index in the direction $\mathbf{b}$ if and only if there exists a number $M > 0$ such that for all $z^0 \in A_0$ the function $g_{z^0}(t)$ is of bounded l_{z^0}-index with $N(g_{z^0}, l_{z^0}) \leq M < +\infty$ and $N_{\mathbf{b}}(F, L) = \max\{N(g_{z^0}, l_{z^0}) : z^0 \in A_0\}$.*

3. Local Behavior of Directional Derivative

The following proposition is crucial in theory of functions of bounded index. It initializes series of propositions which are necessary to prove logarithmic criterion of index boundedness. It was first obtained by G. H. Fricke [17] for entire functions of bounded index. Later the proposition was generalized for entire functions of bounded l-index [18], analytic functions of bounded l-index [19], entire functions of bounded L-index in direction [7], functions analytic in a polydisc [4] or in a ball [5] with bounded **L**-index in joint variables,

Theorem 4. *Let $L \in Q_{\mathbf{b}}^n$. A function $F \in \tilde{\mathcal{H}}_{\mathbf{b}}^n$ is of bounded L-index in the direction $\mathbf{b}$ if and only if for each $\eta > 0$ there exist $n_0 = n_0(\eta) \in \mathbb{Z}_+$ and $P_1 = P_1(\eta) \geq 1$ such that for every $z \in \mathbb{C}^n$ there exists $k_0 = k_0(z) \in \mathbb{Z}_+, 0 \leq k_0 \leq n_0$, and*

$$\max\left\{ \left|\partial_{\mathbf{b}}^{k_0} F(z + t\mathbf{b})\right| : |t| \leq \frac{\eta}{L(z)} \right\} \leq P_1 \left|\partial_{\mathbf{b}}^{k_0} F(z)\right|. \tag{3}$$

Proof. Our proof is based on the proof of appropriate theorem for entire functions of bounded L-index in direction [7].

Necessity. Let $N_{\mathbf{b}}(F; L) \equiv N < +\infty$. Let $[a]$, $a \in \mathbb{R}$, stands for the integer part of the number a in this proof. We denote

$$q(\eta) = [2\eta(N+1)(\lambda_{\mathbf{b}}(\eta))^{2N+1}] + 1.$$

For $z \in \mathbb{C}^n$ and $p \in \{0, 1, \ldots, q(\eta)\}$ we put

$$R_p^{\mathbf{b}}(z, \eta) = \max\left\{ \frac{|\partial_{\mathbf{b}}^k F(z + t\mathbf{b})|}{k! L^k(z + t\mathbf{b})} : |t| \leq \frac{p\eta}{q(\eta)L(z)}, 0 \leq k \leq N \right\},$$

$$\tilde{R}_p^{\mathbf{b}}(z, \eta) = \max\left\{ \frac{|\partial_{\mathbf{b}}^k F(z + t\mathbf{b})|}{k! L^k(z)} : |t| \leq \frac{p\eta}{q(\eta)L(z)}, 0 \leq k \leq N \right\}.$$

However, $|t| \leq \frac{p\eta}{q(\eta)L(z)} \leq \frac{\eta}{L(z)}$, then $\lambda_\mathbf{b}\left(\frac{p\eta}{q(\eta)}\right) \leq \lambda_\mathbf{b}(\eta)$. It is clear that $R_p^\mathbf{b}(z,\eta)$, $\widetilde{R}_p^\mathbf{b}(z,\eta)$ are well-defined. Moreover,

$$\begin{aligned}
R_p^\mathbf{b}(z,\eta) &= \\
=\max&\left\{\frac{|\partial_\mathbf{b}^k F(z+t\mathbf{b})|}{k!L^k(z)}\left(\frac{L(z)}{L(z+t\mathbf{b})}\right)^k : 0 \leq k \leq N, |t| \leq \frac{p\eta}{q(\eta)L(z)}\right\} \leq \\
\leq \max&\left\{\frac{|\partial_\mathbf{b}^k F(z+t\mathbf{b})|}{k!L^k(z)}\left(\lambda_\mathbf{b}\left(\frac{p\eta}{q(\eta)}\right)\right)^k : |t| \leq \frac{p\eta}{q(\eta)L(z)}, 0 \leq k \leq N\right\} \leq \\
\leq \max&\left\{\frac{|\partial_\mathbf{b}^k F(z+t\mathbf{b})|}{k!L^k(z)}(\lambda_\mathbf{b}(\eta))^k : |t| \leq \frac{p\eta}{q(\eta)L(z)}, 0 \leq k \leq N\right\} \leq \\
\leq (\lambda_\mathbf{b}(\eta))^N&\max\left\{\frac{|\partial_\mathbf{b}^k F(z+t\mathbf{b})|}{k!L^k(z)} : |t| \leq \frac{p\eta}{q(\eta)L(z)}, 0 \leq k \leq N\right\} = \\
&= \widetilde{R}_p^\mathbf{b}(z,\eta)(\lambda_\mathbf{b}(\eta))^N,
\end{aligned} \quad (4)$$

$$\begin{aligned}
\widetilde{R}_p^\mathbf{b}(z,\eta) &= \\
=\max&\left\{\frac{|\partial_\mathbf{b}^k F(z+t\mathbf{b})|}{k!L^k(z+t\mathbf{b})}\left(\frac{L(z+t\mathbf{b})}{L(z)}\right)^k : |t| \leq \frac{p\eta}{q(\eta)L(z)}, 0 \leq k \leq N\right\} \leq \\
\leq \max&\left\{\frac{|\partial_\mathbf{b}^k F(z+t\mathbf{b})|}{k!L^k(z+t\mathbf{b})}\left(\lambda_\mathbf{b}\left(\frac{p\eta}{q(\eta)}\right)\right)^k : |t| \leq \frac{p\eta}{q(\eta)L(z)}, 0 \leq k \leq N\right\} \leq \\
\leq \max&\left\{(\lambda_\mathbf{b}(\eta))^k\frac{|\partial_\mathbf{b}^k F(z+t\mathbf{b})|}{k!L^k(z+t\mathbf{b})} : |t| \leq \frac{p\eta}{q(\eta)L(z)}, 0 \leq k \leq N\right\} \leq \\
\leq (\lambda_\mathbf{b}(\eta))^N &\max\left\{\frac{|\partial_\mathbf{b}^k F(z+t\mathbf{b})|}{k!L^k(z+t\mathbf{b})} : |t| \leq \frac{p\eta}{q(\eta)L(z)}, 0 \leq k \leq N\right\} = \\
&= R_p^\mathbf{b}(z,\eta)(\lambda_\mathbf{b}(\eta))^N.
\end{aligned} \quad (5)$$

Let $k_p^z \in \mathbb{Z}$, $0 \leq k_p^z \leq N$, and $t_p^z \in \mathbb{C}$, $|t_p^z| \leq \frac{p\eta}{q(\eta)L(z)}$, be such that

$$\widetilde{R}_p^\mathbf{b}(z,\eta) = \frac{|\partial_\mathbf{b}^{k_p^z} F(z+t_p^z\mathbf{b})|}{k_p^z!L^{k_p^z}(z)}. \quad (6)$$

However, for every given $z \in \mathbb{C}^n$ the function $F(z+t\mathbf{b})$ and its derivative are entire as functions of variables t. Then by the maximum modulus principle, equality (6) holds for t_p^z such that $|t_p^z| = \frac{p\eta}{q(\eta)L(z)}$. We set $\widetilde{t}_p^z = \frac{p-1}{p}t_p^z$. Then

$$|\widetilde{t}_p^z| = \frac{(p-1)\eta}{q(\eta)L(z)}, \quad (7)$$

$$|\widetilde{t}_p^z - t_p^z| = \frac{|t_p^z|}{p} = \frac{\eta}{q(\eta)L(z)}. \quad (8)$$

It follows from (7) and the definition of $\widetilde{R}_{p-1}^\mathbf{b}(z,\eta)$ that

$$\widetilde{R}_{p-1}^\mathbf{b}(z,\eta) \geq \frac{|\partial_\mathbf{b}^{k_p^z} F(z+\widetilde{t}_p^z\mathbf{b})|}{k_p^z!L^{k_p^z}(z)}.$$

Therefore,

$$\begin{aligned}
0 \leq \widetilde{R}_p^\mathbf{b}(z,\eta) - \widetilde{R}_{p-1}^\mathbf{b}(z,\eta) &\leq \frac{\left|\partial_\mathbf{b}^{k_p^z}F(z+t_p^z\mathbf{b})\right| - \left|\partial_\mathbf{b}^{k_p^z}F(z+\widetilde{t}_p^z\mathbf{b})\right|}{k_p^z!L^{k_p^z}(z)} = \\
&= \frac{1}{k_p^z!L^{k_p^z}(z)}\int_0^1 \frac{d}{ds}\left|\partial_\mathbf{b}^{k_p^z}F(z+(\widetilde{t}_p^z+s(t_p^z-\widetilde{t}_p^z))\mathbf{b})\right|ds.
\end{aligned} \quad (9)$$

For every analytic complex-valued function of real variable $\varphi(s)$, $s \in \mathbb{R}$, the inequality $\frac{d}{ds}|\varphi(s)| \leq \left|\frac{d}{ds}\varphi(s)\right|$ holds, where $\varphi(s) \neq 0$. Applying this inequality to (9) and using the mean value theorem we obtain

$$\widetilde{R}_p^{\mathbf{b}}(z, t_0, \eta) - \widetilde{R}_{p-1}^{\mathbf{b}}(z, t_0, \eta) \leq$$

$$\leq \frac{|t_p^z - \widetilde{t}_p^z|}{k_p^z! L^{k_p^z}(z)} \int_0^1 \left|\partial_{\mathbf{b}}^{k_p^z+1} F(z + (\widetilde{t}_p^z + s(t_p^z - \widetilde{t}_p^z))\mathbf{b})\right| ds =$$

$$= \frac{|t_p^z - \widetilde{t}_p^z|}{k_p^z! L^{k_p^z}(z)} \left|\partial_{\mathbf{b}}^{k_p^z+1} F(z + (\widetilde{t}_p^z + s^*(t_p^z - \widetilde{t}_p^z))\mathbf{b})\right| =$$

$$= L(z)(k_p^z+1)|t_p^z - \widetilde{t}_p^z| \frac{|\partial_{\mathbf{b}}^{k_p^z+1} F(z + (\widetilde{t}_p^z + s^*(t_p^z - \widetilde{t}_p^z))\mathbf{b})|}{(k_p^z+1)! L^{k_p^z+1}(z)},$$

where $s^* \in [0, 1]$. The point $\widetilde{t}_p^z + s^*(t_p^z - \widetilde{t}_p^z)$ belongs to the set

$$\left\{t \in \mathbb{C}: |t| \leq \frac{p\eta}{q(\eta)L(z)} \leq \frac{\eta}{L(z)}\right\}.$$

Using the definition of boundedness of L-index in direction, the definition of $q(\eta)$, inequalities (4) and (8), for $k_p^z \leq N$ we have

$$\widetilde{R}_p^{\mathbf{b}}(z, \eta) - \widetilde{R}_{p-1}^{\mathbf{b}}(z, \eta) \leq \frac{|\partial_{\mathbf{b}}^{k_p^z+1} F(z + (\widetilde{t}_p^z + s^*(t_p^z - \widetilde{t}_p^z))\mathbf{b})|}{(k_p^z+1)! L^{k_p^z+1}(z + (\widetilde{t}_p^z + s^*(t_p^z - \widetilde{t}_p^z))\mathbf{b})} \times$$

$$\times \left(\frac{L(z + (\widetilde{t}_p^z + s^*(t_p^z - \widetilde{t}_p^z))\mathbf{b})}{L(z)}\right)^{k_p^z+1} L(z)(k_p^z+1)|t_p^z - \widetilde{t}_p^z| \leq \eta \frac{N+1}{q(\eta)} (\lambda_{\mathbf{b}}(\eta))^{N+1} \times$$

$$\times \max\left\{\frac{|\partial_{\mathbf{b}}^k F(z + (\widetilde{t}_p^z + s^*(t_p^z - \widetilde{t}_p^z))\mathbf{b})|}{k! L^k(z + (\widetilde{t}_p^z + s^*(t_p^z - \widetilde{t}_p^z))\mathbf{b})} : 0 \leq k \leq N\right\} \leq \eta \frac{N+1}{q(\eta)} (\lambda_{\mathbf{b}}(\eta))^{N+1} R_p^{\mathbf{b}}(z, \eta) \leq$$

$$\leq \frac{\eta(N+1)(\lambda_{\mathbf{b}}(\eta))^{2N+1}}{[2\eta(N+1)(\lambda_{\mathbf{b}}(\eta))^{2N+1}]+1} \widetilde{R}_p^{\mathbf{b}}(z, \eta) \leq \frac{1}{2} \widetilde{R}_p^{\mathbf{b}}(z, \eta).$$

It follows that $\widetilde{R}_p^{\mathbf{b}}(z, \eta) \leq 2\widetilde{R}_{p-1}^{\mathbf{b}}(z, \eta)$. Using inequalities (4) and (5), we obtain for $R_p^{\mathbf{b}}(z, \eta)$

$$R_p^{\mathbf{b}}(z, \eta) \leq 2(\lambda_{\mathbf{b}}(\eta))^N \widetilde{R}_{p-1}^{\mathbf{b}}(z, \eta) \leq 2(\lambda_{\mathbf{b}}(\eta))^{2N} R_{p-1}^{\mathbf{b}}(z, \eta).$$

Hence,

$$\max\left\{\frac{|\partial_{\mathbf{b}}^k F(z+t\mathbf{b})|}{k! L^k(z+t\mathbf{b})} : |t| \leq \frac{\eta}{L(z)}, 0 \leq k \leq N\right\} = R_{q(\eta)}^{\mathbf{b}}(z, \eta) \leq$$
$$\leq 2(\lambda_{\mathbf{b}}(\eta))^{2N} R_{q(\eta)-1}^{\mathbf{b}}(z, \eta) \leq (2(\lambda_{\mathbf{b}}(\eta))^{2N})^2 R_{q(\eta)-2}^{\mathbf{b}}(z, \eta) \leq \quad (10)$$
$$\leq \cdots \leq (2(\lambda_{\mathbf{b}}(\eta))^{2N})^{q(\eta)} R_0^{\mathbf{b}}(z, \eta) =$$
$$= (2(\lambda_{\mathbf{b}}(\eta))^{2N})^{q(\eta)} \max\left\{\frac{|\partial_{\mathbf{b}}^k F(z)|}{k! L^k(z)} : 0 \leq k \leq N\right\}.$$

Let $k_z \in \mathbb{Z}$, $0 \leq k_z \leq N$, and $\widetilde{t}_z \in \mathbb{C}$, $|\widetilde{t}_z| = \frac{\eta}{L(z)}$ be such that

$$\frac{|\partial_{\mathbf{b}}^{k_z} F(z)|}{k_z! L^{k_z}(z)} = \max_{0 \leq k \leq N} \frac{|\partial_{\mathbf{b}}^k F(z)|}{k! L^k(z)},$$

and

$$|\partial_{\mathbf{b}}^{k_z} F(z + \widetilde{t}_z \mathbf{b})| = \max\{|\partial_{\mathbf{b}}^{k_z} F(z + t\mathbf{b})| : |t| \leq \eta/L(z)\}.$$

Inequality (10) implies

$$\frac{|\partial_{\mathbf{b}}^{k_z}F(z+\tilde{t}_z\mathbf{b})|}{k_z!L^{k_z}(z+\tilde{t}_z\mathbf{b})} \leq \max\left\{\frac{|\partial_{\mathbf{b}}^{k_z}F(z+t\mathbf{b})|}{k_z!L^{k_z}(z+t\mathbf{b})} : |t| = \frac{\eta}{L(z)}\right\} \leq$$

$$\leq \max\left\{\frac{|\partial_{\mathbf{b}}^{k}F(z+t\mathbf{b})|}{k!L^{k}(z+t\mathbf{b})} : |t| = \frac{\eta}{L(z)}, 0 \leq k \leq N\right\} \leq$$

$$\leq (2(\lambda_{\mathbf{b}}(\eta))^{2N})^{q(\eta)}\frac{|\partial_{\mathbf{b}}^{k_z}F(z)|}{k_z!L^{k_z}(z)}.$$

Hence,

$$\max\left\{|\partial_{\mathbf{b}}^{k_z}F(z+t\mathbf{b})| : |t| \leq \eta/L(z)\right\} \leq$$

$$\leq (2(\lambda_{\mathbf{b}}(\eta))^{2N})^{q(\eta)}\frac{L^{k_z}(z+\tilde{t}_z\mathbf{b})}{L^{k_z}(z)}|\partial_{\mathbf{b}}^{k_z}F(z)| \leq$$

$$\leq (2(\lambda_{\mathbf{b}}(\eta))^{2N})^{q(\eta)}(\lambda_{\mathbf{b}}(\eta))^{N}|\partial_{\mathbf{b}}^{k_z}F(z)| \leq$$

$$\leq (2(\lambda_{\mathbf{b}}(\eta))^{2N})^{q(\eta)}(\lambda_{\mathbf{b}}(\eta))^{N}|\partial_{\mathbf{b}}^{k_z}F(z)|.$$

Thus, we obtain (3) with $n_0 = N_{\mathbf{b}}(F, L)$ and

$$P_1(\eta) = (2(\lambda_{\mathbf{b}}(\eta))^{2N})^{q(\eta)}(\lambda_{\mathbf{b}}(\eta))^{N} > 1.$$

Sufficiency. Suppose that for each $\eta > 0$ there exist $n_0 = n_0(\eta) \in \mathbb{Z}_+$ and $P_1 = P_1(\eta) \geq 1$ such that for every $z \in \mathbb{C}^n$ there exists $k_0 = k_0(z) \in \mathbb{Z}_+, 0 \leq k_0 \leq n_0$, for which inequality (3) holds. We choose $\eta > 1$ and $j_0 \in \mathbb{N}$ such that $P_1 \leq \eta^{j_0}$. For given $z \in \mathbb{C}^n$, $k_0 = k_0(z)$ and $j \geq j_0$ by Cauchy's formula for $F(z + t\mathbf{b})$ as a function of one variable t

$$\partial_{\mathbf{b}}^{k_0+j}F(z) = \frac{j!}{2\pi i}\int_{|t|=\eta/L(z)}\frac{\partial_{\mathbf{b}}^{k_0}F(z+t\mathbf{b})}{t^{j+1}}dt.$$

Therefore, in view of (3) we have

$$\frac{|\partial_{\mathbf{b}}^{k_0+j}F(z)|}{j!} \leq \frac{L^{j}(z)}{\eta^j}\max\left\{|\partial_{\mathbf{b}}^{k_0}F(z+t\mathbf{b})| : |t| = \frac{\eta}{L(z)}\right\} \leq P_1\frac{L^{j}(z)}{\eta^j}|\partial_{\mathbf{b}}^{k_0}F(z)|,$$

Hence, for all $j \geq j_0, z \in \mathbb{C}^n$

$$\frac{|\partial_{\mathbf{b}}^{k_0+j}F(z)|}{(k_0+j)!L^{k_0+j}(z)} \leq \frac{j!k_0!}{(j+k_0)!}\frac{P_1}{\eta^j}\frac{|\partial_{\mathbf{b}}^{k_0}F(z)|}{k_0!L^{k_0}(z)} \leq \eta^{j_0-j}\frac{|\partial_{\mathbf{b}}^{k_0}F(z)|}{k_0!L^{k_0}(z)} \leq \frac{|\partial_{\mathbf{b}}^{k_0}F(z)|}{k_0!L^{k_0}(z)}.$$

Since $k_0 \leq n_0$, the numbers $n_0 = n_0(\eta)$ and $j_0 = j_0(\eta)$ are independent of z and t_0, this inequality means that a function F has bounded L-index in the direction $\mathbf{b}$ and $N_{\mathbf{b}}(F, L) \leq n_0 + j_0$. The proof of Theorem 4 is complete. □

Theorem 4 implies the next proposition that describes the boundedness of L-index in direction for an equivalent function to L. Let $L^*(z)$ be a positive continuous function in $\mathbb{C}^n$. We denote $L \asymp L^*$, if for some $\theta_1, \theta_2, 0 < \theta_1 \leq \theta_2 < +\infty$, and for all $z \in \mathbb{C}^n$ the following inequalities hold $\theta_1 L(z) \leq L^*(z) \leq \theta_2 L(z)$.

Proposition 5. *Let $L \in Q_{\mathbf{b}}^n$, $L \asymp L^*$. A function $F \in \widetilde{\mathcal{H}}_{\mathbf{b}}^n$ has bounded L^*-index in the direction $\mathbf{b}$ if and only if F is of bounded L-index in the direction $\mathbf{b}$.*

Proof. First, it is not easy to check that the function L^* also belongs to the class $Q_{\mathbf{b}}^n$. Let $N_{\mathbf{b}}(F, L^*) < +\infty$. Then by Theorem 4 for every $\eta^* > 0$ there exist $n_0(\eta^*) \in \mathbb{Z}_+$ and $P_1(\eta^*) \geq 1$ such that for every $z \in \mathbb{C}^n$ and some k_0, $0 \leq k_0 \leq n_0$, inequality (3) holds with L^* and η^* instead of L and η. But the condition $L \asymp L^*$ means that for some $\theta_1, \theta_2 \in \mathbb{R}_+$, $0 < \theta_1 \leq \theta_2 < +\infty$ and for all $z \in \mathbb{C}^n$ the double inequality holds $\theta_1 L(z) \leq L^*(z) \leq \theta_2 L(z)$. Taking $\eta* = \theta_2 \eta$ we obtain

$$P_1|\partial_{\mathbf{b}}^{k_0} F(z)| \geq \max\left\{|\partial_{\mathbf{b}}^{k_0} F(z+t\mathbf{b})| : |t| \leq \eta^*/L^*(z)\right\} \geq$$

$$\geq \max\left\{|\partial_{\mathbf{b}}^{k_0} F(z+t\mathbf{b})| : |t| \leq \eta/L(z)\right\}.$$

Thus, by Theorem 4 in view of arbitrariness of η^* the function $F(z)$ has bounded L-index in the direction $\mathbf{b}$. We can obtain the converse proposition by replacing L with L^*. □

Please note that Proposition 5 can be slightly refined. The following proposition is easy deduced from (1).

Proposition 6. *Let $L_1(z)$, $L_2(z)$ be positive continuous functions, $F \in \widetilde{\mathcal{H}}_{\mathbf{b}}^n$ be a function of bounded L_1-index in the direction $\mathbf{b}$, for all $z \in \mathbb{C}^n$ the inequality $L_1(z) \leq L_2(z)$ holds. Then $N_{\mathbf{b}}(L_2, F) \leq N_{\mathbf{b}}(L_1, F)$.*

Using Fricke's idea [20], we obtain modification of Theorem 4.

Theorem 5. *Let $L \in Q_{\mathbf{b}}^n$. If there exist $\eta > 0$, $n_0 = n_0(\eta) \in \mathbb{Z}_+$ and $P_1 = P_1(\eta) \geq 1$ such that for all $z \in \mathbb{C}^n$ there exists $k_0 = k_0(z) \in \mathbb{Z}_+$, $0 \leq k_0 \leq n_0$, for which the inequality holds*

$$\max\{|\partial_{\mathbf{b}}^{k_0} F(z+t\mathbf{b})| : |t| \leq \eta/L(z)\} \leq P_1|\partial_{\mathbf{b}}^{k_0} F(z)|,$$

then the function $F \in \widetilde{\mathcal{H}}_{\mathbf{b}}^n$ has bounded L-index in the direction $\mathbf{b} \in \mathbb{C}^n \setminus \{0\}$.

Proof. Our proof is based on the proof of appropriate theorem for entire functions of bounded L-index in direction [21].

Assume that there exist $\eta > 0$, $n_0 = n_0(\eta) \in \mathbb{Z}_+$ and $P_1 = P_1(\eta) \geq 1$ such that for every $z \in \mathbb{C}^n$ there exists $k_0 = k_0(z) \in \mathbb{Z}_+$, $0 \leq k_0 \leq n_0$, for which

$$\max\{|\partial_{\mathbf{b}}^{k_0} F(z+t\mathbf{b})| : |t| \leq \frac{\eta}{L(z)}\} \leq P_1|\partial_{\mathbf{b}}^{k_0} F(z)|. \tag{11}$$

If $\eta > 1$, then we choose $j_0 \in \mathbb{N}$ such that $P_1 \leq \eta^{j_0}$. And for $\eta \in (0; 1]$ we choose $j_0 \in \mathbb{N}$ obeying the inequality $\frac{j_0! k_0!}{(j_0+k_0)!} P_1 < 1$. This j_0 exists because

$$\frac{j_0! k_0!}{(j_0+k_0)!} P_1 = \frac{k_0!}{(j_0+1)(j_0+2)\cdot \ldots \cdot (j_0+k_0)} P_1 \to 0, \ j_0 \to \infty.$$

Applying Cauchy's formula to the function $F(z+t\mathbf{b})$ as function of complex variable t for $j \geq j_0$ we obtain that for every $z \in \mathbb{C}^n$ there exists integer $k_0 = k_0(z)$, $0 \leq k_0 \leq n_0$, and

$$\partial_{\mathbf{b}}^{k_0+j} F(z) = \frac{j!}{2\pi i} \int_{|t|=\frac{\eta}{L(z)}} \frac{\partial_{\mathbf{b}}^{k_0} F(z+t\mathbf{b})}{t^{j+1}} dt.$$

Taking into account (11), one has

$$\frac{|\partial_{\mathbf{b}}^{k_0+j} F(z)|}{j!} \leq \frac{L^j(z)}{\eta^j} \max\left\{|\partial_{\mathbf{b}}^{k_0} F(z+t\mathbf{b})| : |t| = \frac{\eta}{L(z)}\right\} \leq P_1 \frac{L^j(z)}{\eta^j} |\partial_{\mathbf{b}}^{k_0} F(z)|. \tag{12}$$

In view of choice j_0 for $\eta > 1$ and for all $j \geq j_0$ we deduce

$$\frac{|\partial_{\mathbf{b}}^{k_0+j}F(z)|}{(k_0+j)!L^{k_0+j}(z)} \leq \frac{j!k_0!}{(j+k_0)!}\frac{P_1}{\eta^j}\frac{|\partial_{\mathbf{b}}^{k_0}F(z)|}{k_0!L^{k_0}(z+t_0\mathbf{b})} \leq \eta^{j_0-j}\frac{|\partial_{\mathbf{b}}^{k_0}F(z)|}{k_0!L^{k_0}(z)} \leq \frac{|\partial_{\mathbf{b}}^{k_0}F(z)|}{k_0!L^{k_0}(z)}.$$

Since $k_0 \leq n_0$, the numbers $n_0 = n_0(\eta)$ and $j_0 = j_0(\eta)$ are independent of z, and $z \in \mathbb{C}^n$ is arbitrary, the last inequality means that the function F is of bounded L-index in the direction $\mathbf{b}$ and $N_{\mathbf{b}}(F, L) \leq n_0 + j_0$.

If $\eta \in (0, 1)$, then (12) implies for all $j \geq j_0$

$$\frac{|\partial_{\mathbf{b}}^{k_0+j}F(z)|}{(k_0+j)!L^{k_0+j}(z)} \leq \frac{j!k_0!P_1}{(j+k_0)!}\frac{|\partial_{\mathbf{b}}^{k_0}F(z)|}{\eta^j k_0!L^{k_0}(z)} \leq \frac{|\partial_{\mathbf{b}}^{k_0}F(z)|}{\eta^j k_0!L^{k_0}(z)}$$

or in view of the choice of j_0

$$\frac{|\partial_{\mathbf{b}}^{k_0+j}F(z)|}{(k_0+j)!}\frac{\eta^{k_0+j}}{L^{k_0+j}(z)} \leq \frac{|\partial_{\mathbf{b}}^{k_0}F(z)|}{k_0!}\frac{\eta^{k_0}}{L^{k_0}(z)}.$$

Thus, the function F has bounded $\tilde{L}$-index in the direction $\mathbf{b}$, where $\tilde{L}(z) = \frac{L(z)}{\eta}$. Then by Proposition 5 the function F is of bounded L-index in the direction $\mathbf{b}$. Theorem is proved. □

4. Bounded Index in Direction in Bounded Domain

Let D be a bounded domain in $\mathbb{C}^n$. If inequality (1) is fulfilled for all $z \in D$ instead $\mathbb{C}^n$, then F is called *function of bounded L-index in the direction $\mathbf{b}$ in the domain D*. The least such integer m_0 is called *the L-index in the direction $\mathbf{b}$ in the domain D* and is denoted by $N_{\mathbf{b}}(F, L, D) = m_0$. By $\overline{D}$ we denote a closure of domain D.

Theorem 6. *Let D be an arbitrary bounded domain in $\mathbb{C}^n$, $\mathbf{b} \in \mathbb{C}^n \setminus \{0\}$ be arbitrary direction. If $L: \mathbb{C}^n \to \mathbb{R}_+$ is continuous function, $F \in \mathcal{H}_{\mathbf{b}}^n$ and $(\forall p \in \mathbb{N})\ \partial_{\mathbf{b}}^p F \in \mathcal{H}_{\mathbf{b}}^n$ and $(\forall z^0 \in \overline{D})$: $F(z^0 + t\mathbf{b}) \not\equiv 0$, then $N_{\mathbf{b}}(F, L, \overline{D}) < \infty$.*

Proof. Proof of this theorem is similar to proof of corresponding lemma in [13]. For every given $z^0 \in \overline{D}$ we develop the entire function $F(z^0 + t\mathbf{b})$ in power series by powers t

$$F(z^0 + t\mathbf{b}) = \sum_{m=0}^{\infty} \frac{\partial_{\mathbf{b}}^m F(z^0)}{m!} t^m \tag{13}$$

in the disc $\{t \in \mathbb{C}: |t| \leq \frac{1}{L(z^0)}\}$.

The quantity $\frac{|\partial_{\mathbf{b}}^m F(z^0)|}{m!}$ is the modulus of coefficient of power series (13) at the point $t = 0$. Substitute $t = \frac{1}{L(z^0)}$. Since $F \in \mathcal{H}_{\mathbf{b}}^n$, for every $z_0 \in \overline{D}$

$$\frac{|\partial_{\mathbf{b}}^m F(z^0)|}{m!L^m(z^0)} \to 0 \quad (m \to \infty),$$

i.e., there exists $m_0 = m(z^0, \mathbf{b})$ such that inequality (1) holds at the point $z = z^0$ for all $m \in \mathbb{Z}_+$.

We will show that $\sup\{m_0: z^0 \in \overline{D}\} < +\infty$. On the contrary, we suppose that the set of all values m_0 is unbounded in z^0, that is $\sup\{m_0: z^0 \in \overline{D}\} = +\infty$. Hence, for every $m \in \mathbb{Z}_+$ there exists $z^{(m)} \in \overline{D}$ and $p_m > m$

$$\frac{|\partial_{\mathbf{b}}^{p_m} F(z^{(m)})|}{p_m!L^{p_m}(z^{(m)})} > \max\left\{\frac{|\partial_{\mathbf{b}}^k F(z^{(m)})|}{k!L^k(z^{(m)})}: 0 \leq k \leq m\right\}. \tag{14}$$

Since $\{z^{(m)}\} \subset \overline{D}$, there exists subsequence $z'^{(m)} \to z' \in \overline{G}$ as $m \to +\infty$. By Cauchy's formula one has

$$\frac{\partial_\mathbf{b}^p F(z)}{p!} = \frac{1}{2\pi i} \int_{|t|=r} \frac{F(z+t\mathbf{b})}{t^{p+1}} dt$$

for any $p \in \mathbb{N}$, $z \in D$. Rewrite (14) in the form

$$\max\left\{\frac{|\partial_\mathbf{b}^k F(z^{(m)})|}{k! L^k(z^{(m)})} : 0 \leq k \leq m\right\} < \\ < \frac{1}{L^{p_m}(z^{(m)})} \int_{|t|=r/L(z^{(m)})} \frac{|F(z^{(m)}+t\mathbf{b})|}{|t|^{p_m+1}} |dt| \leq \frac{1}{r^{p_m}} \max\{|F(z)| : z \in D_r\}, \quad (15)$$

where $D_r = \bigcup_{z^* \in \overline{D}} \{z \in \mathbb{C}^n : |z - z^*| \leq \frac{|\mathbf{b}|r}{L(z^*)}\}$. We can choose $r > 1$, because $F \in \mathcal{H}_\mathbf{b}^n$. Evaluating limit for every directional derivative of fixed order in (15) as $m \to \infty$ we obtain

$$(\forall k \in \mathbb{Z}_+) : \quad \frac{|\partial_\mathbf{b}^k F(z')|}{k! L^k(z')} \leq \varlimsup_{m \to \infty} \frac{1}{r^{p_m}} \max\{|F(z)| : z \in D_r\} \leq 0.$$

The passing to the limit is possible because $\partial_\mathbf{b}^k F$ is joint continuous. Thus, all derivatives in the direction $\mathbf{b}$ of the function F at the point z' equal 0 and $F(z') = 0$. In view of (13) $F(z' + t\mathbf{b}) \equiv 0$. This is a contradiction. □

5. Conclusions

The proposed approach can be applied in analytic theory of differential equations. It is known that concept of bounded index allows the investigation of properties of analytic solutions of linear higher-order differential equations with analytic coefficients. Therefore, it leads to the question of what the additional conditions are, providing index boundedness of every slice holomorphic solutions for linear higher-order directional derivative equations with slice holomorphic coefficients? In other words, is joint continuity a sufficient condition?

Since there are known analogs of Cauchy's formula for quaternionic variables and for Clifford algebras, the authors assume that The results in this paper can be generalized in these cases, i.e., in the case of slice holomorphic functions of quaternionic variable.

Author Contributions: These authors contributed equally to this work.

Funding: This research received no external funding.

Conflicts of Interest: The authors declare no conflict of interest.

References

1. Bandura, A.; Skaskiv, O. Asymptotic estimates of entire functions of bounded **L**-index in joint variables. *Novi Sad J. Math.* **2018**, *48*, 103–116. [CrossRef]
2. Nuray, F.; Patterson, R.F. Multivalence of bivariate functions of bounded index. *Le Matematiche* **2015**, *70*, 225–233. [CrossRef]
3. Nuray, F.; Patterson, R.F. Vector-valued bivariate entire functions of bounded index satisfying a system of differential equations. *Mat. Stud.* **2018**, *49*, 67–74. [CrossRef]
4. Bandura, A.; Petrechko, N.; Skaskiv, O. Maximum modulus in a bidisc of analytic functions of bounded L-index and an analogue of Hayman's theorem. *Mat. Bohem.* **2018**, *143*, 339–354. [CrossRef]
5. Bandura, A.; Skaskiv, O. Sufficient conditions of boundedness of **L**-index and analog of Hayman's Theorem for analytic functions in a ball. *Stud. Univ. Babeş-Bolyai Math.* **2018**, *63*, 483–501. [CrossRef]
6. Bandura, A.I.; Skaskiv, O.B.; Tsvigun, V.L. Some characteristic properties of analytic functions in $\mathbb{D} \times \mathbb{C}$ of bounded **L**-index in joint variables. *Bukovyn. Mat. Zh.* **2018**, *6*, 21–31.
7. Bandura, A.I.; Skaskiv, O.B. Entire functions of bounded L-index in direction. *Mat. Stud.* **2007**, *27*, 30–52. (In Ukrainian)

8. Kuzyk, A.D.; Sheremeta, M.N. Entire functions of bounded l-distribution of values. *Math. Notes* **1986**, *39*, 3–8. [CrossRef]
9. Kuzyk, A.D.; Sheremeta, M.N. On entire functions, satisfying linear differential equations. *Differ. Equ.* **1990**, *26*, 1716–1722.
10. Lepson, B. Differential equations of infinite order, hyperdirichlet series and entire functions of bounded index. *Proc. Symp. Pure Math.* **1968**, *2*, 298–307.
11. Macdonnell, J.J. Some Convergence Theorems for Dirichlet-Type Series Whose Coefficients are Entire Functions of Bounded Index. Ph.D. Thesis, Catholic University of America, Washington, DC, USA, 1957.
12. Strelitz, S. Asymptotic properties of entire transcendental solutions of algebraic differential equations. *Contemp. Math.* **1983**, *25*, 171. [CrossRef]
13. Bandura, A.I. Sum of entire functions of bounded L-index in direction. *Mat. Stud.* **2016**, *45*, 149–158. [CrossRef]
14. Kosanyak, M. Slice Entire Functions of Bounded Index in Direction, Master's Thesis, Ivan Franko National University of Lviv, Lviv, Ukraine, 2018.
15. Sheremeta, M. *Analytic Functions of Bounded Index*; VNTL Publishers: Lviv, Ukraine, 1999; 141p.
16. Bandura, A.I.; Skaskiv, O.B. Sufficient sets for boundedness L-index in direction for entire functions. *Mat. Stud.* **2008**, *30*, 177–182.
17. Fricke, G.H. Functions of bounded index and their logarithmic derivatives. *Math. Ann.* **1973**, *206*, 215–223. [CrossRef]
18. Sheremeta, M.N.; Kuzyk, A.D. Logarithmic derivative and zeros of an entire function of bounded l-index. *Sib. Math. J.* **1992**, *33*, 304–312. [CrossRef]
19. Kushnir, V.O.; Sheremeta, M.M. Analytic functions of bounded l-index. *Mat. Stud.* **1999**, *12*, 59–66.
20. Fricke, G.H. Entire functions of locally slow growth. *J. Anal. Math.* **1975**, *28*, 101–122. [CrossRef]
21. Bandura, A.I. A modified criterion of boundedness of L-index in direction. *Mat. Stud.* **2013**, *39*, 99–102.

© 2019 by the authors. Licensee MDPI, Basel, Switzerland. This article is an open access article distributed under the terms and conditions of the Creative Commons Attribution (CC BY) license (http://creativecommons.org/licenses/by/4.0/).

Article

On a New Class of Laplace-Type Integrals Involving Generalized Hypergeometric Functions

Wolfram Koepf [1], Insuk Kim [2,*] and Arjun K. Rathie [3]

[1] Department of Mathematics, University of Kassel, Heinrich-Plett-Str. 40, D-34132 Kassel, Germany
[2] Department of Mathematics Education, Wonkwang University, Iksan 570-749, Korea
[3] Department of Mathematics, Vedant College of Engineering and Technology (Rajasthan Technical University), Bundi 323021, Rajasthan, India
* Correspondence: iki@wku.ac.kr

Received: 30 June 2019; Accepted: 20 July 2019; Published: 26 July 2019

Abstract: In the theory of generalized hypergeometric functions, classical summation theorems for the series $_2F_1$, $_3F_2$, $_4F_3$, $_5F_4$ and $_7F_6$ play a key role. Very recently, Masjed-Jamei and Koepf established generalizations of the above-mentioned summation theorems. Inspired by their work, the main objective of the paper is to provide a new class of Laplace-type integrals involving generalized hypergeometric functions $_pF_p$ for $p = 2, 3, 4, 5$ and 7 in the most general forms. Several new and known cases have also been obtained as special cases of our main findings.

Keywords: generalized hypergeometric functions; classical summation theorems; generalization; laplace transforms

MSC: 33C20; 33C05; 33C90

---

1. Introduction

The generalized hypergeometric function with p numerator and q denominator parameters is defined [1–4] as

$$_pF_q\left[\begin{matrix} a_1, & \cdots, & a_p \\ b_1, & \cdots, & b_q \end{matrix}; z\right] = \sum_{n=0}^{\infty} \frac{(a_1)_n \cdots (a_p)_n}{(b_1)_n \cdots (b_q)_n} \frac{z^n}{n!}, \quad (1)$$

in which no denominator parameters b_j is allowed to be zero or a negative integer. If any numerator parameter a_j in Equation (1) is zero or a negative integer, the series terminates.

In addition, here, $(a)_n$ is the well known Pochhammer symbol [5] for any complex number a defined as

$$(a)_n = \frac{\Gamma(a+n)}{\Gamma(a)} \quad (2)$$

$$= \begin{cases} 1, & (n=0, a \in \mathbb{C} \setminus \{0\}) \\ a(a+1)\cdots(a+n-1), & (n \in \mathbb{N}, a \in \mathbb{C}), \end{cases}$$

where $\Gamma(z)$ is the well known gamma function defined by

$$\Gamma(z) = \int_0^{\infty} e^{-x} x^{z-1} dx \quad (3)$$

for $\operatorname{Re}(z) > 0$.

Further, application of the ratio test shows that the series in Equation (1):

(i) converges for all finite z if $p \leq q$;
(ii) converges for $|z| < 1$ if $p = q+1$; and
(iii) diverges for all z, $z \neq 0$ if $p > q+1$.

In addition, following Bromwich [6] (p. 41 and 241), Knopp [7] (p. 401) or Luke [8], it can be shown that the $_{q+1}F_q$ series is

(i) absolutely convergent for $|z| = 1$ if $\text{Re}(\eta) < 0$;
(ii) conditionally convergent for $|z| = 1$, $z \neq 1$ if $0 \leq \text{Re}(\eta) < 1$; and
(iii) divergent for $|z| = 1$ if $1 \leq \text{Re}(\eta)$, where

$$\eta = \sum_{j=1}^{p} a_j - \sum_{j=1}^{q} b_j.$$

It is not out of place to mention here that, whenever a generalized hypergeometric function reduces to products and quotients of gamma functions, the results are very useful from the point of view of applications. For $p = 2, 3, 4, 5$ and 7 of the generalized hypergeometric function in Equation (1) with proper choice of parameters, the results in the form of summation theorems are available in the literature in terms of gamma function. However, for $p = 6$, we do not have any summation theorem available. Thus, in this paper, we do not consider the case for $p = 6$. Here, we mention the following classical summation theorems [1,2], so that the paper may be self contained.

- Gauss Theorem for $\text{Re}(c - a - b) > 0$

$$_2F_1 \left[\begin{array}{c} a, \ b \\ c \end{array} ; 1 \right] = \frac{\Gamma(c)\Gamma(c-a-b)}{\Gamma(c-a)\Gamma(c-b)} \tag{4}$$

- Kummer's Theorem

$$_2F_1 \left[\begin{array}{c} a, \ b \\ 1+a-b \end{array} ; -1 \right] = \frac{\Gamma(1+a-b)\Gamma(1+\frac{1}{2}a)}{\Gamma(1-b+\frac{1}{2}a)\Gamma(1+a)} \tag{5}$$

- Second Gauss Theorem

$$_2F_1 \left[\begin{array}{c} a, \ b \\ \frac{1}{2}(a+b+1) \end{array} ; \frac{1}{2} \right] = \frac{\sqrt{\pi}\,\Gamma(\frac{1}{2}(a+b+1))}{\Gamma(\frac{1}{2}(a+1))\Gamma(\frac{1}{2}(b+1))} \tag{6}$$

- Bailey's Theorem

$$_2F_1 \left[\begin{array}{c} a, \ 1-a \\ b \end{array} ; \frac{1}{2} \right] = \frac{\Gamma(\frac{1}{2}b)\Gamma(\frac{1}{2}(b+1))}{\Gamma(\frac{1}{2}(a+b))\Gamma(\frac{1}{2}(b-a+1))} \tag{7}$$

- Dixon's Theorem for $\text{Re}(a - 2b - 2c) > -2$

$$_3F_2 \left[\begin{array}{c} a, \ b, \ c \\ 1+a-b, \ 1+a-c \end{array} ; 1 \right] \tag{8}$$

$$= \frac{\Gamma(1+\frac{1}{2}a)\Gamma(1+a-b)\Gamma(1+a-c)\Gamma(1-b-c+\frac{1}{2}a)}{\Gamma(1+a)\Gamma(1-b+\frac{1}{2}a)\Gamma(1-c+\frac{1}{2}a)\Gamma(1+a-b-c)}$$

- Watson's Theorem for $\text{Re}(2c - a - b) > 1$

$$_3F_2 \left[\begin{array}{c} a, \ b, \ c \\ \frac{1}{2}(a+b+1), \ 2c \end{array} ; 1 \right] \tag{9}$$

$$= \frac{\sqrt{\pi}\,\Gamma(c+\frac{1}{2})\Gamma(\frac{1}{2}(a+b+1))\Gamma(c-\frac{1}{2}(a+b-1))}{\Gamma(\frac{1}{2}(a+1))\Gamma(\frac{1}{2}(b+1))\Gamma(c-\frac{1}{2}(a-1))\Gamma(c-\frac{1}{2}(b-1))}$$

- Whipple's Theorem for $\operatorname{Re}(b) > 0$

$$_3F_2\left[\begin{array}{ccc} a, & 1-a, & b \\ c, & 2b-c+1 \end{array};1\right] \qquad (10)$$
$$= \frac{\pi\, 2^{1-2b}\,\Gamma(c)\Gamma(2b-c+1)}{\Gamma(\frac{1}{2}(a+c))\Gamma(b+\frac{1}{2}(a-c+1))\Gamma(\frac{1}{2}(1-a+c))\Gamma(b+1-\frac{1}{2}(a+c))}$$

- Pfaff-Saalschütz Theorem

$$_3F_2\left[\begin{array}{ccc} a, & b, & -n \\ c, & 1+a+b-c-n \end{array};1\right] = \frac{(c-a)_n(c-b)_n}{(c)_n(c-a-b)_n} \qquad (11)$$

- Second Whipple's Theorem

$$_4F_3\left[\begin{array}{cccc} a, & 1+\frac{1}{2}a, & b, & c \\ \frac{1}{2}a, & a-b+1, & a-c+1 \end{array};-1\right] = \frac{\Gamma(a-b+1)\Gamma(a-c+1)}{\Gamma(a+1)\Gamma(a-b-c+1)} \qquad (12)$$

- Dougall's Theorem for $\operatorname{Re}(a-c-d-e) > -1$

$$_5F_4\left[\begin{array}{ccccc} a, & 1+\frac{1}{2}a, & c, & d, & e \\ \frac{1}{2}a, & a-c+1, & a-d+1, & a-e+1 \end{array};1\right] \qquad (13)$$
$$= \frac{\Gamma(a-c+1)\Gamma(a-d+1)\Gamma(a-e+1)\Gamma(a-c-d-e+1)}{\Gamma(a+1)\Gamma(a-d-e+1)\Gamma(a-c-e+1)\Gamma(a-c-d+1)}$$

- Second Dougall's Theorem

$$_7F_6\left[\begin{array}{ccccccc} a, & 1+\frac{1}{2}a, & b, & c, & d, & 1+2a-b-c-d+n, & -n \\ \frac{1}{2}a, & a-b+1, & a-c+1, & a-d+1, & b+c+d-a-n, & a+1+n \end{array};1\right] \qquad (14)$$
$$= \frac{(a+1)_n(a-b-c+1)_n(a-b-d+1)_n(a-c-d+1)_n}{(a+1-b)_n(a+1-c)_n(a+1-d)_n(a+1-b-c-d)_n}$$

For very interesting applications of some of the above-mentioned classical summation theorems, we refer a very popular and useful paper by Bailey [9].

In addition, for finite sums of hypergeometric series, if we use the following symbol

$$_pF_q^{(m)}\left[\begin{array}{ccc} a_1, & \ldots, & a_p \\ b_1, & \ldots, & b_q \end{array};z\right] = \sum_{n=0}^{m} \frac{\prod_{i=1}^{p}(a_i)_n}{\prod_{i=1}^{q}(b_i)_n}\frac{z^n}{n!},$$

where, for instance,

$$_pF_q^{(-1)}(z) = 0, \quad _pF_q^{(0)}(z) = 1, \quad _pF_q^{(1)}(z) = 1 + \frac{a_1\cdots a_p}{b_1\cdots b_q}z,$$

then, by using the following relation [10],

$$_pF_q\left[\begin{array}{c} a_1,\cdots, a_{p-1},\ 1 \\ b_1,\cdots, b_{q-1},\ m \end{array};z\right] \tag{15}$$

$$= \frac{\Gamma(b_1)\cdots\Gamma(b_{q-1})}{\Gamma(a_1)\cdots\Gamma(a_{p-1})}\frac{\Gamma(a_1-m+1)\cdots\Gamma(a_{p-1}-m+1)}{\Gamma(b_1-m+1)\cdots\Gamma(b_{q-1}-m+1)}\frac{(m-1)!}{z^{m-1}}$$

$$\times\left\{{}_{p-1}F_{q-1}\left[\begin{array}{c} a_1-m+1,\cdots, a_{p-1}-m+1 \\ b_1-m+1,\cdots, b_{q-1}-m+1 \end{array};z\right]\right.$$

$$\left.-{}_{p-1}^{(m-2)}F_{q-1}\left[\begin{array}{c} a_1-m+1,\cdots, a_{p-1}-m+1 \\ b_1-m+1,\cdots, b_{q-1}-m+1 \end{array};z\right]\right\},$$

very recently Masjed-Jamei and Koepf [11] established generalizations of the classical summation theorems in Equations (4)–(14) in the following form:

$$_3F_2\left[\begin{array}{c} a,\ b,\ 1 \\ c,\ m \end{array};1\right] \tag{16}$$

$$= \frac{\Gamma(m)\Gamma(c)\Gamma(a-m+1)\Gamma(b-m+1)}{\Gamma(a)\Gamma(b)\Gamma(c-m+1)}$$

$$\times\left\{\frac{\Gamma(c-m+1)\Gamma(c-a-b+m-1)}{\Gamma(c-a)\Gamma(c-b)}-{}_2^{(m-2)}F_1\left[\begin{array}{c} a-m+1,\ b-m+1 \\ c-m+1 \end{array};1\right]\right\}$$

$$=\Omega_1$$

$$_3F_2\left[\begin{array}{c} a,\ b,\ 1 \\ a-b+m,\ m \end{array};-1\right] \tag{17}$$

$$=(-1)^{m-1}\frac{\Gamma(m)\Gamma(a-b+m)\Gamma(a-m+1)\Gamma(b-m+1)}{\Gamma(a)\Gamma(b)\Gamma(a-b+1)}$$

$$\times\left\{\frac{\Gamma(a-b+1)\Gamma(1+\frac{1}{2}(a-m+1))}{\Gamma(2+a-m)\Gamma(m-b+\frac{1}{2}(a-m+1))}-{}_2^{(m-2)}F_1\left[\begin{array}{c} a-m+1,\ b-m+1 \\ a-b+1 \end{array};-1\right]\right\}$$

$$=\Omega_2$$

$$_3F_2\left[\begin{array}{c} a,\ b,\ 1 \\ \frac{1}{2}(a+b+1),\ m \end{array};\frac{1}{2}\right] \tag{18}$$

$$= 2^{m-1}\frac{\Gamma(m)\Gamma(\frac{1}{2}(a+b+1))\Gamma(a-m+1)\Gamma(b-m+1)}{\Gamma(a)\Gamma(b)\Gamma(-m+1+\frac{1}{2}(a+b+1))}$$

$$\times\left\{\frac{\sqrt{\pi}\Gamma(-m+1+\frac{1}{2}(a+b+1))}{\Gamma(1+\frac{1}{2}(a-m))\Gamma(1+\frac{1}{2}(b-m))}-{}_2^{(m-2)}F_1\left[\begin{array}{c} a-m+1,\ b-m+1 \\ -m+1+\frac{1}{2}(a+b+1) \end{array};\frac{1}{2}\right]\right\}$$

$$=\Omega_3$$

$$_3F_2\left[\begin{array}{c} a,\ 2m-a-1,\ 1 \\ b,\ m \end{array};\frac{1}{2}\right] \qquad (19)$$

$$= 2^{m-1}\frac{\Gamma(m)\Gamma(b)\Gamma(a-m+1)\Gamma(m-a)}{\Gamma(a)\Gamma(2m-a-1)\Gamma(b-m+1)}$$

$$\times \left\{\frac{\Gamma(\frac{1}{2}(b-m+1))\Gamma(\frac{1}{2}(b-m+2))}{\Gamma(-m+1+\frac{1}{2}(a+b))\Gamma(\frac{1}{2}(b-a+1))} - \,^{(m-2)}_{2}F_1\left[\begin{array}{c} a-m+1,\ m-a \\ b-m+1 \end{array};\frac{1}{2}\right]\right\}$$

$$= \Omega_4$$

$$_4F_3\left[\begin{array}{c} a,\ b,\ c,\ 1 \\ a-b+m,\ a-c+m,\ m \end{array};1\right] \qquad (20)$$

$$= \frac{\Gamma(m)\Gamma(a-b+m)\Gamma(a-c+m)\Gamma(a+1-m)\Gamma(b+1-m)\Gamma(c+1-m)}{\Gamma(a)\Gamma(b)\Gamma(c)\Gamma(a-b+1)\Gamma(a-c+1)}$$

$$\times \left\{\frac{\Gamma(\frac{1}{2}(a+3-m))\Gamma(a-b+1)\Gamma(a-c+1)\Gamma(-b-c+\frac{1}{2}(a+3m-1))}{\Gamma(a+2-m)\Gamma(-b+\frac{1}{2}(a+m+1))\Gamma(-c+\frac{1}{2}(a+m+1))\Gamma(a-b-c+m)}\right.$$

$$\left. - \,^{(m-2)}_{3}F_2\left[\begin{array}{c} a-m+1,\ b-m+1,\ c-m+1 \\ a-b+1,\ a-c+1 \end{array};1\right]\right\}$$

$$= \Omega_5$$

$$_4F_3\left[\begin{array}{c} a,\ b,\ c,\ 1 \\ \frac{1}{2}(a+b+1),\ 2c+1-m,\ m \end{array};1\right] \qquad (21)$$

$$= \frac{\Gamma(m)\Gamma(\frac{1}{2}(a+b+1))\Gamma(2c+1-m)\Gamma(a+1-m)\Gamma(b+1-m)\Gamma(c+1-m)}{\Gamma(a)\Gamma(b)\Gamma(c)\Gamma(-m+\frac{1}{2}(a+b+3))\Gamma(2c-2m+2)}$$

$$\times \left\{\frac{\sqrt{\pi}\,\Gamma(c-m+\frac{3}{2})\Gamma(-m+\frac{1}{2}(a+b+3))\Gamma(c-\frac{1}{2}(a+b-1))}{\Gamma(1+\frac{1}{2}(a-m))\Gamma(1+\frac{1}{2}(b-m))\Gamma(c+1-\frac{1}{2}(a+m))\Gamma(c+1-\frac{1}{2}(b+m))}\right.$$

$$\left. - \,^{(m-2)}_{3}F_2\left[\begin{array}{c} a-m+1,\ b-m+1,\ c-m+1 \\ -m+1+\frac{1}{2}(a+b+1),\ 2c-2m+2 \end{array};1\right]\right\}$$

$$= \Omega_6$$

$$_4F_3\left[\begin{array}{c} a,\ 2m-1-a,\ b,\ 1 \\ c,\ 2b-c+1,\ m \end{array};1\right] \qquad (22)$$

$$= \frac{\Gamma(m)\Gamma(c)\Gamma(2b-c+1)\Gamma(m-a)\Gamma(a+1-m)\Gamma(b+1-m)}{\Gamma(a)\Gamma(b)\Gamma(2m-1-a)\Gamma(c+1-m)\Gamma(2b-c-m+2)}$$

$$\times \left\{\frac{\pi\, 2^{2m-2b-1}\,\Gamma(c-m+1)}{\Gamma(-m+1+\frac{1}{2}(a+c))\Gamma(-m+1+b+\frac{1}{2}(a-c+1))\Gamma(\frac{1}{2}(1-a+c))}\right.$$

$$\left. \times \frac{\Gamma(2b-c-m+2)}{\Gamma(b+1-\frac{1}{2}(a+c))} - \,^{(m-2)}_{3}F_2\left[\begin{array}{c} a-m+1,\ b-m+1,\ m-a \\ c-m+1,\ 2b-c-m+2 \end{array};1\right]\right\}$$

$$= \Omega_7$$

45

$$_4F_3\left[\begin{array}{c} a,\ b,\ -n+m-1,\ 1 \\ c,\ 1+a+b-c-n,\ m \end{array};1\right] = \frac{(m-1)!\,(1-c)_{m-1}}{(1-a)_{m-1}(1-b)_{m-1}} \quad (23)$$

$$\times \frac{(c-a-b+n)_{m-1}}{(n+2-m)_{m-1}} \times \left\{ \frac{(c-a)_n(c-b)_n}{(c+1-m)_n(c-a-b+m-1)_n} \right.$$

$$\left. - \,_3F_2^{(m-2)}\left[\begin{array}{c} a-m+1,\ b-m+1,\ -n \\ c-m+1,\ 2+a+b-c-m-n \end{array};1\right] \right\}$$

$$= \Omega_8$$

$$_5F_4\left[\begin{array}{c} a,\ \tfrac{1}{2}(a+m+1),\ b,\ c,\ 1 \\ \tfrac{1}{2}(a+m-1),\ a-b+m,\ a-c+m,\ m \end{array};-1\right] = (-1)^{m-1}\Gamma(m) \quad (24)$$

$$\times \frac{\Gamma(\tfrac{1}{2}(a+m-1))\Gamma(a-b+m)\Gamma(a-c+m)\Gamma(\tfrac{1}{2}(a-m+3))\Gamma(a-m+1)}{\Gamma(a)\Gamma(b)\Gamma(c)\Gamma(\tfrac{1}{2}(a+m+1))\Gamma(\tfrac{1}{2}(a-m+1))}$$

$$\times \frac{\Gamma(b+1-m)\Gamma(c+1-m)}{\Gamma(a-b+1)\Gamma(a-c+1)} \times \left\{ \frac{\Gamma(1+a-b)\Gamma(1+a-c)}{\Gamma(2-m+a)\Gamma(m+a-b-c)} \right.$$

$$\left. - \,_4F_3^{(m-2)}\left[\begin{array}{c} a-m+1,\ b-m+1,\ \tfrac{1}{2}(a-m+3),\ c-m+1 \\ \tfrac{1}{2}(a-m+1),\ a-b+1,\ a-c+1 \end{array};-1\right] \right\}$$

$$= \Omega_9$$

$$_6F_5\left[\begin{array}{c} a,\ \tfrac{1}{2}(a+m+1),\ c,\ d,\ e,\ 1 \\ \tfrac{1}{2}(a+m-1),\ a-c+m,\ a-d+m,\ a-e+m,\ m \end{array};1\right] \quad (25)$$

$$= \frac{\Gamma(m)\Gamma(\tfrac{1}{2}(a+m-1))\Gamma(a-c+m)\Gamma(a-d+m)\Gamma(a-e+m)}{\Gamma(a-c+1)\Gamma(a-d+1)\Gamma(a-e+1)}$$

$$\times \frac{\Gamma(a-m+1)\Gamma(\tfrac{1}{2}(a-m+3))\Gamma(c+1-m)\Gamma(d+1-m)\Gamma(e+1-m)}{\Gamma(a)\Gamma(c)\Gamma(d)\Gamma(e)\Gamma(\tfrac{1}{2}(a+m+1))\Gamma(\tfrac{1}{2}(a-m+1))}$$

$$\times \left\{ \frac{\Gamma(a-c+1)\Gamma(a-d+1)\Gamma(a-e+1)\Gamma(a-c-d-e+2m-1)}{\Gamma(2-m+a)\Gamma(a-c-e+m)\Gamma(a-d-e+m)\Gamma(a-c-d+m)} \right.$$

$$\left. - \,_5F_4^{(m-2)}\left[\begin{array}{c} a-m+1,\ c-m+1,\ \tfrac{1}{2}(a-m+3),\ d-m+1,\ e-m+1 \\ \tfrac{1}{2}(a-m+1),\ a-c+1,\ a-d+1,\ a-e+1 \end{array};1\right] \right\}$$

$$= \Omega_{10}$$

$$_8F_7\left[\begin{array}{c} a,\ \tfrac{1}{2}(a+m+1),\ b,\ c,\ d,\ 2a-b-c-d+2m-1+n,\ m-n-1,\ 1 \\ \tfrac{1}{2}(a+m-1),\ a-b+m,\ a-c+m,\ a-d+m,\ b+c+d-a+1-m-n,\ a+n+1,\ m \end{array};1\right] \quad (26)$$

$$= (-1)^{m-1}(m-1)! \times \frac{(\tfrac{1}{2}(3-a-m))_{m-1}(1-a+b-m)_{m-1}}{(\tfrac{1}{2}(1-a-m))_{m-1}(1-a)_{m-1}}$$

$$\times \frac{(1-a+c-m)_{m-1}(1-a+d-m)_{m-1}(m+n+a-b-c-d)_{m-1}(-a-n)_{m-1}}{(1-b)_{m-1}(1-c)_{m-1}(1-d)_{m-1}(b+c+d-2a+2-2m-n)_{m-1}(n+2-m)_{m-1}}$$

$$\times \left\{ \frac{(a-m+2)_n(a-b-c+m)_n(a-b-d+m)_n(a-c-d+m)_n}{(a-b+1)_n(a-c+1)_n(a-d+1)_n(a-b-c-d+2m-1)_n} \right.$$

$$\left. - \,_7F_6^{(m-2)}\left[\begin{array}{c} a-m+1,\ \tfrac{1}{2}(a-m+3),\ b-m+1,\ c-m+1,\ d-m+1,\ 2a-b-c-d+m+n,\ -n \\ \tfrac{1}{2}(a-m+1),\ a-b+1,\ a-c+1,\ a-d+1,\ b+c+d-a+2-2m-n,\ a-m+n+2 \end{array};1\right] \right\}$$

$$= \Omega_{11}$$

Remark 1. *For other generalizations of the results in Equations (5)–(10), we refer to [12–16].*

On the other hand, we define the (direct) Laplace transform of a function $f(t)$ of a real variable t as the integral $g(s)$ over a range of the complex parameter s as

$$g(s) = \mathcal{L}\{f(t); s\} = \int_0^\infty e^{-st} f(t) dt \tag{27}$$

provided the integral exists in the Lebesgue sense. For more details, see, for instance, the works of [17] or [18]. It is interesting to mention here that, in view of the formula

$$\int_0^\infty e^{-st} t^{\alpha-1} dt = \Gamma(\alpha) s^{-\alpha} \tag{28}$$

provided $\operatorname{Re}(s) > 0$ and $\operatorname{Re}(\alpha) > 0$, by using Equation (1), it is a simple exercise to show that the Laplace transform of a generalized hypergeometric function ${}_pF_q$ is obtained as [3,19,20]:

$$\int_0^\infty e^{-st} t^{\nu-1} {}_pF_q \begin{bmatrix} a_1, & \cdots, & a_p \\ b_1, & \cdots, & b_q \end{bmatrix}; wt \end{bmatrix} dt \tag{29a}$$

$$= \Gamma(\nu) s^{-\nu} {}_{p+1}F_q \begin{bmatrix} \nu, a_1, & \cdots, & a_p \\ b_1, & \cdots, & b_q \end{bmatrix}; \frac{w}{s} \end{bmatrix}$$

provided that when $p < q$, $\operatorname{Re}(\nu) > 0$, $\operatorname{Re}(s) > 0$ for w arbitrary, or $p = q > 0$, $\operatorname{Re}(\nu) > 0$ and $\operatorname{Re}(s) > \operatorname{Re}(w)$.

Further, in Equation (29a), if we take $p = q = 1$, $\nu = b$, $a_1 = a$ and $b_1 = c$, we find that (see [20]),

$$\int_0^\infty e^{-st} t^{b-1} {}_1F_1\begin{bmatrix} a \\ c \end{bmatrix}; wt \end{bmatrix} dt = \Gamma(b) s^{-b} {}_2F_1\begin{bmatrix} a, b \\ c \end{bmatrix}; \frac{w}{s} \end{bmatrix}, \tag{29b}$$

provided $\operatorname{Re}(b) > 0$, $\operatorname{Re}(s) > 0$, $\operatorname{Re}(s) > \operatorname{Re}(w)$ and $|s| > |w|$.

Finally, in Equation (29b), if we take $w = \frac{1}{2}s$ and either $c = \frac{1}{2}(a+b+1)$ or $b = 1-a$, then it is easy to see that the ${}_2F_1$ series appearing on the right-hand side of Equation (29b) are summable by known summation theorems in Equations (6) and (7), respectively, and we find that

$$\int_0^\infty e^{-st} t^{b-1} {}_1F_1\begin{bmatrix} a \\ \frac{1}{2}(a+b+1) \end{bmatrix}; \frac{1}{2}st \end{bmatrix} dt = s^{-b} \frac{\Gamma(\frac{1}{2})\Gamma(b)\Gamma(\frac{1}{2}(a+b+1))}{\Gamma(\frac{1}{2}(a+1))\Gamma(\frac{1}{2}(b+1))}, \tag{29c}$$

provided $\operatorname{Re}(b) > 0$ and $\operatorname{Re}(s) > 0$, and that

$$\int_0^\infty e^{-st} t^{-a} {}_1F_1\begin{bmatrix} a \\ c \end{bmatrix}; \frac{1}{2}st \end{bmatrix} dt = s^{a-1} \frac{\Gamma(1-a)\Gamma(\frac{1}{2}c)\Gamma(\frac{1}{2}(c+1))}{\Gamma(\frac{1}{2}(c+a))\Gamma(\frac{1}{2}(c-a+1))}, \tag{29d}$$

provided $\operatorname{Re}(1-a) > 0$ and $\operatorname{Re}(s) > 0$.

The results in Equations (29b) and (29c) are very well-known in the literature and are recorded, for example, in the work of [20].

In addition to Equation (29a), it is interesting to observe here that, when $w = \pm s$ and $q = p$, on similar lines, we can obtain the following result in view of the conditions of convergence of ${}_pF_q$ mentioned in Section 1.

$$\int_0^\infty e^{-st} t^{\nu-1} {}_pF_p \begin{bmatrix} a_1, & \cdots, & a_p \\ b_1, & \cdots, & b_p \end{bmatrix}; \pm st \end{bmatrix} dt \tag{29e}$$

$$= \Gamma(\nu) s^{-\nu} {}_{p+1}F_p \begin{bmatrix} \nu, a_1, & \cdots, & a_p \\ b_1, & \cdots, & b_p \end{bmatrix}; \pm 1 \end{bmatrix}$$

provided $\operatorname{Re}(\nu) > 0$, $\operatorname{Re}(s) > 0$ and $\operatorname{Re}(b_1 + \cdots + b_p - a_1 - \cdots - a_p - \nu) > 0$.

Remark 2. 1. Since there is no summation theorems for the series $_pF_p$ for argument $\frac{1}{2}$, 1 and -1 are available in the literature, at this moment, it is not possible to find the Laplace-type integrals for the generalized hypergeometric function $_pF_p$. We leave this open problem for the readers.
2. Laplace-type integrals in the case $p = 2$ were given by Deepthi et al. [21] and connections with fractional integral operators were recently studied by Parmar and Purohit [22].

The aim of this paper is to provide a new class of Laplace-type integrals involving generalized hypergeometric functions by employing the summation theorems in Equations (16)–(26). Several new and known special cases have also been considered.

2. Laplace-Type Integrals Involving Generalized Hypergeometric Functions

In this section, we establish several new, interesting and elementary Laplace-type integrals in the most general form, involving generalized hypergeometric functions asserted in the following theorems that follow directly from Equations (29a) and (29e) and the known results in Equations (16)–(26). The results presented in this section would serve as key formulas from which, on specializing the parameters, lead to several results, some of which are known and others are believed to be new.

[A] Laplace-type integrals involving generalized hypergeometric function $_2F_2$

The results to be established are asserted in the following theorems.

Theorem 1. *For $m \in \mathbb{N}$, $\operatorname{Re}(s) > 0$, $\operatorname{Re}(a) > 0$ and $\operatorname{Re}(c - a - b + m) > 1$, the following result holds true.*

$$\int_0^\infty e^{-st} t^{a-1} \, _2F_2\left[\begin{matrix} b, & 1 \\ c, & m \end{matrix}; st\right] dt = \Gamma(a)\, s^{-a}\, \Omega_1, \tag{30}$$

where Ω_1 is the same as given in Equation (16).

Theorem 2. *For $m \in \mathbb{N}$, $\operatorname{Re}(s) > 0$ and $\operatorname{Re}(c - a - b + m) > 1$, the following result holds true.*

$$\int_0^\infty e^{-st} \, _2F_2\left[\begin{matrix} a, & b \\ c, & m \end{matrix}; st\right] dt = s^{-1}\Omega_1, \tag{31}$$

where Ω_1 is the same as given in Equation (16).

Theorem 3. *For $m \in \mathbb{N}$, $\operatorname{Re}(s) > 0$ and $\operatorname{Re}(a) > 0$, the following result holds true.*

$$\int_0^\infty e^{-st} t^{a-1} \, _2F_2\left[\begin{matrix} b, & 1 \\ a-b+m, & m \end{matrix}; -st\right] dt = \Gamma(a)\, s^{-a}\, \Omega_2, \tag{32}$$

where Ω_2 is the same as given in Equation (17).

Theorem 4. *For $m \in \mathbb{N}$, $\operatorname{Re}(s) > 0$ and $\operatorname{Re}(b) > 0$, the following result holds true.*

$$\int_0^\infty e^{-st} t^{b-1} \, _2F_2\left[\begin{matrix} a, & 1 \\ a-b+m, & m \end{matrix}; -st\right] dt = \Gamma(b)\, s^{-b}\, \Omega_2, \tag{33}$$

where Ω_2 is the same as given in Equation (17).

Remark 3. *The reader should observe that the results given in Theorems 3 and 4 are different but obtained from the same known result in Equation (17).*

Theorem 5. *For $m \in \mathbb{N}$ and $Re(s) > 0$, the following result holds true.*

$$\int_0^\infty e^{-st} {}_2F_2\left[\begin{matrix} a, & b \\ a-b+m, & m \end{matrix}; -st\right] dt = s^{-1}\Omega_2, \tag{34}$$

where Ω_2 is the same as given in Equation (17).

Theorem 6. *For $m \in \mathbb{N}$, $Re(s) > 0$ and $Re(a) > 0$, the following result holds true.*

$$\int_0^\infty e^{-st} t^{a-1} {}_2F_2\left[\begin{matrix} b, & 1 \\ \frac{1}{2}(a+b+1), & m \end{matrix}; \frac{1}{2}st\right] dt = \Gamma(a)\, s^{-a}\Omega_3, \tag{35}$$

where Ω_3 is the same as given in Equation (18).

Theorem 7. *For $m \in \mathbb{N}$ and $Re(s) > 0$, the following result holds true.*

$$\int_0^\infty e^{-st} {}_2F_2\left[\begin{matrix} a, & b \\ \frac{1}{2}(a+b+1), & m \end{matrix}; \frac{1}{2}st\right] dt = s^{-1}\Omega_3, \tag{36}$$

where Ω_3 is the same as given in Equation (18).

Theorem 8. *For $m \in \mathbb{N}$, $Re(s) > 0$ and $Re(a) > 0$, the following result holds true.*

$$\int_0^\infty e^{-st} t^{a-1} {}_2F_2\left[\begin{matrix} 2m-a-1, & 1 \\ b, & m \end{matrix}; \frac{1}{2}st\right] dt = \Gamma(a)\, s^{-a}\Omega_4, \tag{37}$$

where Ω_4 is the same as given in Equation (19).

Theorem 9. *For $m \in \mathbb{N}$, $Re(s) > 0$ and $Re(2m-a-1) > 0$, the following result holds true.*

$$\int_0^\infty e^{-st} t^{2m-a-2} {}_2F_2\left[\begin{matrix} a, & 1 \\ b, & m \end{matrix}; \frac{1}{2}st\right] dt = \Gamma(2m-a-1)\, s^{a+1-2m}\Omega_4, \tag{38}$$

where Ω_4 is the same as given in Equation (19).

Theorem 10. *For $m \in \mathbb{N}$ and $Re(s) > 0$, the following result holds true.*

$$\int_0^\infty e^{-st} {}_2F_2\left[\begin{matrix} a, & 2m-a-1 \\ b, & m \end{matrix}; \frac{1}{2}st\right] dt = s^{-1}\Omega_4, \tag{39}$$

where Ω_4 is the same as given in Equation (19).

Proof. To establish the result in Equation (30) asserted in Theorem 1, we proceed as follows. In Equation (29e), if we take $p = q = 2$, $\nu = a$, $a_1 = b$, $a_2 = 1$, $b_1 = c$, and $b_2 = m$, considering positive sign, we get

$$\int_0^\infty e^{-st} t^{a-1} {}_2F_2\left[\begin{matrix} b, & 1 \\ c, & m \end{matrix}; st\right] dx = s^{-a}\Gamma(a)\, {}_3F_2\left[\begin{matrix} a, & b, & 1 \\ c, & m \end{matrix}; 1\right]. \tag{40}$$

We now observe that the ${}_3F_2$ appearing on the right-hand side of Equation (40) can be evaluated with the help of the result in Equation (16) and we easily arrive at the right-hand side of Equation (30). This completes the proof of Equation (30) asserted in Theorem 1.

In exactly the same manner, the results in Equations (31)–(39) asserted in Theorems 2–10 can be evaluated. We however omit the details. □

Corollary 1. (a) *In Theorem 1, if we take* $m = 1, 2, 3$, *we get the following results.*

$$\int_0^\infty e^{-st} t^{a-1} {}_1F_1\left[\begin{matrix} b \\ c \end{matrix}; st\right] dt = \frac{\Gamma(a)\Gamma(c)\Gamma(c-a-b)}{s^a \Gamma(c-a)\Gamma(c-b)}, \quad (41)$$

$$\int_0^\infty e^{-st} t^{a-1} {}_2F_2\left[\begin{matrix} b, & 1 \\ c, & 2 \end{matrix}; st\right] dt \quad (42)$$

$$= \frac{(c-1)\Gamma(a-1)}{s^a (b-1)} \left\{\frac{\Gamma(c-1)\Gamma(c-a-b+1)}{\Gamma(c-a)\Gamma(c-b)} - 1\right\}$$

and

$$\int_0^\infty e^{-st} t^{a-1} {}_2F_2\left[\begin{matrix} b, & 1 \\ c, & 3 \end{matrix}; st\right] dt \quad (43)$$

$$= \frac{2\Gamma(a)(c-2)_2}{s^a (a-2)_2(b-2)_2} \left\{\frac{\Gamma(c-2)\Gamma(c-a-b+2)}{\Gamma(c-a)\Gamma(c-b)} - \frac{ab+c-2a-2b+2}{c-2}\right\}.$$

(b) *In Theorem 4, if we take* $m = 1, 2, 3$, *we get the following results.*

$$\int_0^\infty e^{-st} t^{b-1} {}_1F_1\left[\begin{matrix} a \\ 1+a-b \end{matrix}; -st\right] dt = \frac{2^{-a}\Gamma(b)\Gamma(\frac{1}{2})\Gamma(1+a-b)}{s^b \Gamma(\frac{1}{2}a+\frac{1}{2})\Gamma(1+\frac{1}{2}a-b)}, \quad (44)$$

$$\int_0^\infty e^{-st} t^{b-1} {}_2F_2\left[\begin{matrix} a, & 1 \\ 2+a-b, & 2 \end{matrix}; -st\right] dt \quad (45)$$

$$= \frac{(a-b+1)\Gamma(b)}{s^b (a-1)(b-1)} \left\{1 - \frac{\Gamma(1+a-b)\Gamma(\frac{1}{2}a+\frac{1}{2})}{\Gamma(a)\Gamma(\frac{1}{2}a-b+\frac{3}{2})}\right\}$$

and

$$\int_0^\infty e^{-st} t^{a-1} {}_2F_2\left[\begin{matrix} b, & 1 \\ 3+a-b, & 3 \end{matrix}; -st\right] dt \quad (46)$$

$$= \frac{2(a-b+1)_2 \Gamma(a)}{s^b (a-2)_2(b-2)_2} \left\{\frac{\Gamma(\frac{1}{2}a)\Gamma(1+a-b)}{\Gamma(a-1)\Gamma(\frac{1}{2}a-b+2)} - \frac{3a+b-ab-3}{1+a-b}\right\}.$$

(c) *In Theorem 7, if we take* $m = 1, 2, 3$, *we get the following results.*

$$\int_0^\infty e^{-st} t^{a-1} {}_1F_1\left[\begin{matrix} b \\ \frac{1}{2}(a+b+1) \end{matrix}; \frac{1}{2}st\right] dt = \frac{\sqrt{\pi}\Gamma(a)\Gamma(\frac{1}{2}(a+b+1))}{s^a \Gamma(\frac{1}{2}(a+1))\Gamma(\frac{1}{2}(b+1))}, \quad (47)$$

$$\int_0^\infty e^{-st} t^{a-1} {}_2F_2\left[\begin{matrix} b, & 1 \\ \frac{1}{2}(a+b+1), & 2 \end{matrix}; \frac{1}{2}st\right] dt \quad (48)$$

$$= \frac{(a+b-1)\Gamma(a-1)}{s^a (b-1)} \left\{\frac{\sqrt{\pi}\Gamma(\frac{1}{2}(a+b-1))}{\Gamma(\frac{1}{2}a)\Gamma(\frac{1}{2}b)} - 1\right\}$$

and

$$\int_0^\infty e^{-st} t^{a-1} {}_2F_2\left[\begin{matrix} b, & 1 \\ \frac{1}{2}(a+b+1), 3 \end{matrix}; \frac{1}{2}st\right] dt = \frac{2\Gamma(a)(a+b-1)(a+b-3)}{s^a (a-2)_2(b-2)_2} \quad (49)$$
$$\times \left\{ \frac{\sqrt{\pi}\Gamma(\frac{1}{2}(a+b-3))}{\Gamma(\frac{1}{2}(a-1))\Gamma(\frac{1}{2}(b-1))} - \frac{ab-a-b+1}{a+b-3} \right\}.$$

(d) In Theorem 10, if we take $m = 1, 2, 3$, we get the following results.

$$\int_0^\infty e^{-st} t^{a-1} {}_1F_1\left[\begin{matrix} 1-a \\ b \end{matrix}; \frac{1}{2}st\right] dt = \frac{\Gamma(a)\Gamma(\frac{1}{2}b)\Gamma(\frac{1}{2}(b+1))}{s^a \Gamma(\frac{1}{2}(a+b))\Gamma(\frac{1}{2}(b-a+1))}, \quad (50)$$

$$\int_0^\infty e^{-st} t^{a-1} {}_2F_2\left[\begin{matrix} 3-a, & 1 \\ b, & 2 \end{matrix}; \frac{1}{2}st\right] dt \quad (51)$$
$$= \frac{2(1-b)\Gamma(a)}{s^a (1-a)_2} \left\{ \frac{\Gamma(\frac{1}{2}(b-1))\Gamma(\frac{1}{2}b)}{\Gamma(\frac{1}{2}(a+b)-1)\Gamma(\frac{1}{2}(b-a+1))} - 1 \right\}$$

and

$$\int_0^\infty e^{-st} t^{a-1} {}_2F_2\left[\begin{matrix} 5-a, & 1 \\ b, & 3 \end{matrix}; \frac{1}{2}st\right] dt = \frac{8(b-2)_2\Gamma(a)}{s^a (a-4)_4} \quad (52)$$
$$\times \left\{ \frac{\Gamma(\frac{1}{2}(b-1))\Gamma(\frac{1}{2}(b-2))}{\Gamma(\frac{1}{2}(a+b)-2)\Gamma(\frac{1}{2}(b-a+1))} - \frac{5a - a^2 + 2b - 10}{2(b-2)} \right\}.$$

Similarly, other results can be obtained from Theorems 2, 3, 5, 6, 8 and 9.

Remark 4. 1. *The results in Equations (47) and (50) were recorded by [23] as well as [20].*
2. *The proofs of Theorems 11–35 given below are straight forward and can be proven with the help of the result in Equation (29e), thus they are given here without proof.*

[B] Laplace-type integrals involving generalized hypergeometric function ${}_3F_3$

The results to be established are asserted in the following theorems.

Theorem 11. *For $m \in \mathbb{N}$, $Re(s) > 0$, $Re(a) > 0$ and $Re(a - 2b - 2c + 3m) > 1$, the following result holds true.*

$$\int_0^\infty e^{-st} t^{a-1} {}_3F_3\left[\begin{matrix} b, & c, & 1 \\ a-b+m, & a-c+m, & m \end{matrix}; st\right] dt = \Gamma(a) s^{-a} \Omega_5, \quad (53)$$

where Ω_5 is the same as given in Equation (20).

Theorem 12. *For $m \in \mathbb{N}$, $Re(s) > 0$, $Re(b) > 0$ and $Re(a - 2b - 2c + 3m) > 1$, the following result holds true.*

$$\int_0^\infty e^{-st} t^{b-1} {}_3F_3\left[\begin{matrix} a, & c, & 1 \\ a-b+m, & a-c+m, & m \end{matrix}; st\right] dt = \Gamma(b) s^{-b} \Omega_5, \quad (54)$$

where Ω_5 is the same as given in Equation (20).

Theorem 13. For $m \in \mathbb{N}$, $Re(s) > 0$ and $Re(a - 2b - 2c + 3m) > 1$, the following result holds true.

$$\int_0^\infty e^{-st} {}_3F_3\left[\begin{array}{c} a, \quad b, \quad c \\ a-b+m, \; a-c+m, \; m \end{array}; st\right] dt = s^{-1} \Omega_5, \tag{55}$$

where Ω_5 is the same as given in Equation (20).

Theorem 14. For $m \in \mathbb{N}$, $Re(s) > 0$, $Re(a) > 0$ and $Re(2c - a - b) > -1$, the following result holds true.

$$\int_0^\infty e^{-st} t^{a-1} {}_3F_3\left[\begin{array}{c} b, \quad c, \quad 1 \\ \frac{1}{2}(a+b+1), \; 2c+1-m, \; m \end{array}; st\right] dt = \Gamma(a) s^{-a} \Omega_6, \tag{56}$$

where Ω_6 is the same as given in Equation (21).

Theorem 15. For $m \in \mathbb{N}$, $Re(s) > 0$, $Re(c) > 0$ and $Re(2c - a - b) > -1$, the following result holds true.

$$\int_0^\infty e^{-st} t^{c-1} {}_3F_3\left[\begin{array}{c} a, \quad b, \quad 1 \\ \frac{1}{2}(a+b+1), \; 2c+1-m, \; m \end{array}; st\right] dt = \Gamma(c) s^{-c} \Omega_6, \tag{57}$$

where Ω_6 is the same as given in Equation (21).

Theorem 16. For $m \in \mathbb{N}$, $Re(s) > 0$ and $Re(2c - a - b) > -1$, the following result holds true.

$$\int_0^\infty e^{-st} {}_3F_3\left[\begin{array}{c} a, \quad b, \quad c \\ \frac{1}{2}(a+b+1), \; 2c+1-m, \; m \end{array}; st\right] dt = s^{-1} \Omega_6, \tag{58}$$

where Ω_6 is the same as given in Equation (21).

Theorem 17. For $m \in \mathbb{N}$, $Re(s) > 0$, $Re(a) > 0$ and $Re(b - m + 1) > 0$, the following result holds true.

$$\int_0^\infty e^{-st} t^{a-1} {}_3F_3\left[\begin{array}{c} 2m-a-1, \; b, \; 1 \\ c, \quad 2b-c+1, \; m \end{array}; st\right] dt = \Gamma(a) s^{-a} \Omega_7, \tag{59}$$

where Ω_7 is the same as given in Equation (22).

Theorem 18. For $m \in \mathbb{N}$, $Re(s) > 0$, $Re(b) > 0$ and $Re(b - m + 1) > 0$, the following result holds true.

$$\int_0^\infty e^{-st} t^{b-1} {}_3F_3\left[\begin{array}{c} a, \; 2m-a-1, \; 1 \\ c, \quad 2b-c+1, \; m \end{array}; st\right] dt = \Gamma(b) s^{-b} \Omega_7, \tag{60}$$

where Ω_7 is the same as given in Equation (22).

Theorem 19. For $m \in \mathbb{N}$, $Re(s) > 0$, $Re(2m - a - 1) > 0$ and $Re(b - m + 1) > 0$, the following result holds true.

$$\int_0^\infty e^{-st} t^{2m-a-2} {}_3F_3\left[\begin{array}{c} a, \quad c, \quad 1 \\ c, \; 2b-c+1, \; m \end{array}; st\right] dt = \Gamma(2m-a-1) s^{a+1-2m} \Omega_7, \tag{61}$$

where Ω_7 is the same as given in Equation (22).

Theorem 20. For $m \in \mathbb{N}$, $Re(s) > 0$ and $Re(b - m + 1) > 0$, the following result holds true.

$$\int_0^\infty e^{-st} {}_3F_3\left[\begin{matrix} a, & 2m-a-1, & b \\ c, & 2b-c+1, & m \end{matrix}; st\right] dt = s^{-1}\Omega_7, \qquad (62)$$

where Ω_7 is the same as given in Equation (22).

Theorem 21. For $m \in \mathbb{N}$, $n \in \mathbb{N}_0$, $Re(s) > 0$ and $Re(a) > 0$, the following result holds true.

$$\int_0^\infty e^{-st} t^{a-1} {}_3F_3\left[\begin{matrix} b, & -n+m-1, & 1 \\ c, & 1+a+b-c-n, & m \end{matrix}; st\right] dt = \Gamma(a) s^{-a} \Omega_8, \qquad (63)$$

where Ω_8 is the same as given in Equation (23).

Theorem 22. For $m \in \mathbb{N}$, $n \in \mathbb{N}_0$ and $Re(s) > 0$, the following result holds true.

$$\int_0^\infty e^{-st} {}_3F_3\left[\begin{matrix} a, & b, & -n+m-1 \\ c, & 1+a+b-c-n, & m \end{matrix}; st\right] dt = s^{-1}\Omega_8, \qquad (64)$$

where Ω_8 is the same as given in Equation (23).

Corollary 2. *(e)* In Theorem 11, if we take $m = 1, 2, 3$, we get the following results.

$$\int_0^\infty e^{-st} t^{a-1} {}_2F_2\left[\begin{matrix} b, & c \\ 1+a-b, & 1+a-c \end{matrix}; st\right] dt \qquad (65)$$

$$= \frac{\Gamma(\frac{1}{2}a)\Gamma(1+a-b)\Gamma(1+a-c)\Gamma(1+\frac{1}{2}a-b-c)}{2\,s^a\,\Gamma(1+\frac{1}{2}a-b)\Gamma(1+\frac{1}{2}a-c)\Gamma(1+a-b-c)},$$

$$\int_0^\infty e^{-st} t^{a-1} {}_3F_3\left[\begin{matrix} b, & c, & 1 \\ a-b+2, & a-c+2, & 2 \end{matrix}; st\right] dt \qquad (66)$$

$$= \frac{\Gamma(a-1)(1+a-b)(1+a-c)}{s^a\,(b-1)(c-1)}$$

$$\times \left\{ \frac{\Gamma(\frac{1}{2}(a+1))\Gamma(1+a-b)\Gamma(1+a-c)\Gamma(\frac{1}{2}a-b-c+\frac{5}{2})}{\Gamma(a)\Gamma(\frac{1}{2}a-b+\frac{3}{2})\Gamma(\frac{1}{2}a-c+\frac{3}{2})\Gamma(2+a-b-c)} - 1 \right\}$$

and

$$\int_0^\infty e^{-st} t^{a-1} {}_3F_3\left[\begin{matrix} b, & c, & 1 \\ a-b+3, & a-c+3, & 3 \end{matrix}; st\right] dt \qquad (67)$$

$$= \frac{2(a-b+1)_2(a-c+1)_2\,\Gamma(a)}{s^a\,(a-2)_2(b-2)_2(c-2)_2}$$

$$\times \left\{ \frac{\Gamma(\frac{1}{2}a)\Gamma(1+a-b)\Gamma(1+a-c)\Gamma(\frac{1}{2}a-b-c+4)}{\Gamma(a-1)\Gamma(\frac{1}{2}a-b+2)\Gamma(\frac{1}{2}a-c+2)\Gamma(3+a-b-c)} \right.$$

$$\left. - \frac{(a-2)(b-2)(c-2)}{(a-b+1)(a-c+1)} - 1 \right\}.$$

(f) In Theorem 14, if we take $m = 1, 2, 3$, we get the following results.

$$\int_0^\infty e^{-st} t^{a-1} {}_2F_2\left[\begin{array}{c} b, \quad c \\ \frac{1}{2}(a+b+1), \ 2c \end{array}; st\right] dt \qquad (68)$$

$$= \frac{\sqrt{\pi}\Gamma(a)\Gamma(c+\frac{1}{2})\Gamma(\frac{1}{2}(a+b+1))\Gamma(c-\frac{1}{2}(a+b-1))}{s^a \Gamma(\frac{1}{2}(a+1))\Gamma(\frac{1}{2}(b+1))\Gamma(c-\frac{1}{2}(a-1))\Gamma(c-\frac{1}{2}(b-1))},$$

$$\int_0^\infty e^{-st} t^{a-1} {}_3F_3\left[\begin{array}{c} b, \quad c, \quad 1 \\ \frac{1}{2}(a+b+1), \ 2c-1, \ 2 \end{array}; st\right] dt \qquad (69)$$

$$= \frac{(a+b-1)\Gamma(a-1)}{s^a (b-1)} \left\{ \frac{\sqrt{\pi}\Gamma(c-\frac{1}{2})\Gamma(\frac{1}{2}(a+b-1))\Gamma(c-\frac{1}{2}(a+b-1))}{\Gamma(\frac{1}{2}a)\Gamma(\frac{1}{2}b)\Gamma(c-\frac{1}{2}a)\Gamma(c-\frac{1}{2}b)} - 1 \right\}$$

and

$$\int_0^\infty e^{-st} t^{a-1} {}_3F_3\left[\begin{array}{c} b, \quad c, \quad 1 \\ \frac{1}{2}(a+b+1), \ 2c-2, \ 3 \end{array}; st\right] dt \qquad (70)$$

$$= \frac{(2c-3)(a+b-1)(a+b-3)\Gamma(a)}{s^a (c-1)(a-2)_2(b-2)_2}$$

$$\times \left\{ \frac{\sqrt{\pi}\Gamma(c-\frac{3}{2})\Gamma(\frac{1}{2}(a+b-3))\Gamma(c-\frac{1}{2}(a+b-1))}{\Gamma(\frac{1}{2}(a-1))\Gamma(\frac{1}{2}(b-1))\Gamma(c-\frac{1}{2}(a+1))\Gamma(c-\frac{1}{2}(b+1))} \right.$$

$$\left. - \frac{(a-2)(b-2)}{a+b-3} - 1 \right\}.$$

(g) In Theorem 17, if we take $m = 1, 2, 3$, we get the following results.

$$\int_0^\infty e^{-st} t^{a-1} {}_2F_2\left[\begin{array}{c} 1-a, \quad b \\ c, \ 2b-c+1 \end{array}; st\right] dt \qquad (71)$$

$$= \frac{\pi 2^{1-2b} \Gamma(a)\Gamma(c)\Gamma(2b-c+1)}{s^a \Gamma(\frac{1}{2}(a+c))\Gamma(b+\frac{1}{2}(a-c+1))\Gamma(\frac{1}{2}(1-a+c))\Gamma(b+1-\frac{1}{2}(a+c))},$$

$$\int_0^\infty e^{-st} t^{a-1} {}_3F_3\left[\begin{array}{c} 3-a, \quad b, \quad 1 \\ c, \ 2b-c+1, \ 2 \end{array}; st\right] dt = \frac{(c-1)(c-2b)\Gamma(a)}{s^a (a-2)_2(b-1)} \qquad (72)$$

$$\times \left\{ \frac{\pi 2^{3-2b}\Gamma(c-1)\Gamma(2b-c)}{\Gamma(\frac{1}{2}(a+c)-1)\Gamma(b+\frac{1}{2}(a-c-1))\Gamma(\frac{1}{2}(1-a+c))\Gamma(b+1-\frac{1}{2}(a+c))} - 1 \right\}$$

and

$$\int_0^\infty e^{-st} t^{a-1} {}_3F_3\left[\begin{array}{c} 5-a, \quad b, \quad 1 \\ c, \ 2b-c+1, \ 3 \end{array}; st\right] dt = \frac{2(c-2)_2(2b-c-1)_2\Gamma(a)}{s^a (a-4)_4(b-2)_2} \qquad (73)$$

$$\times \left\{ \frac{\pi 2^{5-2b} \Gamma(c-2)\Gamma(2b-c+1)}{\Gamma(\frac{1}{2}(a+c)-2)\Gamma(b+\frac{1}{2}(a-c-3))\Gamma(\frac{1}{2}(1-a+c))\Gamma(b+1-\frac{1}{2}(a+c))} \right.$$

$$\left. - \frac{(a-2)(3-a)(b-2)}{(c-2)(2b-c-1)} - 1 \right\}.$$

(h) In Theorem 21, if we take $m = 1, 2, 3$, we get the following results.

$$\int_0^\infty e^{-st} t^{a-1} {}_2F_2 \left[\begin{array}{c} -n, \quad b \\ 1+a+b-c-n, \quad c \end{array} ; st \right] dt = \frac{\Gamma(a)(c-a)_n(c-b)_n}{s^a (c)_n (c-a-b)_n}, \quad (74)$$

$$\int_0^\infty e^{-st} t^{a-1} {}_3F_3 \left[\begin{array}{c} -n+1, \quad b, \quad 1 \\ 1+a+b-c-n, \quad c, \quad 2 \end{array} ; st \right] dt \quad (75)$$

$$= \frac{(1-c)(c-a-b+n)\Gamma(a-1)}{n(1-b) s^a} \left\{ \frac{(c-a)_n(c-b)_n}{(c)_n(c-a-b+1)_n} - 1 \right\}$$

and

$$\int_0^\infty e^{-st} t^{a-1} {}_3F_3 \left[\begin{array}{c} -n+2, \quad b, \quad 1 \\ 1+a+b-c-n, \quad c, \quad 3 \end{array} ; st \right] dt \quad (76)$$

$$= \frac{2(1-c)_2(c-a-b+n)_2 \Gamma(a)}{s^a (1-a)_2(1-b)_2}$$

$$\times \left\{ \frac{(c-a)_n(c-b)_n}{(c-2)_n(c-a-b+2)_n} + \frac{n(a-2)(b-2)}{(c-2)(a+b-c-n-1)} - 1 \right\}.$$

Remark 5. *The results in Equations* (65), (68) *and* (71) *are known results due to Kim et al.* [24].

[C] Laplace-type integrals involving generalized hypergeometric function ${}_4F_4$

The results to be established are asserted in the following theorems.

Theorem 23. *For $m \in \mathbb{N}$, $Re(s) > 0$, $Re(a) > 0$ and $Re(a - 2b - 2c + 3m) > 2$, the following result holds true.*

$$\int_0^\infty e^{-st} t^{a-1} {}_4F_4 \left[\begin{array}{c} \frac{1}{2}(a+m+1), \quad b, \quad c, \quad 1 \\ \frac{1}{2}(a+m-1), \quad a-b+m, \quad a-c+m, \quad m \end{array} ; -st \right] dt = \Gamma(a) s^{-a} \Omega_9, \quad (77)$$

where Ω_9 is the same as given in Equation (24).

Theorem 24. *For $m \in \mathbb{N}$, $Re(s) > 0$, $Re(c) > 0$ and $Re(a - 2b - 2c + 3m) > 2$, the following result holds true.*

$$\int_0^\infty e^{-st} t^{c-1} {}_4F_4 \left[\begin{array}{c} a, \quad \frac{1}{2}(a+m+1), \quad b, \quad 1 \\ \frac{1}{2}(a+m-1), \quad a-b+m, \quad a-c+m, \quad m \end{array} ; -st \right] dt = \Gamma(c) s^{-c} \Omega_9, \quad (78)$$

where Ω_9 is the same as given in Equation (24).

Theorem 25. *For $m \in \mathbb{N}$, $Re(s) > 0$, $Re(a + m + 1) > 0$ and $Re(a - 2b - 2c + 3m) > 2$, the following result holds true.*

$$\int_0^\infty e^{-st} t^{\frac{1}{2}(a+m-1)} {}_4F_4 \left[\begin{array}{c} a, \quad b, \quad c, \quad 1 \\ \frac{1}{2}(a+m-1), \quad a-b+m, \quad a-c+m, \quad m \end{array} ; -st \right] dt \quad (79)$$

$$= \Gamma(\tfrac{1}{2}(a+m+1)) s^{-\frac{1}{2}(a+m+1)} \Omega_9,$$

where Ω_9 is the same as given in Equation (24).

Theorem 26. For $m \in \mathbb{N}$, $Re(s) > 0$ and $Re(a - 2b - 2c + 3m) > 2$, the following result holds true.

$$\int_0^\infty e^{-st} {}_4F_4\left[\begin{matrix} a, & \frac{1}{2}(a+m+1), & b, & c \\ \frac{1}{2}(a+m-1), & a-b+m, & a-c+m, & m \end{matrix}; -st\right] dt = s^{-1}\Omega_9, \tag{80}$$

where Ω_9 is the same as given in Equation (24).

Corollary 3. (i) In Theorem 24, if we take $m = 1, 2, 3$, we get the following results.

$$\int_0^\infty e^{-st} t^{c-1} {}_3F_3\left[\begin{matrix} a, & \frac{1}{2}(a+2), & b \\ \frac{1}{2}a, & a-b+1, & a-c+1 \end{matrix}; -st\right] dt \tag{81}$$
$$= \frac{\Gamma(1+a-b)\Gamma(1+a-c)\Gamma(c)}{s^c \Gamma(1+a)\Gamma(1+a-b-c)},$$

$$\int_0^\infty e^{-st} t^{c-1} {}_4F_4\left[\begin{matrix} a, & \frac{1}{2}(a+3), & b, & 1 \\ \frac{1}{2}(a+1), & a-b+2, & a-c+2, & 2 \end{matrix}; -st\right] dt \tag{82}$$
$$= \frac{(1+a-b)(1+a-c)\Gamma(c)}{s^c (a+1)(b-1)(c-1)}\left\{1 - \frac{\Gamma(1+a-b)\Gamma(1+a-c)}{\Gamma(a)\Gamma(2+a-b-c)}\right\}$$

and

$$\int_0^\infty e^{-st} t^{c-1} {}_4F_4\left[\begin{matrix} a, & \frac{1}{2}(a+4), & b, & 1 \\ \frac{1}{2}(a+3), & a-b+3, & a-c+3, & 3 \end{matrix}; -st\right] dt \tag{83}$$
$$= \frac{2(1+a-b)_2(1+a-c)_2 \Gamma(c)}{s^c (a+2)(a-1)(b-2)_2(c-2)_2}$$
$$\times \left\{\frac{\Gamma(1+a-b)\Gamma(1+a-c)}{\Gamma(a-1)\Gamma(3+a-b-c)} + \frac{a(b-2)(c-2)}{(1+a-b)(1+a-c)} - 1\right\}.$$

Similarly, other results can be obtained from Theorems 23, 25 and 26.

[D] Laplace-type integrals involving generalized hypergeometric function ${}_5F_5$

The results to be established are asserted in the following theorems.

Theorem 27. For $m \in \mathbb{N}$, $Re(s) > 0$, $Re(a) > 0$ and $Re(2a - 2c - 2d - 2e + 3m) > 2$, the following result holds true.

$$\int_0^\infty e^{-st} t^{a-1} {}_5F_5\left[\begin{matrix} \frac{1}{2}(a+m+1), & c, & d, & e, & 1 \\ \frac{1}{2}(a+m-1), & a-c+m, & a-d+m, & a-e+m, & m \end{matrix}; st\right] dt \tag{84}$$
$$= \Gamma(a) s^{-a} \Omega_{10},$$

where Ω_{10} is the same as given in Equation (25).

Theorem 28. For $m \in \mathbb{N}$, $Re(s) > 0$, $Re(c) > 0$ and $Re(2a - 2c - 2d - 2e + 3m) > 2$, the following result holds true.

$$\int_0^\infty e^{-st} t^{c-1} {}_5F_5\left[\begin{matrix} a, & \frac{1}{2}(a+m+1), & d, & e, & 1 \\ & \frac{1}{2}(a+m-1), & a-c+m, & a-d+m, & a-e+m, & m \end{matrix}; st\right] dt \tag{85}$$
$$= \Gamma(c) s^{-c} \Omega_{10},$$

where Ω_{10} is the same as given in Equation (25).

Theorem 29. For $m \in \mathbb{N}$, $Re(s) > 0$, $Re(a + m + 1) > 0$ and $Re(2a - 2c - 2d - 2e + 3m) > 2$, the following result holds true.

$$\int_0^\infty e^{-st} t^{\frac{1}{2}(a+m-1)} {}_5F_5 \left[\begin{array}{c} a, \quad c, \quad d, \quad e, \quad 1 \\ \frac{1}{2}(a+m-1), \, a-c+m, \, a-d+m, \, a-e+m, \, m \end{array} ; st \right] dt \quad (86)$$

$$= \Gamma(\frac{1}{2}(a+m+1)) \, s^{-\frac{1}{2}(a+m+1)} \, \Omega_{10},$$

where Ω_{10} is the same as given in Equation (25).

Theorem 30. For $m \in \mathbb{N}$, $Re(s) > 0$ and $Re(2a - 2c - 2d - 2e + 3m) > 2$, the following result holds true.

$$\int_0^\infty e^{-st} {}_5F_5 \left[\begin{array}{c} a, \quad \frac{1}{2}(a+m-1), \quad c, \quad d, \quad e \\ \frac{1}{2}(a+m-1), \, a-c+m, \, a-d+m, \, a-e+m, \, m \end{array} ; st \right] dt \quad (87)$$

$$= s^{-1} \Omega_{10},$$

where Ω_{10} is the same as given in Equation (25).

Corollary 4. (j) In Theorem 27, if we take $m = 1, 2, 3$, we get the following results.

$$\int_0^\infty e^{-st} t^{a-1} {}_4F_4 \left[\begin{array}{c} c, \quad \frac{1}{2}(a+2), \quad d, \quad e \\ \frac{1}{2}a, \, a-c+1, \, a-d+1, \, a-e+1 \end{array} ; st \right] dt \quad (88)$$

$$= \frac{\Gamma(1+a-c)\Gamma(1+a-d)\Gamma(1+a-e)\Gamma(1+a-c-d-e)}{s^a \, \Gamma(1+a-d-e)\Gamma(1+a-c-e)\Gamma(1+a-c-d)},$$

$$\int_0^\infty e^{-st} t^{a-1} {}_5F_5 \left[\begin{array}{c} c, \quad \frac{1}{2}(a+3), \quad d, \quad e, \quad 1 \\ \frac{1}{2}(a+1), \, a-c+2, \, a-d+2, \, a-e+2, \, 2 \end{array} ; st \right] dt \quad (89)$$

$$= \frac{(1+a-c)(1+a-d)(1+a-e)\,\Gamma(a)}{s^a \, (1+a)(c-1)(d-1)(e-1)}$$

$$\times \left\{ \frac{\Gamma(1+a-c)\Gamma(1+a-d)\Gamma(1+a-e)\Gamma(3+a-c-d-e)}{\Gamma(a)\Gamma(2+a-d-e)\Gamma(2+a-c-e)\Gamma(2+a-c-d)} - 1 \right\}$$

and

$$\int_0^\infty e^{-st} t^{a-1} {}_5F_5 \left[\begin{array}{c} c, \quad \frac{1}{2}(a+4), \quad d, \quad e, \quad 1 \\ \frac{1}{2}(a+2), \, a-c+3, \, a-d+3, \, a-e+3, \, 3 \end{array} ; st \right] dt \quad (90)$$

$$= \frac{2(1+a-c)_2(1+a-d)_2(1+a-e)_2 \, \Gamma(a)}{s^a \, (a-1)(a+2)(c-2)_2(d-2)_2(e-2)_2}$$

$$\times \left\{ \frac{\Gamma(1+a-c)\Gamma(1+a-d)\Gamma(1+a-e)\Gamma(5+a-c-d-e)}{\Gamma(a-1)\Gamma(3+a-d-e)\Gamma(3+a-c-e)\Gamma(3+a-c-d)} \right.$$

$$\left. - \frac{a(c-2)(d-2)(e-2)}{(1+a-c)(1+a-d)(1+a-e)} \right\}.$$

Similarly, other results can be obtained from Theorems 28–30.

[E] Laplace-type integrals involving generalized hypergeometric function ${}_6F_6$

The results to be established are asserted in the following theorems.

Theorem 31. For $m \in \mathbb{N}$, $Re(s) > 0$ and $Re(a) > 0$, the following result holds true.

$$\int_0^\infty e^{-st} t^{a-1} \times \tag{91}$$
$$\,_7F_7\left[\begin{array}{c}\frac{1}{2}(a+m+1),\ b,\ c,\ d,\ 2a-b-c-d+2m-1+n,\ m-n-1,\ 1\\ \frac{1}{2}(a+m-1), a-b+m, a-c+m, a-d+m, b+c+d-a+1-m-n, a+n+1, m'\end{array}; st\right] dt$$
$$= \Gamma(a)\, s^{-a}\, \Omega_{11},$$

where Ω_{11} is the same as given in Equation (26).

Theorem 32. For $m \in \mathbb{N}$, $Re(s) > 0$ and $Re(a+m+1) > 0$, the following result holds true.

$$\int_0^\infty e^{-st} t^{\frac{1}{2}(a+m-1)} \times \tag{92}$$
$$\,_7F_7\left[\begin{array}{c}a,\ b,\ c,\ d,\ 2a-b-c-d+2m-1+n,\ m-n-1,\ 1\\ \frac{1}{2}(a+m-1), a-b+m, a-c+m, a-d+m, b+c+d-a+1-m-n, a+n+1, m'\end{array}; st\right] dt$$
$$= \Gamma(\tfrac{1}{2}(a+m+1))\, s^{-\frac{1}{2}(a+m+1)}\, \Omega_{11},$$

where Ω_{11} is the same as given in Equation (26).

Theorem 33. For $m \in \mathbb{N}$, $Re(s) > 0$ and $Re(b) > 0$, the following result holds true.

$$\int_0^\infty e^{-st} t^{b-1} \times \tag{93}$$
$$\,_7F_7\left[\begin{array}{c}\frac{1}{2}(a+m+1),\ b,\ c,\ d,\ 2a-b-c-d+2m-1+n,\ m-n-1,\ 1\\ \frac{1}{2}(a+m-1), a-b+m, a-c+m, a-d+m, b+c+d-a+1-m-n, a+n+1, m'\end{array}; st\right] dt$$
$$= \Gamma(b)\, s^{-b}\, \Omega_{11},$$

where Ω_{11} is the same as given in Equation (26).

Theorem 34. For $m \in \mathbb{N}$, $Re(s) > 0$ and $Re(2a-b-c-d+2m-1+n) > 0$, the following result holds true.

$$\int_0^\infty e^{-st} t^{2a-b-c-d+2m+n-2} \times \tag{94}$$
$$\,_7F_7\left[\begin{array}{c}\frac{1}{2}(a+m+1),\ a,\ b,\ c,\ d,\ m-n-1,\ 1\\ \frac{1}{2}(a+m-1), a-b+m, a-c+m, a-d+m, b+c+d-a+1-m-n, a+n+1, m'\end{array}; st\right] dt$$
$$= \Gamma(2a-b-c-d+2m-1+n)\, s^{-(2a-b-c-d+2m-1+n)}\, \Omega_{11},$$

where Ω_{11} is the same as given in Equation (26).

Theorem 35. For $m \in \mathbb{N}$ and $Re(s) > 0$, the following result holds true.

$$\int_0^\infty e^{-st}\, _7F_7\left[\begin{array}{c}\frac{1}{2}(a+m+1),\ a,\ b,\ c,\ d,\ m-n-1,\ 1\\ \frac{1}{2}(a+m-1), a-b+m, a-c+m, a-d+m, b+c+d-a+1-m-n, a+n+1, m'\end{array}; st\right] dt \tag{95}$$
$$= s^{-1}\Omega_{11},$$

where Ω_{11} is the same as given in Equation (26).

Corollary 5. (k) *In Theorem* 31, *if we take* $m = 1, 2, 3$, *we get the following results.*

$$\int_0^\infty e^{-st} t^{a-1}$$ (96)

$$\times {}_6F_6\left[\begin{matrix} b, & \frac{1}{2}(a+2), & c, & d, & 2a-b-c-d+n+1, & -n \\ \frac{1}{2}a, 1+a-b, 1+a-c, 1+a-d, b+c+d-a-n, a+n+1 \end{matrix}; st\right] dt$$

$$= \frac{\Gamma(a)(1+a)_n(a-b-c+1)_n(a-b-d+1)_n(a-c-d+1)_n}{s^a (1+a-b)_n(1+a-c)_n(1+a-d)_n(1+a-b-c-d)_n},$$

$$\int_0^\infty e^{-st} t^{a-1}$$ (97)

$$\times {}_7F_7\left[\begin{matrix} b, & \frac{1}{2}(a+3), & c, & d, & 2a-b-c-d+n+3, & 1-n, & 1 \\ \frac{1}{2}(a+1), 1+a-b, 1+a-c, 1+a-d, b+c+d-a-n-1, a+n+1, 2 \end{matrix}; st\right] dt$$

$$= \frac{(b-a-1)(c-a-1)(d-a-1)(n+2+a-b-c-d)(a+n)\Gamma(a)}{n\,s^a (1+a)(1-b)(1-c)(1-d)(b+c+d-2a-2-n)}$$

$$\times \left\{1 - \frac{(a)_n(a-b-c+2)_n(a-b-d+2)_n(a-c-d+2)_n}{(1+a-b)_n(1+a-c)_n(1+a-d)_n(3+a-b-c-d)_n}\right\}$$

and

$$\int_0^\infty e^{-st} t^{a-1}$$ (98)

$$\times {}_7F_7\left[\begin{matrix} b, & \frac{1}{2}(a+4), & c, & d, & 2a-b-c-d+n+5, & 2-n, & 1 \\ \frac{1}{2}(a+2), 3+a-b, 3+a-c, 3+a-d, b+c+d-a-n-2, a+n+1, 3 \end{matrix}; st\right] dt$$

$$= \frac{(a-2)(b-a-2)_2(c-a-2)_2(d-a-2)_2}{s^a (a+2)(1-a)_2(1-b)_2(1-c)_2(1-d)_2}$$

$$\times \frac{(-a-n)_n(3+n+a-b-c-d)_2\,\Gamma(a)}{(n-1)_2(b+c+d-2a-4-n)_2}$$

$$\times \left\{\frac{(a-1)_n(a-b-c+3)_n(a-b-d+3)_n(a-c-d+3)_n}{(a-b+1)_n(a-c+1)_n(a-d+1)_n(a-b-c-d+5)_n}\right.$$

$$\left. + \frac{na(b-2)(c-2)(d-2)(2a-b-c+n+3)}{(a-b+1)(a-c+1)(a-d+1)(b+c+d-a-n-4)(n+a-1)} - 1\right\}.$$

Similarly, other results can be obtained from Theorems 32–35.

Remark 6. *For evaluation of Eulerian's type integrals involving generalized hypergeometric functions by employing the summation theorems, Equations* (16)–(26), *we refer an interesting paper by Jun et al.* [25].

3. Concluding Remark

In the theory of generalized hypergeometric functions, classical summation theorems such as those of Gauss, Gauss second, Kummer, Bailey, Dixon, Watson, Whipple, Saalschütz and Dougall play a key role. Applications of the above-mentioned classical summation theorems are well-known. Very recently, Masjed-Jamei and Koepf established interesting and useful generalizations of the above-mentioned classical summation theorems in the most general form.

In this paper, an attempt has been made for providing a list of several Laplace-type integrals involving generalized hypergeometric functions ${}_pF_p$ for $p = 2, 3, 4, 5$ and 7 in the most general forms which would serve as key formulas from which, on specializing the parameters, lead to several results, some of which are known and others are believed to be new. The results established in this paper are simple, interesting, easily proven and may be potentially useful.

We conclude this section by remarking that other applications of the generalized summation theorems due to Masjed-Jamei and Koepf are under investigations and the same will form a part of the subsequent paper in this direction.

Author Contributions: All authors contributed equally to writing of this paper. All authors read and approved the final manuscript.

Funding: The research work of Insuk Kim is supported by Wonkwang University research fund in 2019.

Acknowledgments: The authors are grateful to the referees for making certain very useful suggestions which led to a better presentation of the paper.

Conflicts of Interest: The authors declare that they have no competing interests.

References

1. Andrews, G.E.; Askey, R.; Roy, R. Special Functions. In *Encyclopedia of Mathematics and Its Applications*; Cambridge University Press: Cambridge, UK, 1999; Volume 71.
2. Bailey, W.N. *Generalized Hypergeometric Series*; Cambridge University Press: Cambridge, UK, 1935; Reprinted by Stechert-Hafner, New York, NY, USA, 1964.
3. Oberhettinger, F.; Badi, L. *Tables of Laplace Transforms*; Springer: Berlin, Germany, 1973.
4. Rainville, E.D. *Special Functions*; The Macmillan Company: New York, NY, USA, 1960; Reprinted by Chelsea Publishing Company, Bronx, NY, USA, 1971.
5. Koepf, W. *Hypergeometric Summation: An Algorithmic Approach to Summation and Special Function Identities*, 2nd ed.; Springer: London, UK, 2014.
6. Bromwich, T.J. *An Introduction to the Theory of Infinite Series*; Macmillan: New York, NY, USA, 1948.
7. Knopp, K. *Theory and Applications of Infinite Series*: Hafner: New York, NY, USA, 1949; Reprinted by Dover, 1990.
8. Luke, Y.L. *The Special Functions and Their Approximations*; Academic Press: New York, NY, USA, 1969; Volume 1.
9. Bailey, W.N. Products of Generalized Hypergeometric Series. *Proc. Lond. Math. Soc.* **1928**, *28*, 242–254. [CrossRef]
10. Prudnikov, A.P.; Brychkov, Yu.A.; Marichev, O.I. *More Special Functions*; Integrals and Series; Gordon and Breach Science Publishers: Amsterdam, The Netherlands, 1990; Volume 3.
11. Masjed-Jamei, M.; Koepf, W. Some Summation Theorems for Generalized Hypergeometric Functions. *Axioms* **2018**, *7*, 38. [CrossRef]
12. Kim, Y.S.; Rakha, M.A.; Rathie, A.K. Extensions of Certain Classical Summation Theorems for the Series $_2F_1$, $_3F_2$ and $_4F_3$ with Applications in Ramanujan's Summations. *Int. J. Math. Math. Sci.* **2010**, *2010*, 309503. [CrossRef]
13. Lavoie, J.L.; Grondin, F.; Rathie, A.K. Generalizations of Watson's theorem on the sum of a $_3F_2$. *Indian J. Math.* **1992**, *34*, 23–32.
14. Lavoie, J.L.; Grondin, F.; Rathie, A.K.; Arora, K. Generalizations of Dixon's Theorem on the sum of a $_3F_2(1)$. *Math. Comp.* **1994**, *62*, 267–276.
15. Lavoie, J.L.; Grondin, F.; Rathie, A.K. Generalizations of Whipple's theorem on the sum of a $_3F_2$. *J. Comput. Appl. Math.* **1996**, *72*, 293–300. [CrossRef]
16. Rakha, M.A.; Rathie, A.K. Generalizations of classical summation theorems for the series $_2F_1$ and $_3F_2$ with applications. *Integral Transform. Spec. Funct.* **2011**, *22*, 823–840. [CrossRef]
17. Davis, B. *Integral Transforms and Their Applications*, 3rd ed.; Springer: New York, NY, USA, 2002.
18. Doetsch, G. *Introduction to the Theory and Applications of the Laplace Transformation*; Springer: New York, NY, USA, 1974.
19. Erdelyi, A.; Magnus, W.; Oberhettinger, F.; Tricomi, F.G. *Tables of Integral Transforms*; McGraw Hill: New York, NY, USA, 1954; Volume I–II.
20. Slater, L.J. *Confluent Hypergeometric Functions*; Cambridge University Press: Cambridge, UK, 1960.
21. Deepthi, P.; Prajapati, J.C.; Rathie, A.K. New Laplace transforms of the $_2F_2$ hypergeometric function. *J. Fract. Calc. Appl.* **2017**, *8*, 150–155.
22. Parmar, R.K.; Purohit, S.D. Certain integral transforms and fractional integral formulas for the extended hypergeometric functions. *TWMS J. Appl. Eng. Math.* **2017**, *7*, 74–81.

23. Kim, Y.S.; Rathie, A.K.; Civijovic, D. New Laplace transforms of Kummer's confluent hypergeometric functions. *Math. Comput. Model.* **2012**, *55*, 1068–1071. [CrossRef]
24. Kim, Y.S.; Rathie, A.K.; Lee, C.H. New Laplace transforms for the generalized hypergeometric functions $_2F_2$. *Honam Math. J.* **2015**, *37*, 245–252. [CrossRef]
25. Jun, S.; Kim, I.; Rathie, A.K. On a new class of Eulerian's type integrals involving generalized hypergeometric functions. *Aust. J. Math. Anal. Appl.* **2019**, *16*, 10.

© 2019 by the authors. Licensee MDPI, Basel, Switzerland. This article is an open access article distributed under the terms and conditions of the Creative Commons Attribution (CC BY) license (http://creativecommons.org/licenses/by/4.0/).

Article

Generalized Hyers–Ulam Stability of the Additive Functional Equation

Yang-Hi Lee [1] and Gwang Hui Kim [2],*

[1] Department of Mathematics Education, Gongju National University of Education, Gongju 32553, Korea; yanghi2@hanmail.net
[2] Department of Mathematics, Kangnam University, Yongin 16979, Korea
* Correspondence: ghkim@kangnam.ac.kr

Received: 2 May 2019; Accepted: 21 June 2019; Published: 25 June 2019

Abstract: We will prove the generalized Hyers–Ulam stability and the hyperstability of the additive functional equation $f(x_1 + y_1, x_2 + y_2, \ldots, x_n + y_n) = f(x_1, x_2, \ldots, x_n) + f(y_1, y_2, \ldots, y_n)$. By restricting the domain of a mapping f that satisfies the inequality condition used in the assumption part of the stability theorem, we partially generalize the results of the stability theorems of the additive function equations.

Keywords: additive (Cauchy) equation; additive mapping; Hyers–Ulam stability; generalized Hyers–Ulam stability; hyperstability

MSC: 39B82; 39B5

1. Introduction

In 1940, Ulam [1] gave the question concerning the stability of homomorphisms in a conference of the mathematics club of the University of Wisconsin as follows:

Let $(G, \cdot)$ be a group, and let $(G', \cdot, d)$ be a metric group with the metric d. Given $\delta > 0$, does there exist $\epsilon > 0$ such that if a mapping $h : G \to G'$ satisfies the inequality

$$d(h(xy), h(x)h(y)) \leq \delta$$

for all $x, y \in G$, then there is a homomorphism $H : G \to H$ with

$$d(h(x), H(x)) \leq \epsilon$$

for all $x \in G$?

Next year, the Ulam's conjecture was partially solved by Hyers [2] for the additive functional equation.

Theorem 1. [2], Let X and Y be Banach spaces. Suppose that the mapping $f : X \to Y$ satisfies the inequality

$$\|f(x+y) - f(x) - f(y)\| \leq \varepsilon, \quad \forall x, y \in X, \quad \varepsilon : constant.$$

Then, there exists a unique additive mapping

$$A(x+y) = A(x) + A(y),$$

such that $\|f(x) - A(x)\| \leq \varepsilon$, where the limit $A(x) = \lim_{n \to \infty} 2^{-n} f(2^n x)$.

Thereafter, this phenomenon has been called the Hyers–Ulam stability.

Theorem 2. Let X and Y be Banach spaces. Suppose that the mapping $f : X \to Y$ satisfies the inequality

$$\|f(x+y) - f(x) - f(y)\| \le \theta(\|x\|^p + \|y\|^p) \tag{1}$$

for all $x, y \in X\setminus\{0\}$, where θ and p are constants with $\theta > 0$ and $p \ne 1$. Then, there exists a unique additive mapping $T : X \to Y$ such that

$$\|f(x) - T(x)\| \le \frac{\theta}{|1 - 2^{p-1}|} \|x\|^p \tag{2}$$

for all $x \in X\setminus\{0\}$.

Theorem 2 is due to Aoki [3] and Rassias [4] for $0 < p < 1$, Gajda [5] for $p > 1$, Hyers [2] for $p = 0$, and Rassias [6] for $p < 0$.

In 1994, Găvruta [7] generalized these results for additive mapping by replacing $\theta(\|x\|^p + \|y\|^p)$ in (1) by a general function $\varphi(x, y)$, which is called the 'generalized Hyers–Ulam stability' in this paper.

In 2001, the term hyperstability was used for the first time probably by G. Maksa and Z. Páles in [8]. However, in 1949, it seems to have created by D. G. Bourgin [9] that the first hyperstability result concerned the ring homomorphisms.

We say that a functional equation $\mathfrak{D}(f) = 0$ is hyperstable if any function f satisfying the equation $\mathfrak{D}(f) = 0$ approximately is a true solution of $\mathfrak{D}(f) = 0$, which is a phenomenon called hyperstability. The hyperstability results for the additive (Cauchy) equation were investigated by Brzdęk [10,11].

In this paper, let V and W be vector spaces, X be a real normed space, and Y be a real Banach space. We denote the set of natural numbers by $\mathbb{N}$ and the set of real numbers by $\mathbb{R}$.

For a given mapping $f : V^n \to W$, where V^n denotes $V \times V \times \cdots \times V$, let us consider the additive functional equation

$$f(x_1 + y_1, x_2 + y_2, \ldots, x_n + y_n) = f(x_1, x_2, \ldots, x_n) + f(y_1, y_2, \ldots, y_n), \tag{3}$$

for all $x_i, y_i \in V$ ($i = 1, 2, \ldots, n$).

Each solution of the additive functional Equation (3) is called an n-variable additive mapping. A typical example for the solutions of Equation (3) is the mapping $f : \mathbb{R}^n \to \mathbb{R}^l$ given by $f(x_1, x_2, \ldots, x_n) = (\sum_{i=1}^{n} a_{1i} x_i, \sum_{i=1}^{n} a_{2i} x_i, \ldots, \sum_{i=1}^{n} a_{li} x_i)$ with real constants a_{ij}.

In this paper, we will prove the generalized Hyers–Ulam stability of the additive functional Equation (3) in the spirit of Găvruta [7], and the hyperstability of the additive functional Equation (3).

2. Main Results

For a given mapping $f : V^n \to W$, we use the following abbreviation:

$$Df(x_1, y_1, x_2, y_2, \ldots, x_n, y_n) := f(x_1 + y_1, x_2 + y_2, \ldots, x_n + y_n)$$
$$- f(x_1, x_2, \ldots, x_n) - f(y_1, y_2, \ldots, y_n)$$

for all $x_1, y_1, x_2, y_2, \ldots, x_n, y_n \in V$. We need the following lemma to prove main theorems.

Lemma 1. If a mapping $f : V^n \to W$ satisfies (3) for all $x_1, y_1, x_2, y_2, \ldots, x_n, y_n \in V\setminus\{0\}$, then f satisfies (3) for all $x_1, y_1, x_2, y_2, \ldots, x_n, y_n \in V$.

Proof. Let $x \in V\setminus\{0\}$ be a fixed element, and let $i \in \{1,2,\ldots,n\}$. For given $x_i, y_i \in V$, let $x_i^{(1)}, x_i^{(2)}, y_i^{(1)}, y_i^{(2)}$ be

$$
\begin{aligned}
&x_i^{(1)} = x, \; x_i^{(2)} = -x, \; y_i^{(1)} = x, \; y_i^{(2)} = -x && \text{if } x_i = 0 \text{ and } y_i = 0, \\
&x_i^{(1)} = y_i, \; x_i^{(2)} = -y_i, \; y_i^{(1)} = \frac{y_i}{2}, \; y_i^{(2)} = \frac{y_i}{2} && \text{if } x_i = 0 \text{ and } y_i \neq 0, \\
&x_i^{(1)} = \frac{x_i}{2}, \; x_i^{(2)} = \frac{x_i}{2}, \; y_i^{(1)} = x_i, \; y_i^{(2)} = -x_i && \text{if } x_i \neq 0 \text{ and } y_i = 0, \\
&x_i^{(1)} = \frac{x_i}{2}, \; x_i^{(2)} = \frac{x_i}{2}, \; y_i^{(1)} = (k+1)y_i, \; y_i^{(2)} = -ky_i && \text{if } x_i \neq 0 \text{ and } y_i \neq 0,
\end{aligned}
$$

where k is a fixed integer, such that $\frac{x_i}{2} + (k+1)y_i \neq 0$, $\frac{x_i}{2} - ky_i \neq 0$. Then, $x_i^{(1)}, x_i^{(2)}, y_i^{(1)}, y_i^{(2)}, x_i^{(1)} + y_i^{(1)}, x_i^{(2)} + y_i^{(2)} \in V\setminus\{0\}$ and $x_i^{(1)} + y_i^{(1)} + x_i^{(2)} + y_i^{(2)} = x_i + y_i$ for all $i = 1, 2, \ldots, n$.

Hence, the equalities $Df(x_1^{(1)}, y_1^{(1)}, \ldots, x_n^{(1)}, y_n^{(1)}) = 0$, $Df(x_1^{(2)}, y_1^{(2)}, \ldots, x_n^{(2)}, y_n^{(2)}) = 0$, $Df(x_1^{(1)}, x_1^{(2)}, x_2^{(1)}, x_2^{(2)}, \ldots, x_n^{(1)}, x_n^{(2)}) = 0$, and $Df(y_1^{(1)}, y_1^{(2)}, y_2^{(1)}, y_2^{(2)}, \ldots, y_n^{(1)}, y_n^{(2)}) = 0$ hold for all $x_1, y_1, x_2, y_2, \ldots, x_n, y_n \in V$. Since the equality

$$
\begin{aligned}
Df(&x_1, y_1, x_2, y_2, \ldots, x_n, y_n) \\
= Df(&x_1^{(1)} + y_1^{(1)}, x_1^{(2)} + y_1^{(2)}, x_2^{(1)} + y_2^{(1)}, x_2^{(2)} + y_2^{(2)}, \ldots, x_n^{(1)} + y_n^{(1)}, x_n^{(2)} + y_n^{(2)}) \\
&+ Df(x_1^{(1)}, y_1^{(1)}, x_2^{(1)}, y_2^{(1)}, \ldots, x_n^{(1)}, y_n^{(1)}) + Df(x_1^{(2)}, y_1^{(2)}, x_2^{(2)}, y_2^{(2)}, \ldots, x_n^{(2)}, y_n^{(2)}) \\
&- Df(x_1^{(1)}, x_1^{(2)}, x_2^{(1)}, x_2^{(2)}, \ldots, x_n^{(1)}, x_n^{(2)}) - Df(y_1^{(1)}, y_1^{(2)}, y_2^{(1)}, y_2^{(2)}, \ldots, y_n^{(1)}, y_n^{(2)})
\end{aligned}
$$

holds for all $x_1, y_1, x_2, y_2, \ldots, x_n, y_n \in V$, we conclude that f satisfies $Df(x_1, y_1, \ldots, x_n, y_n) = 0$ for all $x_1, y_1, x_2, y_2, \ldots, x_n, y_n \in V$. □

Thereafter, let $i \in \{1, 2, 3, \ldots, n\}$. For a given element $(x_1, x_2, \ldots, x_n) \neq (0, 0, \ldots, 0)$, we can choose a fixed element $x' \neq 0$, such that $x' \in \{x_1, x_2, \ldots, x_n\}$. Moreover, let $x_i^{(1)}, x_i^{(2)} \in V\setminus\{0\}$ be the elements defined by

$$
\begin{aligned}
&x_i^{(1)} = x_i, \; x_i^{(2)} = x_i && \text{if } x_i \neq 0, \\
&x_i^{(1)} = x', \; x_i^{(2)} = -x' && \text{if } x_i = 0.
\end{aligned} \tag{4}
$$

By using Lemma 1, we can prove the following set of stability theorems.

Theorem 3. *Suppose that $f : V^n \to Y$ is a mapping for which there exists a function $\varphi : (V\setminus\{0\})^{2n} \to [0, \infty)$, such that*

$$\sum_{m=0}^{\infty} \frac{\varphi(2^m x_1, 2^m y_1, 2^m x_2, 2^m y_2, \ldots, 2^m x_n, 2^m y_n)}{2^m} < \infty \tag{5}$$

and

$$\|Df(x_1, y_1, x_2, y_2, \ldots, x_n, y_n)\| \leq \varphi(x_1, y_1, x_2, y_2, \ldots, x_n, y_n) \tag{6}$$

for all $x_1, y_1, x_2, y_2, \ldots, x_n, y_n \in V\setminus\{0\}$. Then, there exists a unique mapping $F : V^n \to Y$ that satisfies

$$DF(x_1, y_1, x_2, y_2, \ldots, x_n, y_n) = 0 \tag{7}$$

for all $x_1, y_1, x_2, y_2, \ldots, x_n, y_n \in V$ and

$$\|f(x_1, x_2, \ldots, x_n) - F(x_1, x_2, \ldots, x_n)\| \leq \sum_{m=0}^{\infty} \frac{\mu(2^m x_1, 2^m x_2, \ldots, 2^m x_n)}{2^{m+1}} \tag{8}$$

for all $(x_1, x_2, \ldots, x_n) \in V^n \setminus \{(0, 0, \ldots, 0)\}$, where the function $\mu : V^n \to \mathbb{R}$ is defined by

$$\mu(x_1, x_2, \ldots, x_n)$$
$$:= \varphi\left(x_1^{(1)}, x_1^{(2)}, x_2^{(1)}, x_2^{(2)}, \ldots, x_n^{(1)}, x_n^{(2)}\right) + 2\varphi\left(\frac{x_1^{(1)}}{2}, \frac{x_1^{(2)}}{2}, \frac{x_2^{(1)}}{2}, \frac{x_2^{(2)}}{2}, \ldots, \frac{x_n^{(1)}}{2}, \frac{x_n^{(2)}}{2}\right)$$
$$+ \varphi\left(\frac{x_1^{(1)}}{2}, \frac{x_1^{(1)}}{2}, \frac{x_2^{(1)}}{2}, \frac{x_2^{(1)}}{2}, \ldots, \frac{x_n^{(1)}}{2}, \frac{x_n^{(1)}}{2}\right) + \varphi\left(\frac{x_1^{(2)}}{2}, \frac{x_1^{(2)}}{2}, \frac{x_2^{(2)}}{2}, \frac{x_2^{(2)}}{2}, \ldots, \frac{x_n^{(2)}}{2}, \frac{x_n^{(2)}}{2}\right)$$

for all $(x_1, x_2, \ldots, x_n) \in V^n \setminus \{(0, 0, \ldots, 0)\}$.

Proof. From the inequality (6) and the equalities

$$f(2x_1, 2x_2, \ldots, 2x_n) - 2f(x_1, x_2, \ldots, x_n) \tag{9}$$
$$= f(2x_1, 2x_2, \ldots, 2x_n) - f\left(x_1^{(1)}, x_2^{(1)}, \ldots, x_n^{(1)}\right) - f\left(x_1^{(2)}, x_2^{(2)}, \ldots, x_n^{(2)}\right)$$
$$- 2f(x_1, x_2, \ldots, x_n) + 2f\left(\frac{x_1^{(1)}}{2}, \frac{x_2^{(1)}}{2}, \ldots, \frac{x_n^{(1)}}{2}\right) + 2f\left(\frac{x_1^{(2)}}{2}, \frac{x_2^{(2)}}{2}, \ldots, \frac{x_n^{(2)}}{2}\right)$$
$$+ f\left(x_1^{(1)}, x_2^{(1)}, \ldots, x_n^{(1)}\right) - 2f\left(\frac{x_1^{(1)}}{2}, \frac{x_2^{(1)}}{2}, \ldots, \frac{x_n^{(1)}}{2}\right)$$
$$+ f\left(x_1^{(2)}, x_2^{(2)}, \ldots, x_n^{(2)}\right) - 2f\left(\frac{x_1^{(2)}}{2}, \frac{x_2^{(2)}}{2}, \ldots, \frac{x_n^{(2)}}{2}\right)$$
$$= Df\left(x_1^{(1)}, x_1^{(2)}, x_2^{(1)}, x_2^{(2)}, \ldots, x_n^{(1)}, x_n^{(2)}\right) - 2Df\left(\frac{x_1^{(1)}}{2}, \frac{x_1^{(2)}}{2}, \frac{x_2^{(1)}}{2}, \frac{x_2^{(2)}}{2}, \ldots, \frac{x_n^{(1)}}{2}, \frac{x_n^{(2)}}{2}\right)$$
$$+ Df\left(\frac{x_1^{(1)}}{2}, \frac{x_1^{(1)}}{2}, \frac{x_2^{(1)}}{2}, \frac{x_2^{(1)}}{2}, \ldots, \frac{x_n^{(1)}}{2}, \frac{x_n^{(1)}}{2}\right)$$
$$+ Df\left(\frac{x_1^{(2)}}{2}, \frac{x_1^{(2)}}{2}, \frac{x_2^{(2)}}{2}, \frac{x_2^{(2)}}{2}, \ldots, \frac{x_n^{(2)}}{2}, \frac{x_n^{(2)}}{2}\right)$$

for all $(x_1, x_2, \ldots, x_n) \in V^n \setminus \{(0, 0, \ldots, 0)\}$, we have

$$\left\| f(x_1, x_2, \ldots, x_n) - \frac{f(2x_1, 2x_2, \ldots, 2x_n)}{2} \right\|$$
$$\leq \left\| Df\left(x_1^{(1)}, x_1^{(2)}, x_2^{(1)}, x_2^{(2)}, \ldots, x_n^{(1)}, x_n^{(2)}\right) \right\| + 2 \left\| Df\left(\frac{x_1^{(1)}}{2}, \frac{x_1^{(2)}}{2}, \frac{x_2^{(1)}}{2}, \frac{x_2^{(2)}}{2}, \ldots, \frac{x_n^{(1)}}{2}, \frac{x_n^{(2)}}{2}\right) \right\|$$
$$+ \left\| Df\left(\frac{x_1^{(1)}}{2}, \frac{x_1^{(1)}}{2}, \frac{x_2^{(1)}}{2}, \frac{x_2^{(1)}}{2}, \ldots, \frac{x_n^{(1)}}{2}, \frac{x_n^{(1)}}{2}\right) \right\|$$
$$+ \left\| Df\left(\frac{x_1^{(2)}}{2}, \frac{x_1^{(2)}}{2}, \frac{x_2^{(2)}}{2}, \frac{x_2^{(2)}}{2}, \ldots, \frac{x_n^{(2)}}{2}, \frac{x_n^{(2)}}{2}\right) \right\|$$
$$\leq \frac{1}{2} \mu(x_1, x_2, \ldots, x_n)$$

for all $(x_1, x_2, \ldots, x_n) \in V^n \setminus \{(0,0,\ldots,0)\}$. From the above inequality, we get the (following-4 palces) inequality

$$\left\| \frac{f(2^m x_1, \ldots, 2^m x_n)}{2^m} - \frac{f(2^{m+m'} x_1, \ldots, 2^{m+m'} x_n)}{2^{m+m'}} \right\|$$

$$\leq \sum_{k=m}^{m+m'-1} \left\| \frac{f(2^k x_1, \ldots, 2^k x_n)}{2^k} - \frac{f(2^{k+1} x_1, \ldots, 2^{k+1} x_n)}{2^{k+1}} \right\|$$

$$\leq \sum_{k=m}^{m+m'-1} \frac{\mu(2^k x_1, 2^k x_2, \ldots, 2^k x_n)}{2^{k+1}} \tag{10}$$

for all $(x_1, x_2, \ldots, x_n) \in V^n \setminus \{(0,0,\ldots,0)\}$ and all positive integers m, m'. Thus, the sequence $\{\frac{f(2^n x_1, \ldots, 2^n x_n)}{2^n}\}_{m \in \mathbb{N}}$ is a Cauchy sequence for all $(x_1, x_2, \ldots, x_n) \in V^n \setminus \{(0,0,\ldots,0)\}$. Since Y is a real Banach space and $\lim_{m \to \infty} \frac{f(2^m 0, 2^m 0, \ldots, 2^m 0)}{2^m} = 0$, we can define a mapping $F : V^n \to Y$ by

$$F(x_1, x_2, \ldots, x_n) = \lim_{m \to \infty} \frac{f(2^m x_1, 2^m x_2, \ldots, 2^m x_n)}{2^m}$$

for all $x_1, x_2, \ldots, x_n \in V$. By putting $m = 0$ and by letting $m' \to \infty$ in the inequalities (10), we can obtain the inequalities (8) for all $(x_1, x_2, \ldots, x_n) \in V^n \setminus \{(0,0,\ldots,0)\}$.

From the inequality (6), we can obtain

$$\left\| \frac{Df(2^m x_1, 2^m y_1, 2^m x_2, 2^m y_2, \ldots, 2^m x_n, 2^m y_n)}{2^m} \right\| \leq \frac{\varphi(2^m x_1, 2^m y_1, 2^m x_2, \ldots, 2^m x_n, 2^m y_n)}{2^m}$$

for all $x_1, y_1, x_2, y_2, \ldots, x_n, y_n \in V \setminus \{0\}$. Since the right-hand side in the above equality tends to zero as $m \to \infty$, and the equality

$$DF(x_1, y_1, x_2, y_2, \ldots, x_n, y_n) = \lim_{m \to \infty} \frac{Df(2^m x_1, 2^m y_1, 2^m x_2, 2^m y_2, \ldots, 2^m x_n, 2^m y_n)}{2^m}$$

holds, then F satisfies the equality (7) for all $x_1, y_1, \ldots, x_n, y_n \in V \setminus \{0\}$. By Lemma 1, F satisfies the equality (3) for all $x_1, y_1, x_2, y_2, \ldots, x_n, y_n \in V$. If $G : V^n \to Y$ is another n-variable additive mapping that satisfies (8), then we obtain $G(0, 0, \ldots, 0) = 0 = F(0, 0, \ldots, 0)$ and

$$\| G(x_1, x_2, \ldots, x_n) - F(x_1, x_2, \ldots, x_n) \|$$

$$\leq \left\| \frac{G(2^k x_1, 2^k x_2, \ldots, 2^k x_n)}{2^k} - \frac{f(2^k x_1, 2^k x_2, \ldots, 2^k x_n)}{2^k} \right\|$$

$$+ \left\| \frac{f(2^k x_1, 2^k x_2, \ldots, 2^k x_n)}{2^k} - \frac{F(2^k x_1, 2^k x_2, \ldots, 2^k x_n)}{2^k} \right\|$$

$$\leq \sum_{m=k}^{\infty} \frac{\mu(2^m x_1, 2^m x_2, \ldots, 2^m x_n)}{2^m}$$

for all $(x_1, x_2, \ldots, x_n) \in V^n \setminus \{(0,0,\ldots,0)\}$ and all $k \in \mathbb{N}$. Since $\sum_{m=k}^{\infty} \frac{\mu(2^m x_1, 2^m x_2, \ldots, 2^m x_n)}{2^m} \to 0$ as $k \to \infty$, we have $G(x_1, x_2, \ldots, x_n) = F(x_1, x_2, \ldots, x_n)$ for all $x_1, x_2, \ldots, x_n \in V$. Hence, the mapping F is the unique n-variable additive mapping, as desired. $\square$

The condition $x_1, y_1, x_2, y_2, \ldots, x_n, y_n \in V \setminus \{0\}$ used in the inequality (6) differs from the condition $(x_1, x_2, \ldots, x_n) \neq (0, 0, \ldots, 0)$ and $(y_1, y_2, \ldots, y_n) \neq (0, 0, \ldots, 0)$ handled by the other authors. If the function f satisfies the inequality (3.2) for all $(x_1, x_2, \ldots, x_n) \neq (0, 0, \ldots, 0)$ and $(y_1, y_2, \ldots, y_n) \neq$

$(0,0,\ldots,0)$, then the function f satisfies the inequality (3.2) for all $x_1, y_1, x_2, y_2, \ldots, x_n, y_n \in V \setminus \{0\}$. Therefore, the condition $x_1, y_1, x_2, y_2, \ldots, x_n, y_n \in V \setminus \{0\}$ used in the inequality (3.2) in this paper is a generalization of the conditions used in the inequality (3.2) in the well-known pre-results ([10,11]). This condition will apply until Corollary 1.

Theorem 4. *Suppose that $f : V^n \to Y$ is a mapping for which there exists a function $\varphi : (V \setminus \{0\})^{2n} \to [0, \infty)$ that satisfies*

$$\sum_{i=0}^{\infty} 2^i \varphi\left(\frac{x_1}{2^i}, \frac{y_1}{2^i}, \frac{x_2}{2^i}, \frac{y_2}{2^i}, \ldots, \frac{x_n}{2^i}, \frac{y_n}{2^i}\right) < \infty, \tag{11}$$

and (6) for all $x_1, y_1, x_2, y_2, \ldots, x_n, y_n \in V \setminus \{0\}$. Then, there exists a unique mapping $F : V^n \to Y$ that satisfies (7) for all $x_1, y_1, x_2, y_2, \ldots, x_n, y_n \in V$ and

$$\|f(x_1, x_2, \ldots, x_n) - F(x_1, x_2, \ldots, x_n)\| \leq \sum_{m=0}^{\infty} 2^m \mu\left(\frac{x_1}{2^{m+1}}, \frac{x_2}{2^{m+1}}, \ldots, \frac{x_n}{2^{m+1}}\right) \tag{12}$$

for all $(x_1, x_2, \ldots, x_n) \in V^n \setminus \{(0,0,\ldots,0)\}$, where the function $\mu : V^n \to \mathbb{R}$ is defined as Theorem 3.

Proof. By choosing a fixed element $x \in V \setminus \{0\}$, we can obtain

$$\|f(0,0,\ldots,0)\| = \left\|2Df\left(\frac{x}{2^m}, \frac{-x}{2^m}, \ldots, \frac{x}{2^m}, \frac{-x}{2^m}\right) - Df\left(\frac{x}{2^{m-1}}, \frac{-x}{2^{m-1}}, \ldots, \frac{x}{2^{m-1}}, \frac{-x}{2^{m-1}}\right)\right.$$

$$\left. - Df\left(\frac{x}{2^m}, \frac{x}{2^m}, \ldots, \frac{x}{2^m}, \frac{x}{2^m}\right) - Df\left(\frac{-x}{2^m}, \frac{-x}{2^m}, \ldots, \frac{-x}{2^m}, \frac{-x}{2^m}\right)\right\|$$

$$\leq 2\varphi\left(\frac{x}{2^m}, \frac{-x}{2^m}, \ldots, \frac{x}{2^m}, \frac{-x}{2^m}\right) + \varphi\left(\frac{x}{2^{m-1}}, \frac{-x}{2^{m-1}}, \ldots, \frac{x}{2^{m-1}}, \frac{-x}{2^{m-1}}\right)$$

$$+ \varphi\left(\frac{x}{2^m}, \frac{x}{2^m}, \ldots, \frac{x}{2^m}, \frac{x}{2^m}\right) + \varphi\left(\frac{-x}{2^m}, \frac{-x}{2^m}, \ldots, \frac{-x}{2^m}, \frac{-x}{2^m}\right)$$

$$\to 0 \quad \text{as } m \to \infty,$$

so $f(0,0,\ldots,0) = 0$. Since the equality (9) holds for all $(x_1, x_2, \ldots, x_n) \in V \setminus \{(0,0,\ldots,0)\}$, the inequality (6) implies the inequality

$$\left\|f(x_1, x_2, \ldots, x_n) - 2f\left(\frac{x_1}{2}, \frac{x_2}{2}, \ldots, \frac{x_n}{2}\right)\right\| \leq \mu\left(\frac{x_1}{2}, \frac{x_2}{2}, \ldots, \frac{x_n}{2}\right)$$

for all $(x_1, x_2, \ldots, x_n) \in V^n \setminus \{(0,0,\ldots,0)\}$. From the above inequality, we can also obtain the inequality

$$\left\|2^m f\left(\frac{x_1}{2^m}, \frac{x_2}{2^m}, \ldots, \frac{x_n}{2^m}\right) - 2^{m+m'} f\left(\frac{x_1}{2^{m+m'}}, \frac{x_2}{2^{m+m'}}, \ldots, \frac{x_n}{2^{m+m'}}\right)\right\|$$

$$\leq \sum_{k=m}^{m+m'-1} 2^k \mu\left(\frac{x_1}{2^{k+1}}, \frac{x_2}{2^{k+1}}, \ldots, \frac{x_n}{2^{k+1}}\right) \tag{13}$$

for all $(x_1, x_2, \ldots, x_n) \in V^n \setminus \{(0,0,\ldots,0)\}$ and all positive integers m, m'. Thus, the sequences $\{2^m f(\frac{x_1}{2^m}, \ldots, \frac{x_n}{2^m})\}_{m \in \mathbb{N}}$ is a Cauchy sequence for all $(x_1, \ldots, x_n) \in V^n \setminus \{(0, \ldots, 0)\}$. Since $f(0,0,\ldots,0) = 0$ and Y is a real Banach space, we can define a mapping $F : V^n \to Y$ by

$$F(x_1, x_2, \ldots, x_n) = \lim_{m \to \infty} 2^m f\left(\frac{x_1}{2^m}, \frac{x_2}{2^m}, \ldots, \frac{x_n}{2^m}\right)$$

for all $x_1, x_2, \ldots, x_n \in V$. By putting $m = 0$ and by letting $m' \to \infty$ in the inequality (13), we can obtain the inequality (12) for all $(x_1, x_2, \ldots, x_n) \in V^n \setminus \{(0,0,\ldots,0)\}$.

From the inequality (6), we get

$$\left\|2^m Df\left(\frac{x_1}{2^m}, \frac{y_1}{2^m}, \frac{x_2}{2^m}, \frac{y_2}{2^m}, \ldots, \frac{x_n}{2^m}, \frac{y_n}{2^m}\right)\right\| \leq 2^m \varphi\left(\frac{x_1}{2^m}, \frac{y_1}{2^m}, \frac{x_2}{2^m}, \frac{y_2}{2^m}, \ldots, \frac{x_n}{2^m}, \frac{y_n}{2^m}\right)$$

for all $x_1, y_1, x_2, y_2, \ldots, x_n, y_n \in V\setminus\{0\}$. Since the right-hand side in the above equality tends to zero as $m \to \infty$, then F satisfies the equality (7) for all $x_1, y_1, x_2, y_2, \ldots, x_n, y_n \in V\setminus\{0\}$. By Lemma 1, F satisfies the equality (3) for all $x_1, y_1, x_2, y_2, \ldots, x_n, y_n \in V$. If $G : V^n \to Y$ is another n-variable additive mapping satisfying (12), then we obtain $G(0, 0, \ldots, 0) = 0 = F(0, 0, \ldots, 0)$ and

$$\begin{aligned}
\|G(x_1, x_2, \ldots, x_n) &- F(x_1, x_2, \ldots, x_n)\| \\
&\leq \left\|2^k G\left(\frac{x_1}{2^k}, \frac{x_2}{2^k}, \ldots, \frac{x_n}{2^k}\right) - 2^k f\left(\frac{x_1}{2^k}, \frac{x_2}{2^k}, \ldots, \frac{x_n}{2^k}\right)\right\| \\
&+ \left\|2^k f\left(\frac{x_1}{2^k}, \frac{x_2}{2^k}, \ldots, \frac{x_n}{2^k}\right) - 2^k F\left(\frac{x_1}{2^k}, \frac{x_2}{2^k}, \ldots, \frac{x_n}{2^k}\right)\right\| \\
&\leq \sum_{m=k}^{\infty} 2^m \mu\left(\frac{x_1}{2^{m+1}}, \frac{x_2}{2^{m+1}}, \ldots, \frac{x_n}{2^{m+1}}\right) \\
&\to 0 \text{ as } k \to \infty
\end{aligned}$$

for all $(x_1, x_2, \ldots, x_n) \in V^n \setminus \{(0, 0, \ldots, 0)\}$. Hence, the mapping F is the unique n-variable additive mapping, as desired. □

The following corollary follows from Theorems 3 and 4.

Corollary 1. *Let $(X, \|\|\cdot\|\|)$ be a normed space, $\theta > 0$, and let p be a real number with $p \neq 1$. Suppose that $f : X^n \to Y$ is a mapping that satisfies*

$$\|Df(x_1, y_1, x_2, y_2, \ldots, x_n, y_n)\| \leq \theta(\|\|x_1\|\|^p + \|\|y_1\|\|^p + \|\|x_2\|\|^p + \ldots + \|\|x_n\|\|^p + \|\|y_n\|\|^p) \quad (14)$$

for all $x_1, y_1, x_2, y_2, \ldots, x_n, y_n \in X\setminus\{0\}$. Then, there exists a unique n-variable additive mapping $F : X^n \to Y$, such that

$$\|f(x_1, x_2, \ldots, x_n) - F(x_1, x_2, \ldots, x_n)\| \leq \frac{4(2^p + 4)n\theta}{2^p|2 - 2^p|} \max_{x_i \neq 0}\{\|\|x_i\|\|^p : 1 \leq i \leq n\} \quad (15)$$

for all $(x_1, x_2, \ldots, x_n) \in X^n \setminus \{(0, 0, \ldots, 0)\}$.

Proof. Put $\varphi(x_1, y_1, x_2, y_2, \ldots, x_n, y_n) := \theta(\|\|x_1\|\|^p + \|\|y_1\|\|^p + \|\|x_2\|\|^p + \|\|y_2\|\|^p + \ldots + \|\|x_n\|\|^p + \|\|y_n\|\|^p)$ for all $x_1, y_1, x_2, y_2, \ldots, x_n, y_n \in X\setminus\{0\}$, then $\|\|x_i^{(1)}\|\|, \|\|x_i^{(2)}\|\| \leq \max_{x_i \neq 0}\{\|\|x_i\|\|^p : 1 \leq i \leq n\}$ for all i from (4). Hence, due to μ of Theorems 3 and 4, we obtain that

$$\begin{aligned}
\mu(x_1, x_2, \ldots, x_n) &= \varphi\left(x_1^{(1)}, x_1^{(2)}, x_2^{(1)}, x_2^{(2)}, \ldots, x_n^{(1)}, x_n^{(2)}\right) + 2\varphi\left(\frac{x_1^{(1)}}{2}, \frac{x_1^{(2)}}{2}, \frac{x_2^{(1)}}{2}, \frac{x_2^{(2)}}{2}, \ldots, \frac{x_n^{(1)}}{2}, \frac{x_n^{(2)}}{2}\right) \\
&+ \varphi\left(\frac{x_1^{(1)}}{2}, \frac{x_1^{(1)}}{2}, \frac{x_2^{(1)}}{2}, \frac{x_2^{(1)}}{2}, \ldots, \frac{x_n^{(1)}}{2}, \frac{x_n^{(1)}}{2}\right) + \varphi\left(\frac{x_1^{(2)}}{2}, \frac{x_1^{(2)}}{2}, \frac{x_2^{(2)}}{2}, \frac{x_2^{(2)}}{2}, \ldots, \frac{x_n^{(2)}}{2}, \frac{x_n^{(2)}}{2}\right) \\
&\leq \left(2n + \frac{8n}{2^p}\right) \max_{x_i \neq 0}\{\|\|x_i\|\|^p : 1 \leq i \leq n\}
\end{aligned}$$

for all $(x_1, x_2, \ldots, x_n) \in X^n \setminus \{(0, 0, \ldots, 0)\}$. Therefore, the inequality (15) can be obtained easily from (8) and (12) in Theorems 3 and 4. □

The following theorem for the hyperstability of n-variable additive functional equation follows from Corollary 1.

Theorem 5. *Let $(X, |||\cdot|||)$ be a normed space and p be a real number with $p < 0$. Suppose that $f : X^n \to Y$ is a mapping that satisfies (14) for all $x_1, y_1, x_2, y_2, \ldots, x_n, y_n \in X \backslash \{0\}$. Then, f is an n-variable additive mapping itself.*

Proof. By Corollary 1, there exists a unique n-variable additive mapping $F : X^n \to Y$, such that (15) for all $x_1, x_2, \ldots, x_n \in X^n \backslash \{(0,0,\ldots,0)\}$ and $DF(x_1, y_1, x_2, y_2, \ldots, x_n, y_n) = 0$ for all $x_1, y_1, x_2, y_2, \ldots, x_n, y_n \in X$.

For a given $(x_1, x_2, \ldots, x_n) \neq (0, 0, \ldots, 0)$, let $x' \neq 0$ be a nonzero fixed element in $\{x_1, x_2, \ldots, x_n\}$, and let

$$x_i^{(3)} = (m+1)x_i, \quad x_i^{(4)} = -mx_i \qquad \text{when } x_i \neq 0,$$
$$x_i^{(3)} = mx', \quad x_i^{(4)} = -mx' \qquad \text{when } x_i = 0.$$

Then, we can easily show that $|||x_i^{(3)}|||, |||x_i^{(4)}||| \le m^p \max_{x_i \neq 0}\{|||x_i|||^p : 1 \le i \le n\}$ for all i from (4). If $(x_1, x_2, \ldots, x_n) \in X \backslash \{(0, 0, \ldots, 0)\}$, then the equality $f(x_1, x_2, \ldots, x_n) = F(x_1, x_2, \ldots, x_n)$ follows from the inequalities

$$\|f(x_1, x_2, \ldots, x_n) - F(x_1, x_2, \ldots, x_n)\|$$
$$= \left\| Df\left(x_1^{(3)}, x_1^{(4)}, x_2^{(3)}, x_2^{(4)}, \ldots, x_n^{(3)}, x_n^{(4)}\right) - DF\left(x_1^{(3)}, x_1^{(4)}, x_2^{(3)}, x_2^{(4)}, \ldots, x_n^{(3)}, x_n^{(4)}\right) \right.$$
$$+ f(x_1^{(3)}, x_2^{(3)}, \ldots, (x_n^{(3)}) + f(x_1^{(4)}, x_2^{(4)}, \ldots, x_n^{(4)})$$
$$\left. - F(x_1^{(3)}, x_2^{(3)}, \ldots, (x_n^{(3)}) - F(x_1^{(4)}, x_2^{(4)}, \ldots, x_n^{(4)}) \right\|$$
$$\le m^p \cdot 2n\theta \max_{x_i \neq 0}\{|||x_i|||^p : 1 \le i \le n\} + \left\| f(x_1^{(3)}, x_2^{(3)}, \ldots, x_n^{(3)}) - F(x_1^{(3)}, x_2^{(3)}, \ldots, x_n^{(3)}) \right\|$$
$$+ \left\| f(x_1^{(4)}, x_2^{(4)}, \ldots, x_n^{(4)}) - F(x_1^{(4)}, x_2^{(4)}, \ldots, x_n^{(4)}) \right\|$$
$$\le m^p \left(1 + \frac{4(2^p + 4)}{2^p|2 - 2^p|}\right) 2n\theta \max_{x_i \neq 0}\{|||x_i|||^p : 1 \le i \le n\}$$

as $m \to \infty$. For $(x_1, x_2, \ldots, x_n) = (0, 0, \ldots, 0)$, if we choose a fixed element of $x \in X \backslash \{0\}$, then the equality $f0, 0, \ldots, 0) = F0, 0, \ldots, 0)$ follows from the inequalities

$$\|f(0, 0, \ldots, 0) - F(0, 0, \ldots, 0)\|$$
$$= \|Df(mx, -mx, mx, -mx, \ldots, mx, -mx) - DF(mx, -mx, mx, \ldots, mx, -mx)$$
$$+ f(mx, mx, \ldots, mx) + f(-mx, -mx, \ldots, -mx)$$
$$- F(mx, mx, \ldots, mx) - F(-mx, -mx, \ldots, -mx)\|$$
$$\le m^p \cdot 2n\theta \|x\|^p + \|f(mx, mx, \ldots, mx) - F(mx, mx, \ldots, mx)\|$$
$$+ \|f(-mx, -mx, \ldots, -mx) - F(-mx, -mx, \ldots, -mx)\|$$
$$\le m^p \left(1 + \frac{4(2^p + 4)}{2^p|2 - 2^p|}\right) 2n\theta |||x|||^p$$

as $m \to \infty$. Therefore, f is an n-variable additive mapping itself. □

The following example follows from Theorem 5.

Example 1. Let $(\mathbb{R}, |\cdot|)$ be a normed space with absolute value $|\cdot|$, $(\mathbb{R}^l, \|\cdot\|)$ be a Banach space with Euclid norm $\|\cdot\|$, and $p < 0$ be a real number. Suppose that $f : \mathbb{R}^n \to \mathbb{R}^l$ is a continuous mapping such that

$$\|Df(x_1, y_1, x_2, y_2, \ldots, x_n, y_n)\| \leq \theta(|x_1|^p + |y_1|^p + |x_2|^p + |y_2|^p + \ldots + |x_n|^p + |y_n|^p)$$

for all $x_1, y_1, x_2, y_2, \ldots, x_n, y_n \in \mathbb{R}\backslash\{0\}$. Then, the mapping $f : \mathbb{R}^n \to \mathbb{R}^l$ given by

$$f(x_1, x_2, \ldots, x_n) = \left(\sum_{i=1}^n a_{1i} x_i, \sum_{i=1}^n a_{2i} x_i, \ldots, \sum_{i=1}^n a_{li} x_i \right), \tag{16}$$

where $a_{1i}, a_{2i}, \ldots, a_{li}$ are real constants, indicates that

$$f(1, 0, 0, \ldots, 0) = (a_{11}, a_{21}, \ldots, a_{l1}),$$
$$f(0, 1, 0, \ldots, 0) = (a_{12}, a_{22}, \ldots, a_{l2}),$$
$$\vdots \quad \vdots$$
$$f(0, \ldots, 0, 0, 1) = (a_{1n}, a_{2n}, \ldots, a_{ln}).$$

Proof. Since $f : \mathbb{R}^n \to \mathbb{R}^l$ is a continuous n-variable additive mapping by Theorem 5, then the function $f : \mathbb{R}^n \to \mathbb{R}^l$ is given by (16). □

In the following theorems, we replace the domain $(V\backslash\{0\})^{2n}$ of φ and Df in Theorems 3 and 4 with V^{2n}. Then, we can improve the result inequality (8).

Theorem 6. Suppose that $f : V^n \to Y$ is a mapping for which there exists a function $\varphi : V^{2n} \to [0, \infty)$ satisfying (5) and (6) for all $x_1, y_1, x_2, y_2, \ldots, x_n, y_n \in V$. Then, there exists a unique mapping $F : V^n \to Y$, such that (7) for all $x_1, y_1, x_2, y_2, \ldots, x_n, y_n \in V$ and

$$\|f(x_1, x_2, \ldots, x_n) - F(x_1, x_2, \ldots, x_n)\| \leq \sum_{m=0}^{\infty} \frac{\varphi(2^m x_1, 2^m x_1, 2^m x_2, \ldots, 2^m x_n, 2^m x_n)}{2^{m+1}} \tag{17}$$

for all $x_1, x_2, \ldots, x_n \in V$.

Proof. The equality

$$f(2x_1, 2x_2, \ldots, 2x_n) - 2f(x_1, x_2, \ldots, x_n) = Df(x_1, x_1, x_2, x_2, \ldots, x_n, x_n) \tag{18}$$

for all $x_1, x_2, \ldots, x_n \in V$ and the inequality (6) imply that the inequality

$$\left\| f(x_1, x_2, \ldots, x_n) - \frac{f(2x_1, 2x_2, \ldots, 2x_n)}{2} \right\| \leq \frac{1}{2} \varphi(x_1, x_1, x_2, x_2, \ldots, x_n, x_n)$$

for all $x_1, x_2, \ldots, x_n \in V$. From the above inequality, we can derive the inequalities

$$\left\| \frac{f(2^m x_1, \ldots, 2^m x_n)}{2^m} - \frac{f\left(2^{m+m'} x_1, \ldots, 2^{m+m'} x_n\right)}{2^{m+m'}} \right\|$$
$$\leq \sum_{k=m}^{m+m'-1} \frac{\varphi\left(2^k x_1, 2^k x_1, 2^k x_2, 2^k x_2, \ldots, 2^k x_n, 2^k x_n\right)}{2^{k+1}} \tag{19}$$

for all $x_1, x_2, \ldots, x_n \in V$ and all positive integers m, m'. The remainder of the proof of this theorem developed after inequality (19) is omitted because it is similar to that of Theorem 3. □

Theorem 7. Suppose that $f : V^n \to Y$ is a mapping for which there exists a function $\varphi : V^{2n} \to [0, \infty)$ satisfying (11) and (6) for all $x_1, y_1, x_2, y_2, \ldots, x_n, y_n \in V$. Then, there exists a unique mapping $F : V^n \to Y$ that satisfies (7) for all $x_1, y_1, x_2, y_2, \ldots, x_n, y_n \in V$ and

$$\|f(x_1, \ldots, x_n) - F(x_1, \ldots, x_n)\| \leq \sum_{m=0}^{\infty} 2^m \varphi\left(\frac{x_1}{2^{m+1}}, \frac{x_1}{2^{m+1}}, \frac{x_2}{2^{m+1}}, \ldots, \frac{x_n}{2^{m+1}}, \frac{x_n}{2^{m+1}}\right) \quad (20)$$

for all $x_1, x_2, \ldots, x_n \in V$.

Proof. The equality (18) for all $x_1, x_2, \ldots, x_n \in V$ and the inequality (6) imply that the inequality

$$\left\|f(x_1, x_2, \ldots, x_n) - 2f\left(\frac{x_1}{2}, \frac{x_2}{2}, \ldots, \frac{x_n}{2}\right)\right\| \leq \varphi\left(\frac{x_1}{2}, \frac{x_1}{2}, \frac{x_2}{2}, \ldots, \frac{x_n}{2}, \frac{x_n}{2}\right)$$

for all $x_1, x_2, \ldots, x_n \in V$. From the above inequality, we can derive the inequality

$$\left\|2^m f\left(\frac{x_1}{2^m}, \frac{x_2}{2^m}, \ldots, \frac{x_n}{2^m}\right) - 2^{m+m'} f\left(\frac{x_1}{2^{m+m'}}, \frac{x_2}{2^{m+m'}}, \ldots, \frac{x_n}{2^{m+m'}}\right)\right\|$$
$$\leq \sum_{k=m}^{m+m'-1} 2^k \varphi\left(\frac{x_1}{2^{k+1}}, \frac{x_1}{2^{k+1}}, \frac{x_2}{2^{k+1}}, \ldots, \frac{x_n}{2^k}, \frac{x_n}{2^k}\right) \quad (21)$$

for all $x_1, x_2, \ldots, x_n \in V$ and all positive integers m, m'. The remainder of the proof of this theorem developed after inequality (21) is omitted because it is similar to that of Theorem 4. □

The following corollary follows from Theorems 6 and 7.

Corollary 2. Let $(X, \|\|\cdot\|\|)$ be a normed space and p be a nonnegative real number with $p \neq 1$. Suppose that $f : X^n \to Y$ is a mapping satisfying (14) for all $x_1, y_1, x_2, y_2, \ldots, x_n, y_n \in X$. Then, there exists a unique n-variable additive mapping $F : X^n \to Y$ such that

$$\|f(x_1, x_2, \ldots, x_n) - F(x_1, x_2, \ldots, x_n)\| \leq \frac{2\theta}{|2 - 2^p|}(\||x_1\||^p + \||x_2\||^p + \ldots + \||x_n\||^p) \quad (22)$$

for all $x_1, x_2, \ldots, x_n \in X$.

Proof. By putting $\varphi(x_1, y_1, x_2, y_2, \ldots, x_n, y_n) := \theta(\||x_1\||^p + \||y_1\||^p + \||x_2\||^p + \||y_2\||^p + \cdots + \||x_n\||^p + \||y_n\||^p)$ for all $x_1, y_1, x_2, y_2, \ldots, x_n, y_n \in X$, then we easily obtain (22) from (17) and (20) of Theorems 6 and 7. □

3. Conclusions

We obtained two stability results.

Theorems 3 and 4 are the generalized Hyers–Ulam stability for the additive functional Equation (3) on V^n, which is a generalization for the stability of the Cauchy functional equation in papers of Aoki [3], Rassias [4], Gajda [5], Hyers [2], and Găvruta [7].

Theorems 6 and 7 are the hyperstablity of the additive functional Equation (3) on V^n, which is a generalization of the Brzdęk's results [10,11] for the Cauchy functional equation.

If the function f satisfies the inequality (6) for all $(x_1, x_2, \ldots, x_n) \neq (0, 0, \ldots, 0)$ and $(y_1, y_2, \ldots, y_n) \neq (0, 0, \ldots, 0)$, then the function f satisfies the inequality (6) for all $x_1, y_1, x_2, y_2, \ldots, x_n, y_n \in V \backslash \{0\}$. Therefore, the condition $x_1, y_1, x_2, y_2, \ldots, x_n, y_n \in V \backslash \{0\}$ used in the inequality (3.2) of this paper is a generalization of the conditions used in the inequality (6) in well-known pre-results ([10,11]).

Author Contributions: Conceptualization, Y.-H.L. and G.H.K.; Investigation, Y.-H.L. and G.H.K.

Funding: This research received no external funding.

Conflicts of Interest: The authors declare no conflict of interest.

References

1. Ulam, S.M. *Problems in Modern Mathematics*; Wiley: New York, NY, USA, 1964.
2. Hyers, D.H. On the stability of the linear functional equation. *Proc. Natl. Acad. Sci. USA* **1941**, *27*, 222–224. [CrossRef] [PubMed]
3. Aoki, T. On the stability of the linear transformation in Banach algebras. *J. Math. Soc. Japan* **1950**, *2*, 64–66. [CrossRef]
4. Rassias, T.M. On the stability of the linear mapping in Banach spaces. *Proc. Am. Math. Soc.* **1978**, *72*, 297–300. [CrossRef]
5. Gajda, Z. On stability of additive mappings. *Int. J. Math. Math. Sci.* **1991**, *14*, 431–434. [CrossRef]
6. Rassias, T.M.; Semrl, P. On the behavior of mappings which do not satisfy Hyers-Ulam stability. *Proc. Am. Math. Soc.* **1992**, *114*, 989–993. [CrossRef]
7. Găvruta, P. A generalization of the Hyers-Ulam-Rassias stability of approximately additive mappings. *J. Math. Anal. Appl.* **1994**, *184*, 431–436. [CrossRef]
8. Maksa, G.; Páles, Z. Hyperstability of a class of linear functional equations. *Acta Math. Acad. Paedagog. Nyházi* **2001**, *17*, 107–112.
9. Bourgin, D.G. Approximately isometric and multiplicative transformations on continuous function rings. *Duke Math. J.* **1949**, *16*, 385–397. [CrossRef]
10. Brzdęk, J. Remarks on hyperstability of the the Cauchy equation. *Aequationes Math.* **2013**, *86*, 255–267. [CrossRef]
11. Brzdęk, J. A hyperstability result for the Cauchy equation. *Bull. Aust. Math. Soc.* **2014**, *89*, 33–40. [CrossRef]

© 2019 by the authors. Licensee MDPI, Basel, Switzerland. This article is an open access article distributed under the terms and conditions of the Creative Commons Attribution (CC BY) license (http://creativecommons.org/licenses/by/4.0/).

Article

On Almost b-Metric Spaces and Related Fixed Point Results

Nabil Mlaiki [1], Katarina Kukić [2], Milanka Gardašević-Filipović [3] and Hassen Aydi [4,5]*

1. Department of Mathematics and General Sciences, Prince Sultan University, P. O. Box 66833, Riyadh 11586, Saudi Arabia; nmlaiki@psu.edu.sa
2. Faculty for Traffic and Transport Engineering, University of Belgrade, 11000 Belgrade, Serbia; k.mijailovic@sf.bg.ac.rs
3. School of Computing, Union University, 11000 Belgrade, Serbia; mgardasevic@raf.edu.rs
4. Institut Supérieur d'Informatique et des Techniques de Communication, Université de Sousse, H. Sousse 4000, Tunisia
5. China Medical University Hospital, China Medical University, Taichung 40402, Taiwan
* Correspondence: hassen.aydi@isima.rnu.tn

Received: 15 May 2019; Accepted: 23 May 2019; Published: 1 June 2019

Abstract: In this manuscript, we introduce almost b-metric spaces and prove modifications of fixed point theorems for Reich and Hardy–Rogers type contractions. We present an approach generalizing some fixed point theorems to the case of almost b-metric spaces by reducing almost b-metrics to the corresponding b-metrics. Later, we show that this approach can not work for all kinds of contractions. To confirm this, we present a proof in which the contraction condition is such that it cannot be reduced to corresponding b-metrics.

Keywords: fixed point; Reich contraction; Hardy–Rogers contraction; almost b-metric space

MSC: 46T99; 47H10; 54H25

---

1. Introduction

In [1] Filipović and Kukić considered some classical contraction principles of Kannan [2], Reich [3] and Hardy–Rogers [4] in b-metric spaces and rectangular b-metric spaces without the assumption of continuity of the corresponding metric. The fact that a b-metric d need not be continuous must remind us to use caution in the proofs.

As possibly more general forms of the theorems proven in [1], here we further try to, as many authors before, generalize metric spaces. Plenty of generalizations in previous two decades were done. Starting from 1989, b-metric spaces were introduced in [5]. After, partial b-metric spaces [6], metric-like spaces [7] and b-dislocated metric spaces [8] have been given. For related contraction principles in the setting of above spaces, the readers can see [9–19].

As an attempt to continue in that spirit, we initiate the concept of almost b-metric spaces. The motivation of this initiation comes from [20] where Mitrović, George and Hussain introduced almost rectangular b-metric spaces.

2. Preliminaries

Bakhtin in [5] and Czerwik in [21] introduced b-metric spaces as a generalization of standard metric spaces.

Definition 1 (Ref. [5,21]). *Let X be a nonempty set and $s \geq 1$. The function $d_b : X \times X \to [0, +\infty)$ is a b-metric if and only if, for all $\chi, \zeta, \sigma \in X$, we have*

(bM1) $d_b(\chi,\zeta) = 0$ if and only if $\chi = \zeta$,
(bM2) $d_b(\chi,\zeta) = d_b(\zeta,\chi)$,
(bM3) $d_b(\chi,\sigma) \leq s(d_b(\chi,\zeta) + d_b(\zeta,\sigma))$.

(X, d_b, s) is said a b-metric space and $s \geq 1$ is its coefficient.

In particular, if $s = 1$ then (X, d) is a standard metric space.

Recall that a sequence $\{\chi_n\}$ in X, b-converges to $\chi \in X$ if and only if $d_b(\chi_n, \chi) \to 0$ as $n \to \infty$. $\{\chi_n\}$ is b-Cauchy if and only if $d_b(\chi_n, \chi_m) \to 0$ as $n, m \to \infty$. If each b-Cauchy sequence is b-convergent in X, then (X, d_b, s) is said to be b-complete.

If in previous definition, we assume that only (bM1) and (bM3) hold, then we denote d_b as d_q and we call (X, d_q, s) a quasi-b-metric space.

In next few lines, we make a brief overview of some well known types of contractions. Let (X, d) be a metric space and $T : X \to X$ be such that

- $d(T\chi, T\zeta) \leq \lambda d(\chi, \zeta)$, $\lambda \in [0, 1)$, a Banach type of contraction;
- $d(T\chi, T\zeta) \leq \lambda (d(\chi, T\chi) + d(\zeta, T\zeta))$, $\lambda \in [0, \frac{1}{2})$, a Kannan type of contraction;
- $d(T\chi, T\zeta) \leq \lambda (d(\chi, T\zeta) + d(\zeta, T\chi))$, $\lambda \in [0, \frac{1}{2})$, a Chatterjea type of contraction;
- $d(T\chi, T\zeta) \leq \alpha d(\chi, \zeta) + \beta d(\chi, T\chi) + \gamma d(\zeta, T\zeta)$ where $\alpha, \beta, \gamma \geq 0$ with $\alpha + \beta + \gamma < 1$, a Reich type of contraction;
- $d(T\chi, T\zeta) \leq \alpha d(\chi, \zeta) + \beta d(\chi, T\chi) + \gamma d(\zeta, T\zeta) + \delta d(\chi, T\zeta) + \mu d(\zeta, T\chi)$ where $\alpha, \beta, \gamma, \delta, \mu \geq 0$ with $\alpha + \beta + \gamma + \delta + \mu < 1$, a Hardy–Rogers type of contraction.

In [1] Filipović and Kukić proved new theorems with additional conditions that are necessary to prove the theorems without assumption of continuity of b-metric. Here, we cite only formulations of those theorems and for the proofs, we refer on [1].

Theorem 1. *Ref. [1] let T be a self-mapping on a complete b-metric space $(X, d_b, s \geq 1)$ such that*

$$d_b(T\chi, T\zeta) \leq \lambda d_b(\chi, \zeta) + \mu d_b(\chi, T\chi) + \delta d_b(\zeta, T\zeta),$$

for all $\chi, \zeta \in X$, where $\lambda, \mu, \delta \geq 0$ with $\lambda + \mu + \delta < 1$ and

$$\delta < \frac{1}{s}.$$

Then there is a unique fixed point of T.

Theorem 2. *Ref. [1] let $(X, d_b, s \geq 1)$ be a complete b-metric space and $T : X \to X$ be a mapping satisfying*

$$d_b(T\chi, T\zeta) \leq a_1 d_b(\chi, \zeta) + a_2 d_b(\chi, T\chi) + a_3 d_b(\zeta, T\zeta) + a_4 d_b(\chi, T\zeta) + a_5 d_b(\zeta, T\chi),$$

for all $\chi, \zeta \in X$, where $a_1, a_2, a_3, a_4, a_5 \geq 0$ are such that $a_1 + a_2 + a_3 + s(a_4 + a_5) < 1$ and $a_1 > 1 - \frac{2}{s}$. Then T has a unique fixed point.

In the sequel of this paper, we introduce almost-b-metric spaces and present the related previous theorems in this setting. At the end, we also give some results for different type of contractions, where the proofs cannot be reduced to the corresponding b-metrics.

3. Main Results

In this section, let us firstly introduce the concept of almost-b-metric spaces, as a class of quasi-b-metric spaces with the additional requirement that diminishes a lack of symmetry. We set a demand that existence of the left limit of sequence implies the existence of the right limit (bM2l) or that existence of the right limit of sequence implies the existence of the left limit of the same sequence (bM2r). After that, we introduce a couple of examples of almost-b-metrics and also an example of a quasi-b-metric, which is not an almost-b-metric. Finally, we prove Theorems 1 and 2 with the assumption (bM2left) instead of (bM2).

Definition 2. *Let X be a nonempty set and $s \geq 1$. Let $d_{ab} : X \times X \to [0, +\infty)$ be a function such that for all $\chi, \zeta, \sigma, \chi_n \in X$,*

(bM1) $d_{ab}(\chi, \zeta) = 0$ *iff* $\chi = \zeta$,
(bM2l) $d_{ab}(\chi_n, \chi) \to 0, n \to \infty$ *implies* $d_{ab}(\chi, \chi_n) \to 0, n \to \infty$,
(bM2r) $d_{ab}(\chi, \chi_n) \to 0, n \to \infty$ *implies* $d_{ab}(\chi_n, \chi) \to 0, n \to \infty$,
(bM3) $d_{ab}(\chi, \zeta) \leq s(d_{ab}(\chi, \sigma) + d_{ab}(\sigma, \zeta))$.

Then (X, d_{ab}, s) is called an

1. *l-almost-b-metric space if (bM1), (bM2l) and (bM3) hold;*
2. *r-almost-b-metric space if (bM1), (bM2r) and (bM3) hold;*
3. *almost-b-metric space if (bM1), (bM2l), (bM2r) and (bM3) hold.*

In the next two examples, we present two quasi-b-metrics, which are also almost-b-metrics.

Example 1. *Let $X = \{0, 1, 2\}$. Choose $\alpha \geq 2$. Consider the b-metric $d_{ab} : X \times X \to [0, +\infty)$ defined by*

$$d_{ab}(0,0) = d_{ab}(1,1) = d_{ab}(2,2) = 0,$$

$$d_{ab}(1,0) = 1, \quad d_{ab}(0,1) = \frac{3}{2},$$

$$d_{ab}(2,1) = 1, \quad d_{ab}(1,2) = \frac{3}{2},$$

$$d_{ab}(2,0) = \alpha, \quad d_{ab}(0,2) = \alpha + 1.$$

Note that d_{ab} satisfies (bM1), (bM3), (bM2l) and (bM2r) (but not (bM2)). For $\alpha > 2$, the ordinary triangle inequality is not verified. Indeed,

$$d_{ab}(0,2) = \alpha + 1 > 3 = \frac{3}{2} + \frac{3}{2} = d_{ab}(0,1) + d_{ab}(1,2).$$

However, the following is satisfied for all $x, y, z \in X$,

$$d_{ab}(x,y) \leq \frac{\alpha+2}{2}(d_{ab}(x,z) + d_{ab}(z,y)).$$

Example 2. *Let $X = [0, +\infty)$ and define $d_{ab} : X \times X \to [0, +\infty)$ as*

$$d_{ab}(x,y) = \begin{cases} (x-y)^3, & x \geq y \\ 4(y-x)^3, & x < y \end{cases}$$

Then $(X, d_{ab}, 4)$ is an almost b-metric space. (bM1), (bM2l) and (bM2r) are obvious. It remains to prove that for all $x, y, z, \in X$,

$$d_{ab}(x,y) \leq 4(d_{ab}(x,z) + d_{ab}(z,y)).$$

Case 1. $x \geq y$ and $d_{ab}(x,y) = (x-y)^3$. Starting from the inequality $(\alpha + \beta)^3 \leq 4(\alpha^3 + \beta^3)$, we separate the cases:

$y \leq z \leq x$:

$$d_{ab}(x,y) = (x-y)^3 = (x-z+z-y)^3$$
$$\leq 4((x-z)^3 + (y-z)^3) = 4(d_{ab}(x,z) + d_{ab}(z,y)),$$

$z \leq y \leq x$:
$$d_{ab}(x,y) = (x-y)^3 \leq 4((x-z)^3 + (y-z)^3)$$
$$\leq 4((x-z)^3 + 4(y-z)^3) = 4(d_{ab}(x,z) + d_{ab}(z,y)),$$

$y \leq x \leq z$:
$$d_{ab}(x,y) = (x-y)^3 \leq 4((x-z)^3 + (z-y)^3)$$
$$\leq 4(4(z-x)^3 + (z-y)^3) = 4(d_{ab}(x,z) + d_{ab}(z,y)).$$

Case 2. $x < y$ and $d_{ab}(x,y) = 4(y-x)^3$. Again, we separate the cases:

$x \leq z \leq y$:
$$d_{ab}(x,y) = 4(y-x)^3 = 4(y-z+z-x)^3$$
$$\leq 4(4(y-z)^3 + 4(z-x)^3) = 4(d_{ab}(x,z) + d_{ab}(z,y)),$$

$z \leq x \leq y$:
$$d_{ab}(x,y) = 4(y-x)^3 \leq 4 \cdot 4((y-z)^3 + (z-x)^3)$$
$$= 4(4(y-z)^3 + 4(z-x)^3)$$
$$\leq 4(4(y-z)^3 + (x-z)^3) = 4(d_{ab}(x,z) + d_{ab}(z,y)),$$

$x \leq y \leq z$:
$$d_{ab}(x,y) = 4(y-x)^3 \leq 4 \cdot 4((y-z)^3 + (z-x)^3)$$
$$= 4 \cdot (4(y-z)^3 + 4(z-x)^3)$$
$$\leq 4((z-y)^3 + 4(z-x)^3) = 4(d_{ab}(x,z) + d_{ab}(z,y)).$$

In the two previous examples, we constructed an almost-b-metric, which is also a quasi-b metric. The next example shows that there is a quasi-b-metric d_q, that it is not an almost-b-metric.

Example 3. Let $X = \mathbb{R}$ and define $d_q : X \times X \to [0, \infty)$ as

$$d_q(x,y) = \begin{cases} (x-y)^3, & x \geq y \\ 1, & x < y \end{cases}$$

As in the previous example, (bM3) and (bM1) are obvious. Notice that

$$d_q(\frac{1}{n}, 0) \to 0, \ n \to \infty \quad \text{but} \quad d_q(0, \frac{1}{n}) = 1,$$

so (bM2l) does not hold and it is the same for (bM2r). We conclude that $(X, d_q, 4)$ is a quasi-b-metric space, but it is not an almost-b-metric space.

There are many examples of b-metrics that are not continuous. Here, we modify one of such examples in sense that we do not demand symmetry.

Example 4. Let $A = \mathbb{N} \cup \{\infty\}$ and define $d_q : A \times A \to [0, +\infty)$:

$$d_q(x,y) = \begin{cases} 0, & x = y \\ \frac{1}{x} - \frac{1}{y}, & \text{if } x < y \text{ and one of } x \text{ and } y \text{ is odd and the other} \\ & \text{is odd or } \infty \\ \frac{1}{2}\left(\frac{1}{y} - \frac{1}{x}\right), & \text{if } y < x \text{ and one of } x \text{ and } y \text{ is odd and the other} \\ & \text{is odd or } \infty \\ 3, & \text{if one of } x \text{ and } y \text{ is even and the other is even or } \infty \\ 2, & \text{otherwise.} \end{cases}$$

Then $(A, d_q, \frac{3}{2})$ is a quasi-b-metric space (it is also an almost-b-metric space). Note that d_q is not continuous. Indeed, $d_q(2n+1, \infty) \to 0$, when $n \to \infty$. But, $d_q(2n+1, 2) = 2$, while $d_q(\infty, 2) = 3$.

Here, we introduce some basic concepts for almost-b-metric spaces. The following notions are quite standard and also valid in quasi-b- metric spaces.

Definition 3. *Let (X, d_{ab}, s) be an almost-b-metric space. A sequence $\{\chi_n\}$ in X is said to be*

left-Cauchy *if and only if for each $\varepsilon > 0$ there is an $n_0 \in \mathbb{N}$ such that $d_{ab}(\chi_n, \chi_m) < \varepsilon$ for all $n \geq m > n_0$, which can be written as $\lim\limits_{n \geq m \to \infty} d_{ab}(\chi_n, \chi_m) = 0$,*

right-Cauchy *if and only if for each $\varepsilon > 0$ there is $n_0 \in \mathbb{N}$ so that $d_{ab}(\chi_n, \chi_m) < \varepsilon$ for all $m \geq n > n_0$, which can be written as $\lim\limits_{m \geq n \to \infty} d(\chi_n, \chi_m) = 0$,*

Cauchy *if and only if for each $\varepsilon > 0$, there is $n_0 \in \mathbb{N}$ so that $d_{ab}(\chi_n, \chi_m) < \varepsilon$ for all $n, m > n_0$.*

In a quasi-b-metric space, a sequence is Cauchy if and only if it is left-Cauchy and right-Cauchy. The same is satisfied in almost-b-metric spaces. An almost-b-metric space (X, d_{ab}, s) is left-complete if and only if each left-Cauchy sequence $\{\chi_n\}$ in X satisfies $\lim\limits_{n \to \infty} d_{ab}(\chi_n, \chi) = 0$, right-complete if and only if each right-Cauchy sequence $\{\chi_n\}$ in X satisfies $\lim\limits_{n \to \infty} d_{ab}(\chi, \chi_n) = 0$ and is complete if and only if each Cauchy sequence in X is convergent.

In the next lemma, we will associate a b-metric to a given quasi-b-metric or an almost-b-metric. For some kind of contractions, by virtue of this correlation, the proofs from b-metric spaces can easily be translated into quasi-b-metric spaces and almost-b-metric spaces as their subclass.

Lemma 1. *If (X, d_q, s) is a quasi-b-metric space with $s \geq 1$, then (X, l, s) is a b-metric space, where*

$$l(\chi, \zeta) = \frac{d_q(\chi, \zeta) + d_q(\zeta, \chi)}{2}.$$

Proof. $l(x, y)$ is a b-metric.

(bM1) Suppose that $l(x, y) = 0$. Then $\frac{d_q(x,y) + d_q(y,x)}{2} = 0$ and since $d_q(x, y) \geq 0$, we obtain that $d_q(x, y) = d_q(y, x) = 0$ and that is, $x = y$, so we conclude that $l(x, y)$ satisfies (bM1).

(bM2) $l(x, y)$ is symmetric by definition:

$$l(x, y) = \frac{d_q(x, y) + d_q(y, x)}{2} = \frac{d_q(y, x) + d_q(x, y)}{2} = l(y, x).$$

(bM3) For all $x, y, z, \in X$, the following is satisfied:

$$d_q(x, z) \leq s(d_q(x, y) + d_q(y, z)).$$

Simply, by adding the following inequality to the previous

$$d_q(z, x) \leq s(d_q(z, y) + d_q(y, x))$$

and dividing the resulted sum by two, we obtain

$$l(x, z) \leq s(l(x, y) + l(y, z)).$$

□

Remark 1. *If (X, d_{ab}, s) is a complete almost-b-metric space, then from (bM2l) and (bM2r), we conclude that (X, l, s) is a complete b-metric space.*

The following theorems are modifications of Theorems 1 and 2 for quasi-b metric spaces and almost-b-metric spaces. Since almost-b-metric spaces are contained in quasi-b-metric spaces, we denote a metric by d_q.

Theorem 3. *Let (X, d_q, s) be a b-complete quasi-b-metric space with coefficient $s > 1$ and $T : X \to X$ be a mapping such that*
$$d_q(Tx, Ty) \leq \lambda d_q(x, y) + \mu d_q(x, Tx) + \delta d_q(Ty, y), \tag{1}$$
for all $x, y, z \in X$, where $\lambda, \mu, \delta \geq 0$ and
$$\lambda + 2 \cdot \max\{\mu, \delta\} < 1 \quad \text{and} \quad \max\{\mu, \delta\} < \frac{1}{s}. \tag{2}$$
Then T has a unique fixed point.

Proof. From Lemma 1, we conclude that (X, l, s) is a complete b-metric space. Further, from (1), the b-metric $l(x, y)$ satisfies:

$$l(Tx, Ty) = \frac{d_q(Tx, Ty) + d_Q(Ty, Tx)}{2}$$
$$\leq \frac{1}{2} (\lambda d_q(x, y) + \mu d_q(x, Tx) + \delta d_q(Ty, y))$$
$$+ \frac{1}{2} (\lambda d_q(y, x) + \mu d_q(y, Ty) + \delta d_q(Tx, x))$$
$$= \lambda l(x, y) + \frac{1}{2}(\mu d_q(x, Tx) + \delta d_q(Tx, x)) + \frac{1}{2}(\mu d_q(y, Ty) + \delta d_q(Ty, y))$$
$$\leq \lambda l(x, y) + \frac{1}{2} \cdot \max\{\mu, \delta\}(d_q(x, Tx) + d_q(Tx, x))$$
$$+ \frac{1}{2} \cdot \max\{\mu, \delta\}(d_q(y, Ty) + d_q(Ty, y))$$
$$= \lambda l(x, y) + \max\{\mu, \delta\} l(x, Tx) + \max\{\mu, \delta\} l(y, Ty).$$

Now, from Theorem 1, we conclude that T has a unique fixed point. □

In the next result, we propose a Hardy–Rogers type contraction for quasi-b metric spaces and almost-b-metric spaces.

Theorem 4. *Let (X, d_q, s) be a complete quasi-b-metric space with coefficient $s > 1$ and $T : X \to X$ be a mapping satisfying*
$$d_q(Tx, Ty) \leq a_1 d_q(x, y) + a_2 d_q(x, Tx) + a_3 d_q(Ty, y) + a_4 d_q(x, Ty) + a_5 d_q(Tx, y), \tag{3}$$
for all $x, y \in X$, where $a_1, a_2, a_3, a_4, a_5 \geq 0$ with $a_1 + 2 \cdot \max\{a_2, a_3\} + 2s \cdot \max\{a_4, a_5\} < 1$ and $a_1 > \max\{0, 1 - \frac{2}{s}\}$. Then T has a unique fixed point.

Proof. From Lemma 1, we conclude that (X, l, s) is a complete b-metric space. Starting from (3), we obtain for any $x, y \in X$,

$$2l(Tx, Ty) = d_q(Tx, Ty) + d_q(Ty, Tx)$$
$$\leq a_1 d_q(x, y) + a_1 d_q(y, x) + a_2 d_q(x, Tx) + a_2 d_q(y, Ty) + a_3 d_q(Ty, y)$$
$$+ a_3 d_q(Tx, x) + a_4 d_q(x, Ty) + a_4 d_q(y, Tx) + a_5 d_q(Tx, y) + a_5 d_q(Ty, x).$$

Further, we get

$$l(Tx, Ty) \leq a_1 l(x,y) + \frac{1}{2}\left(a_2 d_q(x, Tx) + a_3 d_q(Tx, x)\right) + \frac{1}{2}\left(a_2 d_q(y, Ty) + a_3 d_q(Ty, y)\right)$$
$$+ \frac{1}{2}\left(a_4 d_q(x, Ty) + a_5 d_q(Ty, x)\right) + \frac{1}{2}\left(a_4 d_q(y, Tx) + a_5 d_q(Tx, y)\right)$$
$$\leq a_1 l(x,y) + \max\{a_2, a_3\} l(x, Tx) + \max\{a_2, a_3\} l(y, Ty)$$
$$+ \max\{a_4, a_5\} l(x, Ty) + \max\{a_4, a_5\} l(y, Tx).$$

From Theorem 2 and conditions from Theorem 4, we conclude that self-mapping T on the complete b-metric space (X, l, s) has an unique fixed point, say x^*. Finally, according to Theorem 2, the result follows. □

It is not difficult to see that Theorems 3 and 4 are also satisfied for $s = 1$. To be specific, then $(X, d, 1)$ is a quasi-metric space, (X, l) is a metric space, while condition (2) reduces to the well known condition $\lambda + \mu + \delta < 1$ for Reich type contractions, and similar for Hardy–Rogers type contractions.

The following results slightly differ from previous in a sense that we use properties (bM2l) and (bM2r). Before we state our result, we prove an auxiliary lemma that we use it in the proof. Since the lemma is satisfied in the quasi-b-metric spaces, it is also valid in almost-b-metric spaces, so again we denote it by d_q (having in mind that it is also valid for d_{ab}).

Lemma 2. *Let $\{\chi_n\}$ be a sequence in a quasi-b-metric space $(X, d_q, s \geq 1)$ such that*

$$d_q(\chi_n, \chi_{n+1}) \leq \lambda \cdot d_q(\chi_{n-1}, \chi_n), \tag{4}$$

for some $\lambda \in [0, \frac{1}{s})$ and each $n \in \mathbb{N}$. Then $\{\chi_n\}$ is a right-Cauchy sequence.

Proof. From (4), we get

$$d_q(\chi_n, \chi_{n+1}) \leq \lambda^n d_q(\chi_0, \chi_1). \tag{5}$$

Let $n, m \in \mathbb{N}$ with $n < m$. Then

$$\begin{aligned}
& d_q(\chi_n, \chi_m) \\
\leq\ & s\left(d_q(\chi_n, \chi_{n+1}) + d_q(\chi_{n+1}, \chi_m)\right) \\
=\ & s d_q(\chi_n, \chi_{n+1}) + s d_q(\chi_{n+1}, \chi_m) \\
\leq\ & s d_q(\chi_n, \chi_{n+1}) + s^2 d_q(\chi_{n+1}, \chi_{n+2}) + s^2 d_q(\chi_{n+2}, \chi_m) \\
\leq\ & s d_q(\chi_n, \chi_{n+1}) + s^2 d_q(\chi_{n+1}, \chi_{n+2}) + s^3 d_q(\chi_{n+2}, \chi_{n+3}) + \ldots \\
+\ & s^{m-n-1} d_q(\chi_{m-2}, \chi_{m-1}) + s^{m-n-1} d_q(\chi_{m-1}, \chi_m) \\
\leq\ & \left[s\lambda^n + s^2 \lambda^{n+1} + s^3 \lambda^{n+2} + \ldots + s^{m-n-1} \lambda^{m-2}\right] d_q(\chi_0, \chi_1) \\
+\ & s^{m-n-1} \lambda^{m-1} d_q(\chi_0, \chi_1) \\
=\ & s\lambda^n \left(1 + (s\lambda) + (s\lambda)^2 + \ldots + (s\lambda)^{m-n-2}\right) d_q(\chi_0, \chi_1) + \frac{(s\lambda)^{m-1}}{s^n} d_q(\chi_0, \chi_1) \\
\leq\ & \left(\frac{s\lambda^n}{1 - s\lambda} + \frac{(s\lambda)^{m-1}}{s^n}\right) d_q(\chi_0, \chi_1) \to 0 \ (m > n \to \infty).
\end{aligned}$$

Since $s\lambda < 1$, we have

$$d_q(\chi_n, \chi_m) \to 0, m > n, n \to \infty \text{ or equivalently } \lim_{m > n \to \infty} d_q(\chi_n, \chi_m) = 0,$$

that is, $\{\chi_n\}$ is right-Cauchy. □

The following result is analogue to Lemma 2 for left- Cauchy sequences.

Lemma 3. *Let $\{\chi_n\}$ be a sequence in a quasi-b-metric space $(X, d_q, s \geq 1)$ such that*

$$d_q(\chi_{n+1}, \chi_n) \leq \lambda \cdot d_q(\chi_n, \chi_{n-1}) \tag{6}$$

for some $\lambda \in [0, \frac{1}{s})$ and each $n \in \mathbb{N}$. Then $\{\chi_n\}$ is a left-Cauchy sequence.

Proof. The proof follows the same steps as in Lemma 2, where, starting from (6), the condition (5) is replaced by

$$d_q(\chi_{n+1}, \chi_n) \leq \lambda^n d_q(\chi_1, \chi_0). \tag{7}$$

Let $n, m \in \mathbb{N}$ with $n > m$. Then

$$\begin{aligned}
& d_q(\chi_n, \chi_m) \\
\leq\ & s\left(d_q(\chi_n, \chi_{m+1}) + d_q(\chi_{m+1}, \chi_m)\right) \\
=\ & sd_q(\chi_{m+1}, \chi_m) + sd_q(\chi_n, \chi_{m+1}) \\
\leq\ & sd_q(\chi_{m+1}, \chi_m) + s^2 d_q(\chi_n, \chi_{m+2}) + s^2 d_q(\chi_{m+2}, \chi_{m+1}) \\
\leq\ & sd_q(\chi_{m+1}, \chi_m) + s^2 d(\chi_{m+2}, \chi_{m+1}) + \ldots \\
+\ & s^{n-m-1}\left(d_q(\chi_n, \chi_{n-1}) + d_q(\chi_{n-1}, \chi_{n-2})\right) \\
\leq\ & \left[s\lambda^m + s^2\lambda^{m+1} + s^3\lambda^{m+2} + \ldots + s^{n-m-1}\lambda^{n-2}\right] d_q(\chi_1, \chi_0) \\
+\ & s^{n-m-1}\lambda^{n-1} d_q(\chi_1, \chi_0) \\
=\ & s\lambda^m \left(1 + (s\lambda) + (s\lambda)^2 + \ldots + (s\lambda)^{n-m-2}\right) d_q(\chi_1, \chi_0) + \frac{(s\lambda)^{n-1}}{s^m} d_q(\chi_1, \chi_0) \\
\leq\ & \left(\frac{s\lambda^m}{1 - s\lambda} + \frac{(s\lambda)^{n-1}}{s^m}\right) d_q(\chi_1, \chi_0) \to 0 \ (n > m \to \infty).
\end{aligned}$$

Since $s\lambda < 1$, we conclude that

$$d_q(\chi_n, \chi_m) \to 0, n > m, m \to \infty \text{ or equivalently } \lim_{n > m \to \infty} d_q(\chi_n, \chi_m) = 0,$$

that is, $\{\chi_n\}$ is left-Cauchy. □

Remark 2. *It is not hard to see that Lemma 2 and Lemma 3 hold if $\lambda \in [\frac{1}{s}, 1)$. For details, see Lemma 5 in [22].*

In the proof of the next theorem, we use the assumption (bM2r), hence we state it an almost-b-metric, and so denote the metric by d_{ab}.

Theorem 5. *Let (X, d_{ab}, s) be a right-complete r-almost b-metric space with coefficient $s > 1$ and $T : X \to X$ be a mapping satisfying*

$$d_{ab}(Tx, Ty) \leq k \cdot \max\{d_{ab}(x, y), d_{ab}(x, Tx), d_{ab}(y, Ty)\}, \tag{8}$$

for all $x, y \in X$, where k is such that $0 \leq k < \frac{1}{s}$. Then T has a unique fixed point.

Proof. At the beginning of the proof, let us consider uniqueness of a possible fixed point. To prove that the fixed point is unique, if it exists, suppose that T has two distinct fixed points $x^*, y^* \in X$. Then we get

$$d_{ab}(x^*, y^*) = d_{ab}(Tx^*, Ty^*)$$
$$\leq k \cdot \max\{d_{ab}(x^*, y^*), d_{ab}(x^*, Tx^*), d_{ab}(y^*, Ty^*)\}$$
$$= k d_{ab}(x^*, y^*) < d_{ab}(x^*, y^*),$$

which is a contradiction.

For an arbitrary $\chi_0 \in X$, consider the sequence $\chi_n = T\chi_{n-1} = T^n\chi_0$, $n \in \mathbb{N}$. If $\chi_n = \chi_{n+1}$ for some n, then χ_n is the unique fixed point of T. We suppose that $d_{ab}(\chi_n, \chi_{n+1}) > 0$ for all $n \in \mathbb{N}$.

We start from (8) for $d_{ab}(\chi_n, \chi_{n+1})$. Then for any $n \in \mathbb{N}$, we get

$$d_{ab}(\chi_n, \chi_{n+1}) = d_{ab}(T\chi_{n-1}, T\chi_n)$$
$$\leq k \cdot \max\{d_{ab}(\chi_{n-1}, \chi_n), d_{ab}(\chi_{n-1}, T\chi_{n-1}), d_{ab}(\chi_n, T\chi_n)\} \quad (9)$$
$$= k \cdot \max\{d_{ab}(\chi_{n-1}, \chi_n), d_{ab}(\chi_{n-1}, \chi_n), d_{ab}(\chi_n, \chi_{n+1})\}$$
$$= k \cdot \max\{d_{ab}(\chi_{n-1}, \chi_n), d_{ab}(\chi_n, \chi_{n+1})\}.$$

If $d_{ab}(\chi_{m-1}, \chi_m) \leq d_{ab}(\chi_m, \chi_{m+1})$ for some $m \in \mathbb{N}$, then from (9) we get

$$d_{ab}(\chi_m, \chi_{m+1}) \leq k \cdot d_{ab}(\chi_m, \chi_{m+1}) < d_{ab}(\chi_m, \chi_{m+1})$$

which is a contradiction. So, we have

$$d_{ab}(\chi_n, \chi_{n+1}) \leq k \cdot d_{ab}(\chi_{n-1}, \chi_n) \quad \text{for all} \quad n \in \mathbb{N}. \quad (10)$$

From (10) and Lemma 2 we can easily conclude that for some $n_0 \in \mathbb{N}$,

$$d_{ab}(\chi_n, \chi_m) < \varepsilon$$

for all $m \geq n > n_0$, so $\{\chi_n\}$ is a right-Cauchy sequence.

Since $(X, d_{ab}, s > 1)$ is a right-complete r-almost-b-metric space, we get that the sequence $\{\chi_n\}$ right converges to $x^* \in X$, i.e., $d_{ab}(x, \chi_n) \to 0$ as $n \to \infty$. (bM2r) implies that $d_{ab}(\chi_n, x^*) \to 0$ as $n \to \infty$.

The end of the proof is analogue to the standard case. From (bM3) and (8), we obtain

$$\frac{1}{s}d_{ab}(x^*, Tx^*) \leq d_{ab}(x^*, \chi_{n+1}) + d_{ab}(\chi_{n+1}, Tx^*)$$
$$= d_{ab}(x^*, \chi_{n+1}) + d_{ab}(T\chi_n, Tx^*)$$
$$\leq d_{ab}(x^*, \chi_{n+1}) + k \cdot \max\{d_{ab}(\chi_n, x^*), d_{ab}(\chi_n, T\chi_n), d_{ab}(x^*, Tx^*)\}$$
$$\to k \cdot d_{ab}(x^*, Tx^*), \, n \to \infty.$$

Finally, $x^* = Tx^*$. In the last inequality, we used property (bM2r) to obtain that $d_{ab}(\chi_n, x^*) \to 0$ as $n \to \infty$ and also that $d_{ab}(\chi_n, T\chi_n) = d_{ab}(\chi_n, \chi_{n+1}) \to 0$ as $n \to \infty$ since $\{\chi_n\}$ is a right-Cauchy sequence. □

From the previous theorem, we can draw several corollaries that are analogous to Banach, Kannan and Reich type contraction principles, respectively.

Corollary 1. *Let (X, d_{ab}, s) be a right-complete r-almost b-metric space with coefficient $s > 1$ and $T : X \to X$ be such that*

Banach contraction:

$$d_{ab}(Tx, Ty) \leq k \cdot d_{ab}(x, y)$$

for all $x, y \in X$ where $0 \leq k < \frac{1}{s}$.
Kannan contraction:
$$d_{ab}(Tx, Ty) \leq k_1 d_{ab}(x, fx) + k_2 d_{ab}(y, fy)$$
for all $x, y \in X$ where $k_1, k_2 \geq 0$ such that $k_1 + k_2 < \frac{1}{s}$.
Reich contraction:
$$d_{ab}(Tx, Ty) \leq k_1 d_{ab}(x, y) + k_2 d_{ab}(x, fx) + k_3 d_{ab}(y, fy),$$
for all $x, y \in X$ where $k_1, k, k_3 \geq 0$ such that $k_1 + k_2 + k_3 < \frac{1}{s}$.

Then T has a unique fixed point.

The next result is analogue to Theorem 5 for left-complete l-almost b-metric spaces.

Theorem 6. *Let (X, d_{ab}, s) be a left-complete l-almost b-metric space with $s > 1$ and $T : X \to X$ be such that*
$$d_{ab}(Tx, Ty) \leq k \cdot max\{d_{ab}(x, y), d_{ab}(Tx, x), d_{ab}(Ty, y)\}, \tag{11}$$
for all $x, y \in X$ where $0 \leq k < \frac{1}{s}$. Then T has a unique fixed point.

Proof. The uniqueness of a possible fixed point is obtained the same way as in proof of Theorem 5.

For arbitrary $\chi_0 \in X$, consider the sequence $\chi_n = T\chi_{n-1} = T^n\chi_0$, $n \in \mathbb{N}$. If $\chi_n = \chi_{n+1}$ for some n, then χ_n is a unique fixed point of T. Hence, we suppose that $d_{ab}(\chi_{n+1}, \chi_n) > 0$ for all $n \in \mathbb{N}$.

We start from (11) for $d_{ab}(\chi_{n+1}, \chi_n)$. Then for any $n \in \mathbb{N}$, using same considerations as in previous proof, we get

$$\begin{aligned} d_{ab}(\chi_{n+1}, \chi_n) &= d_{ab}(T\chi_n, T\chi_{n-1}) \\ &\leq k \cdot max\{d_{ab}(\chi_n, \chi_{n-1}), d_{ab}(T\chi_n, \chi_n), d_{ab}(T\chi_{n-1}, \chi_{n-1})\} \\ &\leq k \cdot d_{ab}(\chi_n, \chi_{n-1}). \end{aligned} \tag{12}$$

From (12) and Lemma 3, we can easily conclude that for some $n_0 \in \mathbb{N}$,

$$d_{ab}(\chi_n, \chi_m) < \varepsilon$$

for all $n \geq m > n_0$, so $\{\chi_n\}$ is a left-Cauchy sequence.

Since $(X, d_{ab}, s > 1)$ is a left-complete l-almost-b-metric space, we get that the sequence $\{\chi_n\}$ left converges to $x^* \in X$, i.e., $d_{ab}(\chi_n, x^*) \to 0$, $n \to \infty$. (bM2l) implies that $d_{ab}(x^*, \chi_n) \to 0$ as $n \to \infty$.

Finally, from (bM3) and (11), we obtain

$$\begin{aligned} \frac{1}{s} d_{ab}(Tx^*, x^*) &\leq d_{ab}(Tx^*, \chi_{n+1}) + d_{ab}(\chi_{n+1}, x^*) \\ &= d_{ab}(Tx^*, T\chi_n) + d_{ab}(\chi_{n+1}, x^*) \\ &\leq k \cdot max\{d_{ab}(x^*, \chi_n), d_{ab}(Tx^*, x^*), d_{ab}(T\chi_n, \chi_n)\} + d_{ab}(\chi_{n+1}, x^*) \\ &\to k \cdot d_{ab}(Tx^*, x^*), \, n \to \infty, \end{aligned}$$

and so $x^* = Tx^*$. In the last inequality, we used property (bM2l) that implies $d_{ab}(x^*, \chi_n) \to 0, n \to \infty$ and also that $d_{ab}(T\chi_n, \chi_n) = d_{ab}(\chi_{n+1}, \chi_n) \to 0, n \to \infty$ since $\{\chi_n\}$ is a left-Cauchy sequence. □

The previous considerations should convince the readers that many generalizations of contraction principles may be obtained in almost-b-spaces, which are introduced here, and present a proper subclass of quasi-b-metric spaces. As another benefit of this paper, we point out the principle applied in Theorems 3 and 4 that elegantly proves some contractions in quasi-b-metric spaces.

Finally, we state some open questions in the context of almost-b-metric spaces (respectively quasi-b-metric spaces). If $s = 1$, we have appropriate unresolved questions in the context of quasi-metric spaces. We present formulations for the case of a right-complete r-almost b-metric space, noting that similar issues remain open in left-complete l-almost b-metric spaces.

Problem 1. *(Generalized Ćirić type contraction of first order) Let $(X, d_{ab}, s \geq 1)$ be a right-complete r-almost b-metric space and $T : X \to X$ be a mapping satisfying*

$$d_{ab}(Tx, Ty) \leq k \max \left\{ d_{ab}(x,y), \frac{d_{ab}(x,Tx) + d_{ab}(y,Ty)}{2s}, \frac{d_{ab}(x,Ty) + d_{ab}(y,Tx)}{2s} \right\},$$

for all $x, y \in X$ where $0 \leq k < \frac{1}{s}$. Then T has a unique fixed point.

Problem 2. *(Generalized Ćirić type contraction of second order) Let $(X, d_{ab}, s \geq 1)$ be a right-complete r-almost b-metric space and $T : X \to X$ be a mapping satisfying*

$$d_{ab}(Tx, Ty) \leq k \max \left\{ d_{ab}(x,y), d_{ab}(x,Tx), d_{ab}(y,Ty), \frac{d_{ab}(x,Ty) + d_{ab}(y,Tx)}{2s} \right\},$$

for all $x, y \in X$ where $0 \leq k < \frac{1}{s}$. Then T has a unique fixed point.

Problem 3. *(Quasicontraction of Ćirić type) Let $(X, d_{ab}, s \geq 1)$ be a right-complete r-almost b-metric space and $T : X \to X$ be such that*

$$d_{ab}(Tx, Ty) \leq k \max \left\{ d_{ab}(x,y), d_{ab}(x,Tx), d_{ab}(y,Ty), d_{ab}(x,Ty), d_{ab}(y,Tx) \right\},$$

for all $x, y \in X$ where $0 \leq k < \frac{1}{s}$. Then T has a unique fixed point.

Author Contributions: All authors contributed equally and significantly in writing this article. All authors read and approved the final manuscript.

Funding: This research received no external funding.

Acknowledgments: The first author would like to thank Prince Sultan University for funding this work through research group Nonlinear Analysis Methods in Applied Mathematics (NAMAM) group number RG-DES-2017-01-17. The research of the second author (K.K.) was partially supported by the Serbian Ministry of Science and Technological Development, Project TR36002.

Conflicts of Interest: The authors declare that they have no competing interests regarding the publication of this paper.

References

1. Filipović, M.G.; Kukić, K. Some results about contraction principles in bMS and RbMS without assumption of b-metric continuity. *Fixed Point Theory* **2019**, submitted.
2. Kannan, R. Some results on fixed points. *Bull. Calcutta Math. Soc.* **1968**, *60*, 71–76.
3. Reich, S. Some remarks concerning contraction mappings. *Can. Math. Bull.* **1971**, *14*, 121–124. [CrossRef]
4. Hardy, G.E.; Rogers, T.D. A generalization of a fixed point theorem of Reich. *Can. Math. Bull.* **1973**, *16*, 201–206. [CrossRef]
5. Bakhtin, I.A. The contraction mapping principle in quasimetric spaces. *Funct. Anal. Ulianowsk Gos. Ped. Inst.* **1989**, *30*, 26–37.
6. Shukla, S. Partial b-metric spaces and fixed point theorems. *Mediterr. J. Math.* **2013**, *11*, 703–711. [CrossRef]
7. Amini-Harandi, A. Metric-like spaces, partial metric spaces and fixed points. *Fixed Point Theory Appl.* **2012**, *2012*, 204. [CrossRef]
8. Hussain, N.; Roshan, J.R.; Parvaneh, V.; Abbas, M. Common fixed point results for weak contractive mappings in ordered b-dislocated metric spaces with applications. *J. Inequal. Appl.* **2013**, *2013*, 486. [CrossRef]

9. Aydi, H.; Bota, M.F.; Karapinar, E.; Moradi, S. A common fixed point for weak ϕ-contractions on b-metric spaces. *Fixed Point Theory* **2012**, *13*, 337–346.
10. Aydi, H.; Felhi, A.; Sahmim, S. Common fixed points via implicit contractions on b-metric-like spaces. *J. Nonlinear Sci. Appl.* **2017**, *10*, 1524–1537. [CrossRef]
11. Aydi, H.; Felhi, A.; Sahmim, S. On common fixed points for (α, ψ)-contractions and generalized cyclic contractions in b-metric-like spaces and consequences. *J. Nonlinear Sci. Appl.* **2016**, *9*, 2492–2510. [CrossRef]
12. Aydi, H.; Karapinar, E.; Bota, M.F.; Mitrović, S. A fixed point theorem for set-valued quasi-contractions in b-metric spaces. *Fixed Point Theory Appl.* **2012**, *2012*, 88. [CrossRef]
13. Faraji, H.; Savić, D.; Radenović, S. Fixed point theorems for Geraghty contraction type mappings in b-metric spaces and applications. *Axioms* **2019**, *8*, 34. [CrossRef]
14. Karapınar, E.; Czerwik, S.; Aydi, H. (α, ψ)-Meir-Keeler contraction mappings in generalized b-metric spaces. *J. Funct. Spaces* **2018**, *2018*, 3264620. [CrossRef]
15. Hussain, A.; Kanwal, T.; Adeel, M.; Radenović, S. Best proximity point results in b-metric spaces and application to nonlinear fractional differential equation. *Mathematics* **2018**, *6*, 221. [CrossRef]
16. Kirk, W.; Shahzad, N. *Fixed Point Theory in Distance Spaces*; Springer International Publishing: Cham, Switzerland, 2014; xii+173p.
17. Patle, P.R.; Vujaković, L.; Radenović, S.; Patel, D.K. Topology induced by θ-metric and multivalued mappings. *Symmetry* **2019**, *7*, 144.
18. Vujaković, J.; Aydi, H.; Radenović, S.; Mukheimer, A. Some remarks and new results in ordered partial b-metric spaces. *Mathematics* **2019**, *7*, 334. [CrossRef]
19. Vujaković, J.; Kishore, G.N.V.; Rao, K.P.R.; Radenoviić, S.; Sadik, S.K. Existence and unique coupled solution in S_b-metric spaces by rational contraction with application. *Mathematics* **2019**, *7*, 313. [CrossRef]
20. Mitrović, Z.D.; George, R.; Hussain, N. Some remarks on contraction mappings in rectangular b-metric spaces. *BSPM* **2018**, in press. [CrossRef]
21. Czerwik, S. Contraction mappings in b-metric spaces. *Acta. Math. Inf. Univ. Ostrav* **1993**, *1*, 5–11.
22. Miculescu, R.; Mihail, A. New fixed points theorems for set-valued contractions in b-metric spaces. *J. Fixed Point Theory Appl.* **2017**, *19*, 2153–2163. [CrossRef]

© 2019 by the authors. Licensee MDPI, Basel, Switzerland. This article is an open access article distributed under the terms and conditions of the Creative Commons Attribution (CC BY) license (http://creativecommons.org/licenses/by/4.0/).

Article

(p, q)-Hermite–Hadamard Inequalities for Double Integral and (p, q)-Differentiable Convex Functions

Julalak Prabseang [1], Kamsing Nonlaopon [1,*] and Jessada Tariboon [2]

1   Department of Mathematics, Faculty of Science, Khon Kaen University, Khon Kaen 40002, Thailand; julalak.pra@kkumail.com
2   Department of Mathematics, Faculty of Applied Science, King Mongkut's University of Technology North Bangkok, Bangkok 10800, Thailand; jessada.t@sci.kmutnb.ac.th
*   Correspondence: nkamsi@kku.ac.th; Tel.: +668-6642-1582

Received: 27 April 2019; Accepted: 24 May 2019; Published: 28 May 2019

Abstract: The aim of this paper is to establish some new (p, q)-calculus of Hermite–Hadamard inequalities for the double integral and refinements of the Hermite–Hadamard inequality for (p, q)-differentiable convex functions.

Keywords: Hermite–Hadamard inequalities; (p, q)-derivative; (p, q)-integral; convex functions

1. Introduction

Quantum calculus is the study of calculus without limits and is sometimes called q-calculus. In q-calculus, we obtain the original mathematical formulas when q tends to one. The beginning of the study of q-calculus can be dated back to the era of Euler (1707–1783), who first launched the q-calculus in the tracks of Newton's work on infinite series. Then, in the early Twentieth Century, Jackson [1] defined an integral, which is known as the q-Jackson integral, and studied it in a systematic way. The subject of q-calculus has many applications in the field of mathematics and other areas such as number theory, special functions, combinatorics, basic hypergeometric functions, orthogonal polynomials, quantum theory, mechanics, and the theory of relativity and physics. In recent years, the topic of q-calculus has increasingly interested many researchers. For more details, see [2–9] and the references therein. Recently, Tunç and Göv [10–12] studied the concept of (p, q)-calculus over the intervals of $[a, b] \subset \mathbb{R}$. The (p, q)-derivative and (p, q)-integral were defined and some basic properties are given. Furthermore, they obtained some new result for the (p, q)-calculus of several important integral inequalities. Currently, the (p, q)-calculus is being investigated extensively by many researchers, and a variety of new results can be found in the literature [13–18] and the references cited therein.

Mathematical inequalities are important to the study of mathematics, as well as in other area of mathematics such as analysis, differential equations, geometry, etcetera.

In 1893, Hadamard [19] investigated one of the fundamental inequalities in analysis as:

$$f\left(\frac{a+b}{2}\right) \leq \frac{1}{b-a}\int_a^b f(x)dx \leq \frac{f(a)+f(b)}{2}, \quad (1)$$

which is now known as the Hermite–Hadamard inequality.

In 2014, Tariboon and Ntouyas [20] studied the extension to q-calculus on the finite interval of (1), which is called the q-Hermite–Hadamard inequality, and some important inequalities. Next, Alp et al. [21] approved the q-Hermite–Hadamard inequality and then obtained generalized q-Hermite–Hadamard inequalities.

In 2018, Mehmet Kunt et al. [22] proved the left-hand side of the (p,q)-Hermite–Hadamard's inequality of (1) through (p,q)-differentiable convex and quasi-convex functions, and then, they gave some new (p,q)-Hermite–Hadamard's inequalities.

In 2019, Prabseang et al. [23] established the q-calculus of Hermite–Hadamard inequalities for the double integral as:

$$f\left(\frac{a+b}{2}\right) \leq \frac{1}{(b-a)^2}\int_a^b\int_a^b f(tx+(1-t)y)dxdy \leq \frac{f(a)+f(b)}{2}, \qquad (2)$$

which was given by Dragomir [24]. Moreover, they obtained refinements of the Hermite–Hadamard inequality for q-differentiable convex functions.

The aim of this paper is to present the (p,q)-calculus of Hermite–Hadamard inequalities for double integrals (2) and refinements of the Hermite–Hadamard inequality. These are obtained as special cases when $p=1$ and $q\to 1$.

Before we proceed to our main theorem, the following definitions and some concepts require some clarifications.

2. Preliminaries

Throughout this paper, let $[a,b] \subseteq \mathbb{R}$ be an interval and $0 < q < p \leq 1$ be a constant. The following definitions for the (p,q)-derivative and (p,q)-integral were given in [10,11].

Definition 1. *Let $f : [a,b] \to \mathbb{R}$ be a continuous function, and let $x \in [a,b]$. Then, the (p,q)-derivative of f on $[a,b]$ at x is defined as:*

$$_aD_{p,q}f(x) = \frac{f(px+(1-p)a) - f(qx+(1-q)a)}{(p-q)(x-a)}, \quad x \neq a \qquad (3)$$

$$_aD_{p,q}f(a) = \lim_{x\to a} {}_aD_{p,q}f(x).$$

Obviously, a function f is (p,q)-differentiable on $[a,b]$ if $_aD_{p,q}f(x)$ exists for all $x \in [a,b]$. In Definition 1, if $a=0$, then $_0D_{p,q}f = D_{p,q}f$, where $D_{p,q}f$ is defined by:

$$D_{p,q}f(x) = \frac{f(px) - f(qx)}{(p-q)x}, \quad x \neq 0. \qquad (4)$$

Furthermore, if $p=1$ in (4), then it reduces to $D_q f$, which is the q-derivative of the function f; see [5].

Example 1. *Define function $f:[a,b]\to\mathbb{R}$ by $f(x) = x^2+1$. Let $0<q<p\leq 1$. Then, for $x\neq a$, we have:*

$$\begin{aligned}_aD_{p,q}(x^2+1) &= \frac{[(px+(1-p)a)^2+1]-[(qx+(1-q)a)^2+1]}{(p-q)(x-a)} \\ &= \frac{(p+q)x^2 + 2ax[1-(p+q)] + a^2[(p+q)-2]}{(x-a)} \\ &= \frac{x(p+q)(x-a) - a(p+q)(x-a) + 2a(x-a)}{(x-a)} \\ &= (p+q)(x-a) + 2a.\end{aligned} \qquad (5)$$

Definition 2. *Let $f:[a,b]\to\mathbb{R}$ be a continuous function. Then, the (p,q)-integral on $[a,b]$ is defined by:*

$$\int_a^x f(t)\, _ad_{p,q}t = (p-q)(x-a)\sum_{n=0}^\infty \frac{q^n}{p^{n+1}} f\left(\frac{q^n}{p^{n+1}}x + \left(1-\frac{q^n}{p^{n+1}}\right)a\right), \qquad (6)$$

for $x\in [a,b]$. If $a=0$ and $p=1$ in (6), then we have the classical q-integral [5].

Example 2. *Define function* $f : [a,b] \to \mathbb{R}$ *by* $f(x) = 2x$. *Let* $0 < q < p \leq 1$. *Then, we have:*

$$\int_a^b f(x) \, _a d_{p,q} x = \int_a^b 2x \, _a d_{p,q} x$$
$$= 2(p-q)(b-a) \sum_{n=0}^{\infty} \frac{q^n}{p^{n+1}} \left(\frac{q^n}{p^{n+1}} b + \left(1 - \frac{q^n}{p^{n+1}}\right) a \right) \quad (7)$$
$$= \frac{2(b-a)(b-a(1-p-q))}{p+q}.$$

Theorem 1. *Let* $f : [a,b] \to \mathbb{R}$ *be a continuous function. Then, we have the following:*

(i) $_a D_{p,q} \int_a^x f(t) \, _a d_{p,q} t = f(x)$;
(ii) $\int_c^x {}_a D_{p,q} f(t) \, _a d_{p,q} t = f(x) - f(c)$ *for* $c \in (a,x)$.

Theorem 2. *Let* $f, g : [a,b] \to \mathbb{R}$ *be continuous functions and* $\alpha \in \mathbb{R}$. *Then, we have the following:*

(i) $\int_a^x [f(t) + g(t)] \, _a d_{p,q} t = \int_a^x f(t) \, _a d_{p,q} t + \int_a^x g(t) \, _a d_{p,q} t$;
(ii) $\int_a^x (\alpha f)(t) \, _a d_{p,q} t = \alpha \int_a^x f(t) \, _a d_{p,q} t$;
(iii) $\int_c^x f(pt + (1-p)a) \, _a D_{p,q} g(t) \, _a d_q t = (fg)|_c^x - \int_c^x g(qt + (1-q)a) \, _a D_{p,q} f(t) \, _a d_{p,q} t$ *for* $c \in (a,x)$.

For the proof properties of Theorems 1 and 2, we refer to [10,11].
The proofs of the following theorems were given in [22].

Theorem 3. *Let* $f : [a,b] \to \mathbb{R}$ *be a convex differentiable function on* (a,b) *and* $0 < q < p \leq 1$. *Then, we have:*

$$f\left(\frac{qa+pb}{p+q}\right) \leq \frac{1}{p(b-a)} \int_a^{pb+(1-p)a} f(x) \, _a d_{p,q} x \leq \frac{qf(a) + pf(b)}{p+q}. \quad (8)$$

Theorem 4. *Let* $f : [a,b] \to \mathbb{R}$ *be a convex differentiable function on* (a,b) *and* $0 < q < p \leq 1$. *Then, we have:*

$$f\left(\frac{pa+qb}{p+q}\right) + \frac{(p-q)(b-a)}{p+q} f'\left(\frac{pa+qb}{p+q}\right) \leq \frac{1}{p(b-a)} \int_a^{pb+(1-p)a} f(x) \, _a d_{p,q} x$$
$$\leq \frac{qf(a)+pf(b)}{p+q}. \quad (9)$$

Theorem 5. *Let* $f : [a,b] \to \mathbb{R}$ *be a convex differentiable function on* (a,b) *and* $0 < q < p \leq 1$. *Then, we have:*

$$f\left(\frac{a+b}{2}\right) + \frac{(p-q)(b-a)}{2(p+q)} f'\left(\frac{a+b}{2}\right) \leq \frac{1}{p(b-a)} \int_a^{pb+(1-p)a} f(x) \, _a d_{p,q} x$$
$$\leq \frac{qf(a)+pf(b)}{p+q}. \quad (10)$$

Lemma 1. *Let* $f : [a,b] \to \mathbb{R}$ *be a convex continuous function on* $[a,b]$ *and* $0 < q < p \leq 1$. *Then, we have:*

$$f\left(\frac{1}{(pb-pa)^2} \int_a^{pb+(1-p)a} \int_a^{pb+(1-p)a} (tx + (1-t)y) \, _a d_{p,q} x \, _a d_{p,q} y\right)$$
$$\leq \frac{1}{(pb-pa)^2} \int_a^{pb+(1-p)a} \int_a^{pb+(1-p)a} f(tx + (1-t)y) \, _a d_{p,q} x \, _a d_{p,q} y. \quad (11)$$

Proof. The proof of this lemma can be obtained by Definition 2 and Jensen's inequality. □

3. Main Results

In this section, we present the (p,q)-Hermite–Hadamard inequality for double integrals and the refinement of Hermite–Hadamard inequalities on the interval $[a,b]$.

Theorem 6. Let $f : [a, b] \to \mathbb{R}$ be a convex continuous function on $[a, b]$ and $0 < q < p \leq 1$. Then, we have:

$$f\left(\frac{qa+pb}{p+q}\right) \leq \frac{1}{(pb-pa)^2} \int_a^{pb+(1-p)a} \int_a^{pb+(1-p)a} f(tx+(1-t)y) \, _ad_{p,q}x \, _ad_{p,q}y$$

$$\leq \frac{1}{p(b-a)} \int_a^{pb+(1-p)a} f(x) \, _ad_{p,q}x \, _ad_{p,q}y \quad (12)$$

$$\leq \frac{qf(a)+pf(b)}{p+q}.$$

Proof. Since f is convex on $[a, b]$, it follows that:

$$f(tx+(1-t)y) \leq tf(x) + (1-t)f(y) \quad (13)$$

for all $x, y \in [a, b]$ and $t \in [0, 1]$. Taking the double (p, q)-integration on both sides for (13) on $[a, pb + (1-p)a] \times [a, pb + (1-p)a]$, we obtain:

$$\int_a^{pb+(1-p)a} \int_a^{pb+(1-p)a} f(tx+(1-t)y) \, _ad_{p,q}x \, _ad_{p,q}y$$

$$\leq \int_a^{pb+(1-p)a} \int_a^{pb+(1-p)a} [tf(x) + (1-t)f(y)] \, _ad_{p,q}x \, _ad_{p,q}y \quad (14)$$

$$= (pb-pa) \int_a^{pb+(1-p)a} f(x) \, _ad_{p,q}x,$$

which show the second part of (12) by using the right-hand side of the (p, q)-Hermite–Hadamard's inequality.

On the other hand, by Lemma 1, we have:

$$f\left(\frac{1}{(pb-pa)^2} \int_a^{pb+(1-p)a} \int_a^{pb+(1-p)a} (tx+(1-t)y) \, _ad_{p,q}x \, _ad_{p,q}y\right)$$

$$\leq \frac{1}{(pb-pa)^2} \int_a^{pb+(1-p)a} \int_a^{pb+(1-p)a} f(tx+(1-t)y) \, _ad_{p,q}x \, _ad_{p,q}y,$$

and since:

$$\frac{1}{(pb-pa)^2} \int_a^{pb+(1-p)a} \int_a^{pb+(1-p)a} (tx+(1-t)y) \, _ad_{p,q}x \, _ad_{p,q}y = \frac{qa+pb}{p+q}.$$

This completes the proof. □

Remark 1. If $p = 1$ and $q \to 1$, then (12) reduces to (2), that is,

$$f\left(\frac{a+b}{2}\right) \leq \frac{1}{(b-a)^2} \int_a^b \int_a^b f(tx+(1-t)y) dx dy \leq \frac{f(a)+f(b)}{2}.$$

Corollary 1. Let $f : [a, b] \to \mathbb{R}$ be a convex continuous function on $[a, b]$ and $0 < q < p \leq 1$. Then, we have:

$$f\left(\frac{qa+pb}{p+q}\right) \leq \frac{1}{(pb-pa)^2} \int_a^{pb+(1-p)a} \int_a^{pb+(1-p)a} f\left(\frac{x+y}{2}\right) \, _ad_{p,q}x \, _ad_{p,q}y$$

$$\leq \frac{1}{p(b-a)} \int_a^{pb+(1-p)a} f(x) \, _ad_{p,q}x$$

$$\leq \frac{qf(a)+pf(b)}{p+q}. \quad (15)$$

Remark 2. If $p = 1$ and $q \to 1$, then (15) reduces to:

$$f\left(\frac{a+b}{2}\right) \leq \frac{1}{(b-a)^2} \int_a^b \int_a^b f\left(\frac{x+y}{2}\right) dx dy \leq \frac{1}{b-a} \int_a^b f(x) dx \leq \frac{f(a)+f(b)}{2},$$

which readily appeared in [25].

Theorem 7. Let $f : [a,b] \to \mathbb{R}$ be a convex continuous function on $[a,b]$ and $0 < q < p \leq 1$. Then, we have:

$$\frac{p}{(pb-pa)^2} \int_a^{pb+(1-p)a} \int_a^{pb+(1-p)a} f\left(\frac{px+qy}{p+q}\right) {}_ad_{p,q}x\, {}_ad_{p,q}y$$

$$\leq \frac{1}{(pb-pa)^2} \int_0^p \int_a^{pb+(1-p)a} \int_a^{pb+(1-p)a} f(tx+(1-t)y) \, {}_ad_{p,q}x\, {}_ad_{p,q}y\, d_{p,q}t \qquad (16)$$

$$\leq \frac{1}{pb-pa} \int_a^{pb+(1-p)a} f(x) \, {}_ad_{p,q}x.$$

Proof. Let $g : [a,b] \to \mathbb{R}$ be given by:

$$g(t) = \frac{1}{(pb-pa)^2} \int_a^{pb+(1-p)a} \int_a^{pb+(1-p)a} f(tx+(1-t)y) \, {}_ad_{p,q}x\, {}_ad_{p,q}y.$$

For all $t_1, t_2 \in [0,1]$ and $\alpha, \beta \geq 0$ with $\alpha + \beta = 1$, we consider:

$$g(\alpha t_1 + \beta t_2) = \frac{1}{(pb-pa)^2} \int_a^{pb+(1-p)a} \int_a^{pb+(1-p)a} f((\alpha t_1 + \beta t_2)x + (1-(\alpha t_1 + \beta t_2))y) \, {}_ad_{p,q}x\, {}_ad_{p,q}y$$

$$\leq \frac{\alpha}{(pb-pa)^2} \int_a^{pb+(1-p)a} \int_a^{pb+(1-p)a} f(t_1 x + (1-t_1)y) \, {}_ad_{p,q}x\, {}_ad_{p,q}y$$

$$+ \frac{\beta}{(pb-pa)^2} \int_a^{pb+(1-p)a} \int_a^{pb+(1-p)a} f(t_2 x + (1-t_2)y) \, {}_ad_{p,q}x\, {}_ad_{p,q}y$$

$$= \alpha g(t_1) + \beta g(t_2),$$

which show that g is convex on $[0,1]$. Using Theorem 3 for the convex function g, we have:

$$\frac{1}{(pb-pa)^2} \int_a^{pb+(1-p)a} \int_a^{pb+(1-p)a} f\left(\frac{px+qy}{p+q}\right) d_q x\, d_q y$$

$$= g\left(\frac{p}{p+q}\right) \leq \frac{1}{p} \int_0^p g(t) \, {}_ad_{p,q}t$$

$$= \frac{1}{p(pb-pa)^2} \int_0^p \int_a^{pb+(1-p)a} \int_a^{pb+(1-p)a} f(tx+(1-t)y) \, {}_ad_{p,q}x\, {}_ad_{p,q}y\, d_{p,q}t$$

$$\leq \frac{qg(0) + pg(1)}{p(p+q)} = \frac{1}{p(pb-pa)} \int_a^{pb+(1-p)a} f(x) \, {}_ad_{p,q}x.$$

This completes the proof. □

Remark 3. If $p = 1$ and $q \to 1$, then (16) reduces to:

$$\frac{1}{(b-a)^2} \int_a^b \int_a^b f\left(\frac{x+y}{2}\right) dx dy \leq \frac{1}{(b-a)^2} \int_0^1 \int_a^b \int_a^b f(tx+(1-t)y) dx dy dt$$

$$\leq \frac{1}{b-a} \int_a^b f(x) dx,$$

which readily appeared in [25].

Theorem 8. Let $f : [a, b] \to \mathbb{R}$ be a (p, q)-differentiable convex continuous function and $0 < q < p \leq 1$, then the following inequalities:

$$0 \leq \frac{p}{b-a} \int_a^{pb+(1-p)a} f(x) \, _a d_{p,q} x$$
$$- \frac{1}{(b-a)^2} \int_a^{pb+(1-p)a} \int_a^{pb+(1-p)a} f(tx + (1-t)y) \, _a d_{p,q} x \, _a d_{p,q} y \quad (17)$$
$$\leq t \left[\frac{p^2 f(a) + pqf(pb+(1-p)a)}{p+q} - \frac{p}{b-a} \int_a^{pb+(1-p)a} f(qx + (1-q)a) \, _a d_{p,q} x \right],$$

are valid for all $t \in [0, 1]$.

Proof. Since f is convex on J, it follows that:

$$f(tx + (1-t)y) \leq tf(x) + (1-t)f(y)$$

for all $x, y \in [a, b]$ and $t \in [0, 1]$. Taking double (p, q)-integration on both sides of the above inequality on $[a, pb + (1-p)a] \times [a, pb + (1-p)a]$, we obtain:

$$\int_a^{pb+(1-p)a} \int_a^{pb+(1-p)a} f(tx + (1-t)y) \, _a d_{p,q} x \, _a d_{p,q} y$$
$$\leq \int_a^{pb+(1-p)a} \int_a^{pb+(1-p)a} [tf(x) + (1-t)f(y)] \, _a d_{p,q} x \, _a d_{p,q} y$$
$$= p(b-a) \int_a^{pb+(1-p)a} f(x) \, _a d_{p,q} x.$$

On the other hand, since f is (p, q)-differentiable convex on $[a, b]$ and $f' \geq \, _a D_{p,q} f$, we have:

$$f(tx + (1-t)y) - f(y) \geq t(x - y) \, _a D_{p,q} f(y)$$

for all $x, y \in [a, b]$ and $t \in [0, 1]$. Taking the double (p, q)-integration on both sides of the above inequality on $[a, pb + (1-p)a] \times [a, pb + (1-p)a]$, we obtain:

$$\int_a^{pb+(1-p)a} \int_a^{pb+(1-p)a} f(tx + (1-t)y) \, _a d_{p,q} x \, _a d_{p,q} y - (pb - pa) \int_a^{pb+(1-p)a} f(x) \, _a d_{p,q} x$$
$$\geq t \int_a^{pb+(1-p)a} \int_a^{pb+(1-p)a} (x - y) \, _a D_{p,q} f(y) \, _a d_{p,q} x \, _a d_{p,q} y. \quad (18)$$

Since,

$$\int_a^{pb+(1-p)a} \int_a^{pb+(1-p)a} (x - y) \, _a D_{p,q} f(y) \, _a d_{p,q} x \, _a d_{p,q} y$$
$$= (pb - pa) \int_a^{pb+(1-p)a} f(qx + (1-q)a) \, _a d_{p,q} x - (b-a)^2 \frac{[p^2 f(a) + pqf(pb+(1-p)a)]}{p+q}.$$

Substituting the above inequality in (18), we have:

$$(pb - pa) \int_a^{pb+(1-p)a} f(x) \, _a d_{p,q} x - \int_a^{pb+(1-p)a} \int_a^{pb+(1-p)a} f(tx + (1-t)y) \, _a d_{p,q} x \, _a d_{p,q} y$$
$$\leq t \left[(b-a)^2 \frac{[p^2 f(a) + pqf(pb+(1-p)a)]}{p+q} - (pb - pa) \int_a^{pb+(1-p)a} f(qx + (1-q)a) \, _a d_{p,q} x \right]$$

for all $t \in [0, 1]$, which completes the proof. □

Remark 4. If $p = 1$ and $q \to 1$, then (17) reduces to:

$$0 \leq \frac{1}{b-a}\int_a^b f(x)dx - \frac{1}{(b-a)^2}\int_a^b\int_a^b f(tx + (1-t)y)dxdy$$
$$\leq t\left[\frac{f(a) + f(b)}{2} - \frac{1}{b-a}\int_a^b f(x)dx\right],$$

which readily appeared in [25,26].

Corollary 2. *Let $f : [a,b] \to \mathbb{R}$ be a (p,q)-differentiable convex continuous function and $0 < q < p \leq 1$. Then, we have:*

$$0 \leq \frac{p}{b-a}\int_a^{pb+(1-p)a} f(x)\,_adp_{,q}x - \frac{1}{(b-a)^2}\int_a^{pb+(1-p)a}\int_a^{pb+(1-p)a} f\left(\frac{x+y}{2}\right)\,_adp_{,q}x\,_adp_{,q}y$$
$$\leq \frac{1}{2}\left[\frac{p^2 f(a) + pqf(pb+(1-p)a)}{p+q} - \frac{p}{b-a}\int_a^{pb+(1-p)a} f(qx + (1-q)a)\,_adp_{,q}x\right]. \quad (19)$$

Remark 5. If $p = 1$ and $q \to 1$, then (19) reduces to:

$$0 \leq \frac{1}{b-a}\int_a^b f(x)dx - \frac{1}{(b-a)^2}\int_a^b\int_a^b f\left(\frac{x+y}{2}\right)dxdy$$
$$\leq \frac{1}{2}\left[\frac{f(a) + f(b)}{2} - \frac{1}{b-a}\int_a^b f(x)dx\right],$$

which readily appeared in [25].

Theorem 9. *Let $f : [a,b] \to \mathbb{R}$ be a (p,q)-differentiable convex continuous function, which is defined at the point $\frac{qa+pb}{p+q} \in (a,b)$ and $0 < q < p \leq 1$. Then, the following inequalities:*

$$0 \leq \frac{1}{b-a}\int_a^{pb+(1-p)a} f(x)\,_adp_{,q}x - \frac{1}{b-a}\int_a^{pb+(1-p)a} f\left(tx + (1-t)\frac{qa+pb}{p+q}\right)\,_adp_{,q}x$$
$$\leq (1-t)\left[\frac{pf(a) + qf(pb+(1-p)a)}{p+q} - \frac{1}{b-a}\int_a^{pb+(1-p)a} f(qx+(1-q)a)\,_adp_{,q}x\right] \quad (20)$$

are valid for all $t \in [0,1]$.

Proof. Since f is convex on $[a,b]$ and using Theorem 3, we have:

$$\frac{1}{p(b-a)}\int_a^{pb+(1-p)a} f\left(tx + (1-t)\frac{qa+pb}{p+q}\right)\,_adp_{,q}x$$
$$\leq \frac{t}{p(b-a)}\int_a^{pb+(1-p)a} f(x)\,_adp_{,q}x + (1-t)f\left(\frac{qa+pb}{p+q}\right)$$
$$\leq \frac{t}{p(b-a)}\int_a^{pb+(1-p)a} f(x)\,_adp_{,q}x + \frac{1-t}{p(b-a)}\int_a^{pb+(1-p)a} f(x)\,_adp_{,q}x$$
$$= \frac{1}{p(b-a)}\int_a^{pb+(1-p)a} f(x)\,_adp_{,q}x$$

for all $t \in [0,1]$.

On the other hand, since f is the (p,q)-differentiable convex on $[a,b]$, we have:

$$f\left(tx + (1-t)\frac{qa+pb}{p+q}\right) - f(x) \geq (1-t)\left(\frac{qa+pb}{p+q} - x\right){}_aD_{p,q}(x).$$

Taking the double (p,q)-integration on both sides of the above inequality on $[a,b]$, we obtain:

$$\frac{1}{p(b-a)} \int_a^{pb+(1-p)a} f\left(tx+(1-t)\frac{qa+pb}{p+q}\right) {}_a d_{p,q} x - \frac{1}{p(b-a)} \int_a^{pb+(1-p)a} f(x) {}_a d_{p,q} x$$
$$\geq \frac{(1-t)}{p(b-a)} \int_a^{pb+(1-p)a} \left(\frac{qa+pb}{p+q}-x\right) {}_a D_{p,q} f(x) {}_a d_{p,q} x. \tag{21}$$

Since,

$$\int_a^{pb+(1-p)a} \left(\frac{qa+pb}{p+q}-x\right) {}_a D_{p,q} f(x) {}_a d_{p,q} x$$
$$= \int_a^{pb+(1-p)a} f(qx+(1-q)a) {}_a d_{p,q} x - (b-a) \frac{pf(a)+qf(pb+(1-p)a)}{p+q}. \tag{22}$$

This completes the proof. □

Corollary 3. *Let $f : [a,b] \to \mathbb{R}$ be a (p,q)-differentiable convex continuous function and $0 < q < p \leq 1$. Then, we have:*

$$0 \leq \frac{1}{b-a} \int_a^{pb+(1-p)a} f(x) {}_a d_{p,q} x - \frac{2}{b-a} \int_{\frac{a(p+2q)+pb}{2(1+q)}}^{\frac{(p^2+pq)(b-a)+(p+2q)a+pb}{2(p+q)}} f(x) {}_a d_{p,q} x$$
$$\leq \frac{1}{2} \left[\frac{pf(a)+qf(pb+(1-p)a)}{p+q} - \frac{1}{b-a} \int_a^{pb+(1-p)a} f(qx+(1-q)a) {}_a d_{p,q} x \right]. \tag{23}$$

Theorem 10. *Let $f : [a,b] \to \mathbb{R}$ be a (p,q)-differentiable convex continuous function, which is defined at the point $\frac{pa+qb}{p+q} \in (a,b)$ and $0 < q < p \leq 1$. Then, the following inequalities:*

$$(1-t) \frac{p(p-q)(b-a)}{p+q} f'\left(\frac{pa+qb}{p+q}\right)$$
$$\leq \frac{1}{b-a} \int_a^{pb+(1-p)a} f(x) {}_a d_{p,q} x - \frac{1}{b-a} \int_a^{pb+(1-p)a} f\left(tx+(1-t)\frac{pa+qb}{p+q}\right) {}_a d_{p,q} x \tag{24}$$
$$\leq (1-t) \left[\frac{qf(a)+pf(pb+(1-p)a)}{p+q} - \frac{1}{b-a} \int_a^{pb+(1-p)a} f(qx+(1-q)a) {}_a d_{p,q} x\right]$$

are valid for all $t \in [0,1]$.

Proof. The proof of this theorem follows a similar procedure as Theorem 9 by using Theorem 4. □

Corollary 4. *Let $f : [a,b] \to \mathbb{R}$ be a (p,q)-differentiable convex continuous function and $0 < q < p \leq 1$. Then, we have:*

$$\frac{p(p-q)(b-a)}{2(p+q)} f'\left(\frac{pa+qb}{p+q}\right)$$
$$\leq \frac{1}{b-a} \int_a^{pb+(1-p)a} f(x) {}_a d_{p,q} x - \frac{2}{b-a} \int_{\frac{2pa+q(a+b)}{2(p+q)}}^{\frac{(p^2+pq)(b-a)+(2p+q)a+qb}{2(1+q)}} f(x) {}_a d_{p,q} x \tag{25}$$
$$\leq \frac{1}{2} \left[\frac{qf(a)+pf(pb+(1-p)a)}{p+q} - \frac{1}{b-a} \int_a^{pb+(1-p)a} f(qx+(1-q)a) {}_a d_{p,q} x\right].$$

Theorem 11. *Let $f : [a,b] \to \mathbb{R}$ be a (p,q)-differentiable convex continuous function, which is defined at the point $\frac{a+b}{2} \in (a,b)$ and $0 < q < p \leq 1$. Then, the following inequalities:*

$$(1-t) \frac{p(p-q)(b-a)}{2(p+q)} f'\left(\frac{a+b}{2}\right)$$
$$\leq \frac{1}{b-a} \int_a^{pb+(1-p)a} f(x) {}_a d_{p,q} x - \frac{1}{b-a} \int_a^{pb+(1-p)a} f\left(tx+(1-t)\frac{a+b}{2}\right) {}_a d_{p,q} x \tag{26}$$
$$\leq (1-t) \left[\frac{f(a)+f(pb+(1-q)a)}{2} - \frac{1}{b-a} \int_a^{pb+(1-p)a} f(qx+(1-q)a) {}_a d_{p,q} x\right]$$

are valid for all $t \in [0,1]$.

Proof. The proof of this theorem follows a similar procedure as Theorem 9 by using Theorem 5. □

Corollary 5. *Let $f : [a,b] \to \mathbb{R}$ be a (p,q)-differentiable convex continuous function and $0 < q < p \le 1$. Then, we have:*

$$\frac{p(p-q)(b-a)}{4(p+q)} f'\left(\frac{a+b}{2}\right)$$

$$\le \frac{1}{b-a} \int_a^{pb+(1-p)a} f(x) \, {}_ad_{p,q}x - \frac{2}{b-a} \int_{\frac{3a+b}{4}}^{\frac{2p(b-a)+3a+b}{4}} f(x) \, {}_ad_{p,q}x \quad (27)$$

$$\le \frac{1}{2}\left[\frac{qf(a)+pf(pb+(1-p)a)}{p+q} - \frac{1}{b-a}\int_a^{pb+(1-p)a} f(qx+(1-q)a) \, {}_ad_{p,q}x\right].$$

Remark 6. *If $p = 1$ and $q \to 1$, then (20), (24), and (26) reduce to:*

$$0 \le \frac{1}{b-a}\int_a^b f(x)dx - \frac{1}{b-a}\int_a^b f\left(tx+(1-t)\frac{a+b}{2}\right)dx$$

$$\le (1-t)\left[\frac{f(a)+f(b)}{2} - \frac{1}{b-a}\int_a^b f(x)dx\right],$$

which readily appeared in [25].

Remark 7. *If $p = 1$ and $q \to 1$, then (23), (25), and (27) reduce to:*

$$0 \le \frac{1}{b-a}\int_a^b f(x)dx - \frac{2}{b-a}\int_{\frac{3a+b}{4}}^{\frac{a+3b}{4}} f(x)dx \le \frac{1}{2}\left[\frac{f(a)+f(b)}{2} - \frac{1}{b-a}\int_a^b f(x)dx\right],$$

which readily appeared in [25].

4. Conclusions

In this paper, we have obtained some new results for the (p,q)-calculus of Hermite–Hadamard inequalities for the double integral and refinements of the Hermite–Hadamard inequality. Our work has improved the results of [23] and can be reduced to the classical inequality formulas in special cases when $p = 1$ and $q \to 1$. It is expected that this paper may stimulate further research in this field.

Author Contributions: The order of the author list reflects the contributions to the paper.

Funding: This research received no external funding.

Conflicts of Interest: The authors declare no conflict of interest.

References

1. Jackson, F.H. On a q-definite integrals. *Quart. J. Pure Appl. Math.* **1910**, *41*, 193–203.
2. Bangerezako, G. Variational q-calculus. *J. Math. Anal. Appl.* **2004**, *289*, 650–665. [CrossRef]
3. Ernst, T. *The History of q-Calculus and a New Method*; Uppsala University: Uppsala, Sweden, 2000.
4. Jackson, F.H. q-Difference equations. *Am. J. Math.* **1910**, *32*, 305–314. [CrossRef]
5. Kac, V.; Cheung, P. *Quantum Calculus*; Springer: New York, NY, USA, 2002.
6. Gauchman, H. Integral inequalities in q-Calculus. *Comput. Math. Appl.* **2004**, *47*, 281–300. [CrossRef]
7. Noor, M.A.; Noor, K.I.; Awan, M.U. Some quantum estimates for Hermite–Hadamard inequalities. *Appl. Math. Comput.* **2015**, *251*, 675–679. [CrossRef]
8. Noor, M.A.; Noor, K.I.; Awan, M.U. Some quantum integral inequalities via preinvex functions. *Appl. Math. Comput.* **2015**, *269*, 242–251. [CrossRef]

9. Sudsutad, W.; Ntouyas, S.K.; Tariboon, J. Quantum integral inequalities for convex functions. *J. Math. Inequal.* **2015**, *9*, 781–793. [CrossRef]
10. Tunç, M.; Göv, E. (p,q)-Integral inequalities. *RGMIA Res. Rep. Coll.* **2016**, *19*, 97.
11. Tunç, M.; Göv, E. Some integral inequalities via (p,q)-calculus on finite intervals. *RGMIA Res. Rep. Coll.* **2016**, *19*, 95.
12. Tunç, M.; Göv, E. (p,q)-integral inequalities for convex functions. *RGMIA Res. Rep. Coll.* **2016**, *19*, 98.
13. Araci, S.; Duran, U.; Acikgoz, M.; Srivastava, H.M. A certain (p,q)-derivative operator and associated divided differences. *J. Inequal. Appl.* **2016**, *1*, 301. [CrossRef]
14. Duran, U.; Acikgoz, M.; Esi, A.; Araci, S. A note on the (p,q) Hermite polynomials. *Appl. Math. Inf. Sci.* **2018**, *12*, 227–231. [CrossRef]
15. Mursaleen, M.; Ansari, K.J.; Khan, A. Some Approximation Results by (p,q)-analogue of Bernstein-Stancu operators. *Appl. Math. Comput.* **2015**, *264*, 392–402. [Corrigendum: *Appl. Math. Comput.* **2015**, *269*, 744–746.] [CrossRef]
16. Sahai, V.; Yadav, S. Representations of two parameter quantum algebras and p,q-special functions. *J. Math. Anal. Appl.* **2007**, *335*, 268–279. [CrossRef]
17. Sadjang, P.N. On the fundamental theorem of (p,q)-calculus and some (p,q)-Taylor formulas. *Results Math.* **2018**, *73*, 39. [CrossRef]
18. Sadjang, P.N. On the (p,q)-Gamma and the (p,q)-Beta functions. *arXiv* **2015**, arXiv:1506.07394.
19. Hadamard, J. Etude sur les propriétés des fonctions entiéres et en particulier d'une fonction considérée par Riemann. *J. Math. Pures Appl.* **1893**, *9*, 171–216.
20. Tariboon, J.; Ntouyas, S.K. Quantum integral inequalities on finite intervals. *J. Inequal. App.* **2014**, *2014*, 121. [CrossRef]
21. Alp, N.; Sarıkaya, M.Z.; Kunt, M.; İşcan, İ. q-Hermite–Hadamard inequalities and quantum estimates for midpoint type inequalities via convex and quasi-convex functions. *J. King Saud Univ. Sci.* **2018**, *30*, 193–203. [CrossRef]
22. Kunt, M.; İşcan, İ.; Alp, N.; Sarikaya, M.Z. (p,q)-Hermite–Hadamard inequalities and (p,q)-estimates for midpoint typeinequalities via convex and quasi-convex functions. *Rev. Real Acad. Cienc. Exactas Fís. Nat. Ser. A Mat.* **2018**, *112*, 969–992. [CrossRef]
23. Prabseang, J.; Nonlaopon, K.; Tariboon, J. Quantum Hermite–Hadamard inequalities for double integral and q-differentiable convex functions. *J. Math. Inequal.* **2019**, *13*, 675–686.
24. Dragomir, S.S. Two refinements of Hadamard's inequalities. *Coll. Sci. Pap. Fac. Kragujevac* **1990**, *11*, 23–26.
25. Pachpatte, B.G. *Mathematical Inequalities*; North-Holland Library, Elsevier Science: Amsterdam, Holland, 2005; Volume 67.
26. Dragomir, S.S. Some integral inequalities for differentiable convex functions. *Contrib. Sec. Math. Technol. Sci.* **1992**, *13*, 13–17.

© 2019 by the authors. Licensee MDPI, Basel, Switzerland. This article is an open access article distributed under the terms and conditions of the Creative Commons Attribution (CC BY) license (http://creativecommons.org/licenses/by/4.0/).

Article

Some New Results Involving the Generalized Bose–Einstein and Fermi–Dirac Functions

Rekha Srivastava [1], Humera Naaz [2], Sabeena Kazi [2] and Asifa Tassaddiq [2,*]

[1] Department of Mathematics and Statistics, University of Victoria, Victoria, BC V8W 3R4, Canada; rekhas@math.uvic.ca

[2] College of Computer and Information Sciences, Majmaah University, Al Majmaah 11952, Saudiarabia; h.naaz@mu.edu.sa (H.N.); s.badesaheb@mu.edu.sa (S.K.)

* Correspondence: a.tassaddiq@mu.edu.sa

Received: 28 March 2019; Accepted: 17 May 2019; Published: 21 May 2019

Abstract: In this paper, we obtain a new series representation for the generalized Bose–Einstein and Fermi–Dirac functions by using fractional Weyl transform. To achieve this purpose, we obtain an analytic continuation for these functions by generalizing the domain of Riemann zeta functions from $(0 < \Re(s) < 1)$ to $(0 < \Re(s) < \mu)$. This leads to fresh insights for a new generalization of the Riemann zeta function. The results are validated by obtaining the classical series representation of the polylogarithm and Hurwitz–Lerch zeta functions as special cases. Fractional derivatives and the relationship of the generalized Bose–Einstein and Fermi–Dirac functions with Apostol–Euler–Nörlund polynomials are established to prove new identities.

Keywords: Fermi–Dirac function; Bose–Einstein function; Weyl transform; series representation

1. Introduction

The importance of the Fermi–Dirac and Bose–Einstein functions emerges from their fundamental presence in quantum physics and related sciences. Unlike the classical mechanics of particles, where the Maxwell distribution is used to study the velocity of classical gas molecules, the quantum gas is analyzed by using the Fermi–Dirac and Bose–Einstein functions. The distinct particles obey Fermi–Dirac statistics, while the indistinct particles follow Bose–Einstein statistics. All particles have a spin in relation to the usual theory. Fermions have half-integer spin and bosons have integer spin. The Fermi–Dirac and Bose–Einstein distribution functions are used to analyze them in the language of mathematics and physics. Indistinguishable particles that are not categorized through either of the aforementioned types are called anyons. The extensions of the Bose–Einstein and Fermi–Dirac functions interpolate between the two. Therefore, Chaudhry et al. [1] proposed that the extensions of the Bose–Einstein and Fermi–Dirac functions may help to describe anyons. In this paper, we generalize the results of Chaudhry and Qadir [2] by proving a general representation theorem to establish a new series representation of the generalized Bose–Einstein and Fermi–Dirac functions. However, we also discuss the fractional derivative, and the relationship of the generalized Bose–Einstein and Fermi–Dirac functions with Apostol–Euler–Nörlund polynomials. Before we provide our research results, it is necessary to enlist all the basic definitions and preliminaries that are required to present and understand this work.

2. Materials and Methods

2.1. Generalized Bose–Einstein and Fermi–Dirac Functions

During the course of our investigation, we consider the subsequent usual notations:

$$\mathbb{N} := \{1,2,3,\ldots\}, \quad \mathbb{N} \cup \{0\} = \mathbb{N}_0; \quad \mathbb{Z}^- = \{-1,-2,-3\ldots\}.$$

In addition, $\mathbb{Z}$ is the set of integers, $\mathbb{R}$ denotes the set of real numbers, $\mathbb{R}^+$ denotes the set of positive numbers, and $\mathbb{C}$ is the set of complex numbers, $s = \sigma + i\tau$. Gamma function $\Gamma(s)$ as a generalization of factorials is also used here as a basic special function. For a detailed study of gamma and related functions, we refer the interested reader to [3,4].

More recently, Bayad and Chikhi [5] introduced and studied the generalized Fermi–Dirac functions given by ([5], (p. 12), Equation (45))

$$\Theta_\nu(s,\mu;x) := \frac{\Gamma(\mu)}{\Gamma(s)} \int_0^\infty \frac{t^{s-1} e^{-\nu(x+t)}}{(e^{x+t}+1)^\mu} dt \tag{1}$$
$$(\mathfrak{R}(x) \geq 0, \ \mathfrak{R}(\nu) > -\mathfrak{R}(\mu) \wedge \mathfrak{R}(s) > \mathfrak{R}(\mu) > 0 \text{ when } e^{-x} \neq -1)$$

and their series representation is given by ([5], (p. 12), Equation (46))

$$\Theta_\nu(s,\mu;x) := \sum_{n=1}^\infty \frac{(-1)^n \Gamma(\mu+n) e^{-(\nu+\mu+n)x}}{n!(\nu+\mu+n)^s}. \tag{2}$$

For $\mu = 1$, in Equation (1) the extended Fermi–Dirac functions ([6], (p. 113), Equation (3.14)) are given here by

$$\Theta_\nu(s;x) := \Theta_\nu(s,1;x) = \frac{1}{\Gamma(s)} \int_0^\infty \frac{t^{s-1} e^{-\nu(t+x)}}{e^{t+x}+1} dt \tag{3}$$
$$(\mathfrak{R}(x) \geq 0, \ \mathfrak{R}(\nu) > -1),$$

and for $\nu = 0$ and $\mu = 1$ in Equation (1), the original Fermi–Dirac function is given by ([6], (p. 109), Equation (1.12))

$$\mathcal{F}_{s-1}(x) := \Theta_0(s,1;x) = \frac{1}{\Gamma(s)} \int_0^\infty \frac{t^{s-1}}{e^{t+x}+1} dt \ (\mathfrak{R}(x) \geq 0; \mathfrak{R}(s) > 0). \tag{4}$$

Similarly, the generalized Bose–Einstein functions $\Psi_\nu(s,\alpha;x)$, which are defined by ([5], (p. 13), Equation (51)), are as follows

$$\Psi_\nu(s,\mu;x) := \frac{\Gamma(\mu)}{\Gamma(s)} \int_0^\infty \frac{t^{s-1} e^{-\nu t}}{(e^{t+x}-1)^\mu} dt \tag{5}$$
$$(\mathfrak{R}(x) \geq 0, \mathfrak{R}(\nu) > -\mathfrak{R}(\mu) \wedge \mathfrak{R}(s) > \mathfrak{R}(\mu) > 0 \text{ when } e^{-x} = 1 \wedge \mathfrak{R}(s) > 0),$$

and their series representation is given by ([5], (p. 13), Equation (52))

$$\Psi_\nu(s,\mu;x) := \sum_{n=1}^\infty \frac{\Gamma(\mu+n) e^{-(\nu+\mu+n)x}}{n!(\nu+\mu+n)^s}. \tag{6}$$

For $\mu = 1$, the extended Bose–Einstein functions ([6], (p. 115), Equation (4.4)) are given here by

$$\Psi_\nu(s;x) := \Psi_\nu(s,1;x) = \frac{1}{\Gamma(s)} \int_0^\infty \frac{t^{s-1} e^{-\nu t}}{e^{t+x}-1} dt \ (\mathfrak{R}(x) \geq 0, \ \mathfrak{R}(\nu) > -1), \tag{7}$$

and the original Bose–Einstein function is given by ([6], (p. 109), Equation (1.13)).

$$\mathcal{B}_{s-1}(x) := \Psi_0(s,1;x) = \frac{1}{\Gamma(s)} \int_0^\infty \frac{t^{s-1}}{e^{t+x}-1} dt \ (\mathfrak{R}(x) \geq 0; \mathfrak{R}(s) > 1). \tag{8}$$

For further study of the Fermi–Dirac and Bose–Einstein functions, we refer the interested reader to [7–9]. The reduction and duality theorems for these functions are given by ([5], (p. 12–13))

$$\Theta_{\nu-M}(s;\ \mu+M;x) = \sum_{m=0}^{M} R_1(M,\ m,-\nu)\Theta_\nu(s-m,\mu;x), \tag{9}$$

$$\Theta_\nu(s-M;\mu;x) = \sum_{m=0}^{M} (-1)^{M-m} R(M,\ m,-\nu)\Theta_{\nu-m}(s,\ \mu+m;x), \tag{10}$$

$$\Psi_{\nu-M}(s;\ \mu+M;x) = \sum_{m=0}^{M} R_1(M,\ m,-\nu)\Psi_\nu(s-m,\ \mu;x), \tag{11}$$

$$\Psi_\nu(s-M;\ \mu;x) = \sum_{m=0}^{M} (-1)^{M-m} R(M,\ m,-\nu)\Psi_{\nu-m}(s,\ \mu+m;x), \tag{12}$$

respectively, where $R_1(M, m, -\nu)$ and $R(M, m, -\nu)$ are the polynomials having explicit representations in terms of Stirling numbers. For examples and details see Carlitz [10,11]. More recently, Tassaddiq [12,13] considered the λ-generalized extended Fermi–Dirac functions and λ-generalized extended Bose–Einstein functions as a transformed form of Srivastava's λ-generalized Hurwitz–Lerch zeta functions ([14], (p. 1487), Equation (1.14)). In this research, we generalize the results of Chaudhry and Qadir [2]. To achieve this goal, it is important to briefly highlight their relationship with the zeta functions. It should be noted that for $x = 0$, the Bose–Einstein and Fermi–Dirac functions are related to the Riemann zeta functions respectively.

$$\zeta(s) := \mathcal{B}_{s-1}(0); \Re(s) > 1 \tag{13}$$

$$\zeta(s)\left(1 - 2^{1-s}\right) := \mathcal{F}_{s-1}(0); \Re(s) > 0. \tag{14}$$

The polylogarithm function is an important function in the study of theory of polymers that was introduced and investigated by Truesdell [15]

$$\text{Li}_s(z) := \sum_{n=1}^{\infty} \frac{z^n}{n^s} \quad (s \in \mathbb{C},\ |z| < 1; \Re(s) > 1,\ |z| = 1). \tag{15}$$

It generalizes the Riemann zeta function, as we have

$$\text{Li}_s(1) = \phi(1,s) = \zeta(s)\ (\Re(s) > 1), \tag{16}$$

and it can also be represented as an integral

$$\text{Li}_s(z) = \frac{z}{\Gamma(s)} \int_0^\infty \frac{t^{s-1}}{e^t - z} dt \quad (s \in \mathbb{C} \text{ when } |z| < 1; \Re(s) > 1 \text{ and when } |z| = 1). \tag{17}$$

In our present analysis, we are especially interested in Lindelöf's representation of these functions given by ([15], (p. 149), Equation (13)),

$$\text{Li}_s(z) = \Gamma(1-s)(\log z)^{s-1} + \sum_{n=0}^{\infty} \zeta(s-n)\frac{(\log z)^n}{n!} \tag{18}$$
$$(|\log z| < 2\pi, s \neq 1, 2, 3, \ldots, \nu \neq 0, -1, -2, \ldots,).$$

The Hurwitz–Lerch zeta function ([16], (p. 27)), as a generalization of the polylogarithm, is given by

$$\Phi(z,s,a) = \sum_{n=0}^{\infty} \frac{z^n}{(n+a)^s} \quad (a \in \mathbb{C}\setminus\mathbb{Z}^-; s \in \mathbb{C} \text{ when } |z| < 1; \mathfrak{R}(s) > 1 \text{ when } |z| = 1). \tag{19}$$

It has a meromorphic extension to the whole complex s-plane, while it has a simple singularity at $s = 1$ of residue 1. It is also represented by ([16], (p. 27), Equation (1.6))

$$\Phi(z,s,a) = \frac{1}{\Gamma(s)} \int_0^\infty \frac{t^{s-1} e^{-at}}{1 - z e^{-t}} dt \quad (|z| < 1 \Rightarrow \mathfrak{R}(s) > 0; \mathfrak{R}(a) > 0; z = 1 \Rightarrow \mathfrak{R}(s) > 1). \tag{20}$$

Apart from other applications, the Hurwitz–Lerch zeta function is the most general function in the original zeta family. For example, different values of the involved parameters in (19–20) yield the following relationships with the polylogarithm, Hurwitz, and Riemann zeta functions, respectively:

$$\text{Li}_s(z) := \sum_{n=1}^{\infty} \frac{z^n}{n^s} = z\Phi(s,z,1), \tag{21}$$

$$\zeta(s,a) := \sum_{n=0}^{\infty} \frac{1}{(n+a)^s} = \Phi(s,1,a), \tag{22}$$

$$\zeta(s) := \sum_{n=1}^{\infty} \frac{1}{n^s} = \Phi(s,1,1) = \zeta(s,1). \tag{23}$$

For our purposes, it is important to note that the Hurwitz–Lerch zeta function has a series representation ([16], (pp. 28–29))

$$\Phi(z,s,\nu) = \frac{\Gamma(1-s)}{z^\nu}\left(\log\tfrac{1}{z}\right)^{s-1} + z^{-\nu} \sum_{n=0}^{\infty} \zeta(s-n,\nu)\frac{(\log z)^n}{n!} \tag{24}$$
$$\left(|\log z| < 2\pi, s \ne 1,2,3,\ldots, \nu \ne 0,-1,-2,\ldots,\right)$$

that generalizes Lindelöf's representation (18).

Further to all of the above discussion, Chaudhry et al. [17] defined a new generalization of the Riemann zeta function in the critical strip by

$$\Xi_a(s;x) := \frac{1}{\Gamma(s)} \int_x^\infty (t-x)^{s-1} \left(\frac{1}{e^t - 1} - \frac{1}{t}\right) e^{-at} dt \quad (0 < \mathfrak{R}(s) < 1 : x \ge 0; a \ge 0). \tag{25}$$

The Riemann hypothesis is a well-known unsolved problem in analytic number theory [18]. It states that "all the non-trivial zeros of the zeta function exist on the line $s = 1/2$". These zeros seem to be complex conjugates and are hence symmetrical on this line. The Riemann zeta function in the critical strip is defined and studied in [18]

$$\zeta(s) := \frac{1}{\Gamma(s)} \int_0^\infty t^{s-1}\left(\frac{1}{e^t - 1} - \frac{1}{t}\right) dt \quad (0 < \mathfrak{R}(s) < 1), \tag{26}$$

which can be obtained as a special case of Equation (25) by putting $x = a = 0$.

2.2. A Class $\mathfrak{R}_\infty(A,P,\delta)$ of Functions and the Representation Theorem

More recently Chaudhry and Qadir [19] discussed some important classes of functions. The statements of this section are taken from [19–21].

We first give a brief introduction to the function spaces $H(\xi;\eta)$ and $H(\infty;\eta)$. The elements of $H(\xi;\eta)$ are particular functions $f \in C^\infty(0,\infty)$ that satisfy the following conditions

1. $\int_0^T f(t)dt$ is well defined for $T \in [0, \infty)$;
2. $f(t) = O(t^{-\eta})$ $(t \to 0^+)$;
3. $f(t) = O(t^{-\xi})$ $(t \to \infty)$.

Furthermore, if $f(t) = O(t^{-\xi})$ $(t \to \infty; \xi \in \mathbb{R}_0^+)$, then $f(t) \in H(\infty; \eta)$. We can note that $H(\infty; \eta) \subset H(\xi; \eta)$ $(\forall \xi \in \mathbb{R}_0^+)$.

Clearly, we have

$$f(t) = e^{-bt} \in H(\infty, 0) \; (b > 0). \tag{27}$$

The Mellin transform of $f \in H(\xi; \eta)$ is defined by

$$f_M(s) = M[f(t); s] := \int_0^\infty f(t) t^{s-1} dt \; (s = \sigma + i\tau, \eta < \mathcal{R}(s) < \xi). \tag{28}$$

The fractional Weyl transform of $f \in H(\xi; 0)$ is defined by

$$\Omega(s; x) := W^{-s}[f(t)](x) := \frac{1}{\Gamma(s)} M[f(t+x); s]$$
$$= \frac{1}{\Gamma(s)} \int_0^\infty f(t+x) t^{s-1} dt = \frac{1}{\Gamma(s)} \int_x^\infty f(t)(t-x)^{s-1} dt; \; (s = \sigma + i\tau, 0 < \mathcal{R}(s) < \xi, x \geq 0). \tag{29}$$

Considering $\mathcal{R}(s) \leq 0$, we define the Weyl transform of $w \in H(\xi; 0)$ as follows,

$$\Omega(s; x) := W^{-s}[f(t)](x) := (-1)^n \frac{d^n}{dx^n}(\Omega(n+s; x)), \; (0 \leq n + \mathcal{R}(s) < \xi), \tag{30}$$

and

$$\Omega(0; x) := w(x). \tag{31}$$

We can rewrite Equation (30) alternately as

$$\Omega(-s; x) := W^s[w(t)](x) = (-1)^n \frac{d^n}{dx^n}\left(W^{-(n-s)}[w(t)](x)\right)$$
$$=: (-1)^n \frac{d^n}{dx^n}(\Omega(n-s; x)) \; (0 \leq n - \mathcal{R}(s) < \xi, \mathcal{R}(s) > 0). \tag{32}$$

For these formulae, $n \geq \mathcal{R}(s)$ where n is the positive and smallest such integer. For $s = n$ in Equation (32), we get

$$\Omega(-n; x) := W^n[w(t)](x) := (-1)^n \frac{d^n}{dx^n}(\Omega(0; x)) = (-1)^n \frac{d^n}{dx^n}(w(x)). \tag{33}$$

Note that $\{W^s\}$ $(s \in \mathbb{C})$ satisfies

$$W^{-(\mu+s)}[w(t)](x) = W^{-\mu}[\Omega(s; t)](x) = \Omega(s + \mu; x) \tag{34}$$

the multiplicative group property. For further detailed study of Weyl and related integral transforms, we refer the interested reader to [22–24].

The space of analytic functions [20,21] as discussed by Hardy is reviewed here as follows: Let $0 < \delta < 1$ and $H(\delta) := \{s = \sigma + i\tau : \mathcal{R}(s) \geq -\delta\}$ be the half space. Further, for an analytic function $\phi(s)$, $s \in H(\delta)$, suppose that $0 < A < \pi$ and

$$\mathcal{R} = \mathcal{R}(A, P, \delta) := \{\phi(s) : |\phi(s)| \leq Ce^{P\sigma + A|\tau|}\} \tag{35}$$

is called the Hardy space of analytic functions that restricts the parameter A to lie in $(0, \pi)$. Consider a function $\phi \in \mathcal{R}$ and define

$$\Phi(x) := \frac{1}{2\pi i} \int_{c-i\infty}^{c+i\infty} \frac{\pi}{\sin \pi s} \phi(-s) x^{-s} ds \quad (0 < c < \delta), \tag{36}$$

such that the kernel is majorized by

$$e^{-(\pi-A)|\tau|} e^{-Pc} x^{-c}, \quad (x > 0). \tag{37}$$

These are uniformly convergent in an interval of $0 < x \leq x_0 \leq X < \infty$. Therefore, the function $\Phi(x)$ is regular, and represented by the integral (36), for all positive x. We combine these classes to define a new class of functions for our purposes. Assume that $w(0) := \Omega(0;0)$ is well defined and $w(x) := \Omega(0;x)$ $(x \geq 0)$. Then, $w \in \mathcal{R}_\infty(A, P, \delta)$ iff $w \in H(\delta; 0)$ and

$$\left| \frac{\Omega(s;0)}{\Gamma(1-s)} \right| \leq C e^{\sigma p + A|\tau|} \quad (0 \leq \Re(s) < \delta). \tag{38}$$

Theorem 1. *Let $\varphi \in H(\delta; 0)$ and $\Phi(s; x)$ $(x > 0)$ be its Weyl transform; then, the series representation is*

$$\Phi(s; x) = \sum_{n=0}^{\infty} \frac{\Phi(s-n; 0)(-x)^n}{n!} \quad (0 \leq \Re(s) < \delta, 0 < x < \infty). \tag{39}$$

Proof. Since $\varphi \in \mathcal{R}_\infty(A, P, \delta)$, the inverse Mellin transform is

$$\begin{aligned}
\Phi(0; x) &:= \frac{1}{2\pi i} \int_{c-i\infty}^{c+i\infty} \varphi_M(s) x^{-s} ds \\
&= \frac{1}{2\pi i} \int_{c-i\infty}^{c+i\infty} \Gamma(s) \Phi(s; 0) x^{-s} ds = \frac{1}{2\pi i} \int_{c-i\infty}^{c+i\infty} \frac{\pi \Phi(s;0)}{\sin(\pi s) \Gamma(1-s)} x^{-s} ds \\
&\quad (0 \leq c < \delta, \, x > 0),
\end{aligned} \tag{40}$$

well defined because the integrand is majorized by a constant multiple of $e^{-(\pi-A)|\tau|} e^{-Pc} x^{-c}$.

Familiar Cauchy's theorem for complex numbers is used to invert the Mellin transform in Equation (40). The integrand has singularities of order 1 at $s = -n$ $(n = 0, 1, 2, 3, \ldots)$ with residues $\frac{\Phi(-n;0)(-x)^n}{n!}$. Therefore,

$$\Phi(0; x) = \sum_{n=0}^{\infty} \frac{(-1)^n \Phi(-n; 0) x^n}{n!} \quad (0 < x < e^{-P}). \tag{41}$$

Because $\varphi \in H(\delta; 0)$, the series (41) extends uniquely for the Weyl transform as

$$(s; x) = \sum_{n=0}^{\infty} \frac{\Phi(s-n; 0)(-x)^n}{n!} \quad (0 < x < e^{-P}, 0 \leq \Re(s) < \delta). \tag{42}$$

□

Theorem 2. *Let $\varphi \in \mathcal{R}_\infty(A, P, \delta)$ and*

$$\psi(t) = \lambda t^{-\mu} + \varphi(t) \quad (\mu > 0). \tag{43}$$

Then, the Weyl transform $\Psi(s; x)$ has a closed form representation

$$\Psi(s; x) = \lambda \frac{\Gamma(\mu - s)}{\mu} x^{s-\mu} + \sum_{n=0}^{\infty} \frac{\Phi(s-n; 0)(-x)^n}{n!} \quad (0 \leq \Re(s) < \min(\delta, \mu); 0 < x < e^{-P}). \tag{44}$$

Proof. An application of the linearity property of Weyl's transform to Equation (43) gives

$$\Psi(s;x) = W^{-s}[\psi(t)](x) = \lambda W^{-s}[t^{-\mu}](x) + \Phi(s;x) \quad (45)$$
$$(0 \leq \Re(s) < \min(\delta,\mu); 0 < x < e^{-P}).$$

However, (see [22], (p. 249)) we have

$$W^{-s}[t^{-\mu}](x) = \frac{\Gamma(\mu-s)}{\mu} x^{s-\mu}; \quad (0 < \Re(s) < \mu; 0 < x < \infty). \quad (46)$$

From Equations (38), (42), and (46) we arrive at Equation (44). □

Example 1. Define

$$\varphi(t) := \frac{1}{e^t - 1} - \frac{1}{t} \quad (t > 0). \quad (47)$$

Note that $\varphi \in \Re_\infty(\frac{\pi}{2}, \ln(1/2\pi), \delta)$ and

$$\Phi(s,0) = \zeta(s) \quad (0 < \Re(s) < 1). \quad (48)$$

Hence, we have an expansion

$$\Phi(0,x) = \frac{1}{e^x - 1} - \frac{1}{x} = \zeta(0) + \sum_{n=1}^{\infty} \frac{(-1)^n \zeta(-n) x^n}{n!} \quad (0 < x < 2\pi), \quad (49)$$

which is the standard result. Using

$$\zeta(-n) = -\frac{B_n}{n+1} \quad (n = 0,1,2,3,\ldots), \quad (50)$$

we can rewrite $\Phi(s;x)$ in terms of Bernoulli numbers.

3. Results

Application of the General Representation Theorem to the Generalized Bose–Einstein and Fermi–Dirac and Related Functions

In this section, we first evaluate the fractional Weyl transform for the function involved in the integrand of generalized Bose–Einstein functions and then analytically continued this function in the interval $(0 < \Re(s) < \mu)$, namely the generalized critical strip.

Remark 1. To apply the general representation theorem, we first discussed analytic continuation of the Bose–Einstein function in the critical strip. The integral representation (5) of the generalized Bose–Einstein function $\Psi_\nu(s,\mu;0)$ can be continued to the domain, $0 < \Re(s) < \mu$, where a particular case of this domain $0 < \Re(s) < 1$ is known as the critical strip for the zeta function. For $\Re(s) > \mu$, we may write in the usual sense as we write for the zeta function ([18], (p. 37))

$$\Gamma(s)\Psi_\nu(s,\mu;0) = \int_0^1 \left(\frac{e^{-\nu t}}{(e^t-1)^\mu} - \frac{1}{t^\mu}\right) t^{s-1} dt + \frac{1}{s-\mu}$$
$$+ \int_1^\infty \frac{e^{-\nu t}}{(e^t-1)^\mu} t^{s-1} dt, \quad (51)$$

which is true by analytic continuation for $\Re(s) > 0$. For these values $0 < \Re(s) < \mu$, we get

$$\frac{1}{s-\mu} = \int_1^\infty \frac{t^{s-1}}{t^\mu} dt, \qquad (52)$$

such that we can write

$$\Gamma(s)\Psi_\nu(s,\mu;0) = \int_0^\infty \left(\frac{e^{-\nu t}}{(e^t-1)^\mu} - \frac{1}{t^\mu}\right) t^{s-1} dt \quad (0 < \Re(s) < \mu; \Re(\nu) > 0). \qquad (53)$$

Putting $\nu = 0$ in Equation (53), we get the representation

$$\Gamma(s)\Psi_0(s,\mu;0) = \int_0^\infty \left(\frac{1}{(e^t-1)^\mu} - \frac{1}{t^\mu}\right) t^{s-1} dt \quad (0 < \Re(s) < \mu). \qquad (54)$$

Putting $\nu = 0$; $\mu = 1$ in Equation (53), the classical representation (26) for the Riemann zeta function is recovered.

Remark 2. *The series representation (24) for the Hurwitz–Lerch function is proved in ([16], (p. 28)) by using the following steps.*

1. Using the contour integral to state the involved function
2. Using the Cauchy residue theorem from complex analysis
3. Using the following identity known as Hurwitz formula [16]

$$\zeta(s,\nu) = 2(2\pi)^{s-1}\Gamma(1-s)\sum_{n=1}^\infty \frac{\sin(2\pi n\nu + \frac{\pi s}{2})}{n^{1-s}} \quad (\Re(s) < 0, \, 0 < \nu \le 1). \qquad (55)$$

In this section, we have obtained a new series representation for the generalized Bose–Einstein and Fermi–Dirac functions. We have shown that the above stated results (18) and (24) for the polylogarithm and Hurwitz–Lerch functions are special cases by using the fractional Weyl transform.

Theorem 3. *Show that the generalized Fermi–Dirac functions have a series representation*

$$\Theta_\nu(s,\mu;x) := \Gamma(\mu)\sum_{M=0}^\infty \frac{\Theta_\nu(s-M,\mu;0)x^M}{M!} = \Gamma(\mu)\sum_{M=0}^\infty \frac{\sum_{m=0}^M (-1)^{M-m} R(M,m,-\nu)\Theta_{\nu-m}(s,\mu+m;0)}{M!} x^M \qquad (56)$$

$$(0 \le \Re(s) < \mu; \nu \ne 0, -1, -2, \ldots).$$

Proof. The generalized Fermi–Dirac function (1) can be written as

$$\begin{aligned}\Theta_\nu(s,\mu;x) &= \frac{\Gamma(\mu)}{\Gamma(s)}\int_x^\infty e^{-\nu t}\frac{(t-x)^{s-1}}{(e^t+1)^\mu} dt \\ &= \frac{\Gamma(\mu)}{\Gamma(s)}\int_x^\infty \left[\frac{e^{-\nu t}}{(e^t+1)^\mu}\right](t-x)^{s-1} dt \\ &= \Gamma(\mu)W^{-s}\left[\frac{e^{-\nu t}}{(e^t+1)^\mu}\right](x) = \Gamma(\mu)\sum_{M=0}^\infty \frac{\Theta_\nu(s-M,\mu;0)}{M!} x^M,\end{aligned} \qquad (57)$$

which leads to the required result by using Equations (10) and (39). □

Corollary 1. *The Fermi–Dirac function has a representation ([2], Equation (4.2)):*

$$F_{s-1}(x) := \sum_{M=0}^\infty \frac{(1-2^{M-s+1})\zeta(s-M)x^M}{M!}. \qquad (58)$$

Proof. This result follows by putting $\mu = 1; \nu = 0$ in Equation (56) and using Equation (14). □

Theorem 4. *Show that the generalized Bose–Einstein functions have a series representation:*

$$\Psi_\nu(s,\mu;x) = \frac{\Gamma(\mu)\Gamma(\mu-s)}{\mu} x^{s-\mu} + \Gamma(\mu) \sum_{M=0}^{\infty} \frac{(-1)^M \Psi_\nu(s-M,\mu;0)}{M!} x^M \quad (59)$$

$$(0 \le \Re(s) < \mu; \nu \ne 0, -1, -2, \ldots).$$

Proof. First, we note that the integral representation (5) can be rewritten as

$$\Psi_\nu(s,\mu;x) := \frac{\Gamma(\mu)}{\Gamma(s)} \int_0^\infty \frac{t^{s-1} e^{-\nu t}}{(e^{t+x}-1)^\mu} dt = \frac{\Gamma(\mu)}{\Gamma(s)} \int_x^\infty e^{-\nu t} \frac{(t-x)^{s-1}}{(e^t-1)^\mu} dt, \quad (60)$$

which can be rearranged as follows

$$= \frac{\Gamma(\mu)}{\Gamma(s)} \int_x^\infty \left[\frac{e^{-\nu t}}{(e^t-1)^\mu} - \frac{1}{t^\mu} + \frac{1}{t^\mu} \right] (t-x)^{s-1} dt. \quad (61)$$

Next, by making use of the definition of Weyl transform, we get

$$\Psi_\nu(s,\mu;x) = \Gamma(\mu) W^{-s}\left[\frac{e^{-\nu t}}{(e^t-1)^\mu} - \frac{1}{t^\mu} \right](x) + \Gamma(\mu) W^{-s}\left[\frac{1}{t^\mu} \right](x). \quad (62)$$

However, an application of the Weyl transform (46) along with an application of the general representation theorem (39) on the left hand side of the above Equation (62) leads to the required series representation. □

Corollary 2. *The Bose–Einstein function has a representation ([2], Equation (4.7)):*

$$\mathcal{B}_{s-1}(x) = \Gamma(1-s) x^{s-1} + \sum_{M=0}^{\infty} \frac{(-1)^n \zeta(s-M) x^M}{M!}. \quad (63)$$

Proof. The result follows by putting $\mu = 1; \nu = 0$ in Equation (59) and using Equation (13). □

Remark 3. *Putting $\mu = 1; x = \log \frac{1}{z} \Rightarrow z = e^{-x}; -x = \log z$, replacing ν by $\nu - 1$ in (59), and using the relation, ([6], Equation (4.5)) $\Psi_\nu(s;x) = e^{-(\nu+1)x} \Phi(e^{-x}, s, \nu+1)$, we obtain*

$$z^\nu \Phi(z,s,\nu) = \Gamma(1-s)(\log \tfrac{1}{z})^{s-1} + \sum_{M=0}^{\infty} \zeta(s-M,\nu) \frac{(\log z)^M}{M!} \quad (|\log z| < 2\pi, s \ne 1,2,3,\ldots, \nu \ne 0,-1,-2,\ldots), \quad (64)$$

which is exactly Equation (24). Further, by putting $\nu = 0$, we deduce Lindelöf's representation (18) for the polylogarithm function.

Remark 4. *The use of fractional derivatives and fractional integrals has become vital to solve many physical problems that were unsolvable otherwise, see for example [25,26]. For our interest, the Riemann–Liouville fractional derivative is defined by ([22], (p. 70)) and [23].*

$$\mathfrak{D}_z^\mu \{f(z)\} := \begin{cases} \frac{1}{\Gamma(-\mu)} \int_0^z (z-t)^{-\mu-1} f(t) dt & \Re(\mu) > 0 \\ \frac{d^m}{dz^m} \left\{ \mathfrak{D}_z^{\mu-m} \{f(z)\} \right\} & (m-1 \le \Re(\mu) < m \, (m \in \mathbb{N})). \end{cases} \quad (65)$$

It is important to notice from integral representations (1) and (5) that the functions $\Theta_\nu(s,\mu;x)$ and $\Psi_\nu(s,\mu;x)$ are in effect a Riemann–Liouville fractional derivative of the Fermi–Dirac and Bose–Einstein functions respectively given by

$$\Theta_\nu(s,\mu;x) = \frac{1}{\Gamma(\mu)} \mathfrak{D}_x^{\mu-1}\left\{e^{-x(\mu-1)}\Theta_\nu(s;x)\right\}; \Re(\mu) > 0, \tag{66}$$

$$\Theta_\nu(s,\mu;x) = \frac{1}{\Gamma(\mu)} \mathfrak{D}_x^{\mu-1}\left\{e^{-x(\mu-1)}\Theta_\nu(s;x)\right\}; \Re(\mu) > 0. \tag{67}$$

Remark 5. *The Apostol–Euler–Nörlund polynomials $E_n^{(\mu)}(x;\lambda)$ [27,28] are defined by the generating function*

$$\left(\frac{2}{\lambda e^t + 1}\right)^\mu e^{vt} = \sum_{n=0}^\infty E_n^{(\mu)}(v;\lambda) \frac{t^n}{n!}; |t| < |\log(-\lambda)|; \lambda \neq -1 \tag{68}$$

and Bernoulli–Nörlund polynomials [27,28] are defined by

$$\left(\frac{t}{e^t - 1}\right)^\mu e^{vt} = \sum_{n=0}^\infty B_n^{(\mu)}(v) \frac{t^n}{n!}; |t| < |2\pi|. \tag{69}$$

It is important to further mention that the relation of the generalized Fermi–Dirac and Bose–Einstein functions with Apostol—Euler–Nörlund [27,28] polynomials can be established in view of integral representations (1) and (5), respectively, as follows.

Consider ([5], Equation (47))

$$\Theta_v(s,\mu;x) = e^{-x(v+\mu)}\zeta(s,\mu;v+\mu,-e^{-x}). \tag{70}$$

Now, replace v by $v - \mu$ and $s = -m$ in Equation (70); we get

$$\Theta_{v-\mu}(-m,\mu;x) = e^{-xv}\zeta(-m,\mu;v,-e^{-x}). \tag{71}$$

Next, by using $\lambda = e^{-x}; \alpha = \mu$ in ([5], Equation (27)), we get

$$\Theta_{v-\mu}(-m,\mu;x) = e^{-xv}\Gamma(\mu)2^{-\mu}E_m^{(\mu)}(v;e^{-x}). \tag{72}$$

Similarly, by considering (5], Equation (53)) and replacing v by $v - \mu$, $s = -m$, we get

$$\Psi_{v-\mu}(-m,\mu;x) = e^{-xv}\zeta(-m,\mu;v,e^{-x}). \tag{73}$$

Next, by using $\lambda = e^{-x}$ in ([5], Equation (27)) and using the result in the above Equation (73), we get

$$\Psi_{v-\mu}(-m,\mu;x) = e^{-xv}\Gamma(\mu)2^{-\mu}E_m^{(\mu)}(v;-e^{-x}). \tag{74}$$

For $x = 0$, in Equations (72) and (74), we get

$$\Theta_{v-\mu}(-m,\mu;0) = \Gamma(\mu)2^{-\mu}E_m^{(\mu)}(v;1)\Psi_{v-\mu}(-m,\mu;0) = \Gamma(\mu)2^{-\mu}E_m^{(\mu)}(v;-1),$$

which can be used in Equations (56) and (59) to obtain the representation in terms of special cases of Apostol–Euler–Nörlund polynomials $E_m^{(\mu)}(v;\mp 1)$. Considering the further restrictions $v = \mu = 1$, we can get these relations in terms of commonly used Bernoulli and Euler numbers.

Remark 6. *One can note that Equation (59) may be stated alternately in terms of Stirling numbers by using Equation (13) as follows*

$$\Psi_\nu(s,\mu;x) = \frac{\Gamma(\mu)\Gamma(\mu-s)}{\mu}x^{s-\mu} + \Gamma(\mu)\sum_{M=0}^{\infty}\frac{(-1)^M \sum_{m=0}^{M}(-1)^{M-m}R(M,m,-\nu)\Psi_{\nu-m}(s,\mu+m;0)}{M!}x^M \quad (75)$$
$$(0 \leq \Re(s) < \mu;\ \nu \neq 0, -1, -2, \ldots).$$

4. Concluding Remarks

One important aspect in relation to the analysis of special functions is to study their representations. These special functions can be studied in different regions by using their series, asymptotic, and integral representations. This fact is also important when writing simpler mathematical proofs of known results. Here, we have provided a new series representation of the generalized Bose–Einstein and Fermi–Dirac functions by using a general representation theorem. To accomplish this work, we discussed an analytic continuation for these functions by generalizing the Riemann zeta function from $(0 < \Re(s) < 1)$ to $(0 < \Re(s) < \mu)$. This gives new insights for a possible generalization of the Rieman zeta function

$$\zeta^*_\mu(s) := \frac{1}{\Gamma(s)}\int_0^\infty t^{s-1}\left(\frac{1}{(e^t-1)^\mu} - \frac{1}{t^\mu}\right)dt \quad (0 < \Re(s) < \mu)$$

and will be discussed in more detail in our future research. Our results were validated by obtaining known series representations for the polylogarithm and the Hurwitz–Lerch zeta functions as special cases. A comparison of the known proof of their series representation was given with this new proof. It is hoped that the general representation theorem can also be applied to analyze other special functions.

Author Contributions: R.S. did project administration and supervision of this research. All the authors (R.S., H.N., S.K., and A.T.) participated equally in the methodology and conceptualization of this research. H.N. and A.T. wrote, reviewed, and edited the manuscript. S.K. checked the validation of the results. All the authors finalized the manuscript after its internal evaluation and contributed substantially to the work reported.

Funding: This research received no external funding.

Acknowledgments: The authors are thankful to the anonymous reviewers for their useful comments. They significantly improved the quality of this manuscript.

Conflicts of Interest: The authors declare no conflict of interest.

References

1. Chaudhry, M.A.; Iqbal, A.; Qadir, A. A representation for the anyon integral function. *arXiv* **2005**, arXiv:math-ph/0504081.
2. Chaudhry, M.A.; Qadir, A. Operator representation of Fermi-Dirac and Bose-Einstein integral functions with applications. *Int. J. Math. Math. Sci.* **2007**, 80515. [CrossRef]
3. Tassaddiq, A. A new representation of the k-gamma functions. *Mathematics* **2019**, *7*, 133. [CrossRef]
4. Tassaddiq, A.; Qadir, A. Fourier transform and distributional representation of the generalized gamma function with some applications. *Appl Math. Comput.* **2011**, *218*, 1084–1088. [CrossRef]
5. Bayad, A.; Chikhi, J. Reduction and duality of the generalized Hurwitz-Lerch zetas. *Fixed Point Theory Appl.* **2013**, *82*, 1–14. [CrossRef]
6. Srivastava, H.M.; Chaudhry, M.A.; Qadir, A.; Tassaddiq, A. Some extensions of the Bose-Einstein and Fermi-Dirac functions with applications to zeta and related functions. *Russ. J. Math. Phys.* **2011**, *18*, 107–121. [CrossRef]
7. Tassaddiq, A.; Qadir, A. Fourier transform representation of the extended Bose-Einstein and Fermi-Dirac functions with applications to the family of the zeta and related functions. *Integral Transforms Spec. Funct.* **2011**, *22*, 453–466. [CrossRef]
8. Tassaddiq, A. Some Representations of the Extended Fermi-Dirac and Bose-Einstein Functions with Applications. Ph.D. Dissertation, National University of Sciences and Technology, Islamabad, Pakistan, 2012.

9. Tassaddiq, A. A New Representation of the Extended Fermi-Dirac and Bose-Einstein Functions. *Int. J. Appl. Math.* **2017**, *5*, 435–446.
10. Carlitz, L. Weighted Stirling numbers of the first and second kind—I. *Fibonacci Q.* **1980**, *18*, 147–162.
11. Carlitz, L. Weighted Stirling numbers of the first and second kind—II. *Fibonacci Q.* **1980**, *18*, 242–257.
12. Tassaddiq, A. A new representation of the Srivastava λ-generalized Hurwitz-Lerch zeta functions. *Symmetry* **2018**, *10*, 733. [CrossRef]
13. Tassaddiq, A. Some difference equations for Srivastava's λ-generalized Hurwitz–Lerch zeta functions with applications. *Symmetry* **2019**, *11*, 311. [CrossRef]
14. Srivastava, H.M. A new family of the λ-generalized Hurwitz-Lerch Zeta functions with applications. *Appl. Math. Inf. Sci.* **2014**, *8*, 1485–1500. [CrossRef]
15. Truesdell, C. On a function which occurs in the theory of the structure of polymers. *Ann. Math.* **1945**, *46*, 144–157. [CrossRef]
16. Erdelyi, A.; Magnus, W.; Oberhettinger, F.; Tricomi, F.G. *Higher Transcendental Functions*; McGraw-Hill Book Company: New York, NY, USA; Toronto, ON, Canada; London, UK, 1953; Volume I.
17. Chaudhry, M.A.; Qadir, A.; Tassaddiq, A. A new generalization of the Riemann zeta function. *Adv. Differ. Equ.* **2011**, *2011*, 20. [CrossRef]
18. Titchmarsh, E.C. *The Theory of the Riemann Zeta Function*; Oxford University Press: Oxford, UK, 1951.
19. Chaudhry, M.A.; Qadir, A. Extension of Hardy's class for Ramanujan's interpolation formula and master theorem with applications. *J Inequal Appl.* **2012**, *2012*, 52. [CrossRef]
20. Hardy, G.H.; Littlewood, J.E. Contributions to the Theory of the Riemann Zeta function and the Theory of the Distribution of Primes. *Acta Math.* **1918**, *41*, 119–196. [CrossRef]
21. Hardy, G.H. *Ramanujan: Twelve Lectures on Subjects Suggested by His Life and Work*; Chelsea Publishing Company: New York, NY, USA, 1959.
22. Kilbas, A.A.; Srivastava, H.M.; Trujillo, J.J. *Theory and Applications of Fractional Differential Equations, North-Holland Mathematical Studies*; Elsevier (North- Holland) Science Publishers: Amsterdam, The Netherlands; London, UK; New York, NY, USA, 2006; Volume 204.
23. Samko, S.G.; Kilbas, A.A.; Marichev, O.I. *Fractional Integrals and Derivatives: Theory and Applications*; Translated from the Russian: Integrals and Derivatives of Fractional Order and Some of Their Applications ("Nauka i Tekhnika", Minsk, 1987); Gordon and Breach Science Publishers: Reading, UK; Tokyo, Japan; Paris, French; Berlin, Germany; Langhorne, PA, USA, 1993.
24. Zayed, A.I. *Handbook of Functions and Generalized Function Transforms*; CRC Press: Boca Raton, FL, USA, 1996.
25. Gomez-Aguilar, J.F.; Abro, K.A.; Kolebaje, O.; Yildirim, A. Chaos in a calcium oscillation model via Atangana-Baleanu operator with strong memory. *Eur. Phys. J. Plus.* **2019**, *134*, 140. [CrossRef]
26. Tassaddiq, A. MHD flow of a fractional second grade fluid over an inclined heated plate. *Chaos Solitons Fractals* **2019**, *123*, 341–346. [CrossRef]
27. He, Y.; Araci, S.; Srivastava, H.M.; Abdel-Aty, M. Higher-order convolutions for apostol-bernoulli, apostol-euler and apostol-genocchi polynomials. *Mathematics* **2018**, *6*, 329. [CrossRef]
28. Liu, G.-D.; Srivastava, H.M. Explicit formulas for the Nörlund polynomials $B_n^{(x)}$ and $b_n^{(x)}$. *Comput. Math. Appl.* **2006**, *51*, 1377–1384.

© 2019 by the authors. Licensee MDPI, Basel, Switzerland. This article is an open access article distributed under the terms and conditions of the Creative Commons Attribution (CC BY) license (http://creativecommons.org/licenses/by/4.0/).

Article

A Short Note on Integral Transformations and Conversion Formulas for Sequence Generating Functions

Maxie D. Schmidt

School of Mathematics, Georgia Institute of Technology, Atlanta, GA 30318, USA; maxieds@gmail.com or mschmidt34@gatech.edu

Received: 23 April 2019; Accepted: 17 May 2019; Published: 19 May 2019

Abstract: The purpose of this note is to provide an expository introduction to some more curious integral formulas and transformations involving generating functions. We seek to generalize these results and integral representations which effectively provide a mechanism for converting between a sequence's ordinary and exponential generating function (OGF and EGF, respectively) and vice versa. The Laplace transform provides an integral formula for the EGF-to-OGF transformation, where the reverse OGF-to-EGF operation requires more careful integration techniques. We prove two variants of the OGF-to-EGF transformation integrals from the Hankel loop contour for the reciprocal gamma function and from Fourier series expansions of integral representations for the Hadamard product of two generating functions, respectively. We also suggest several generalizations of these integral formulas and provide new examples along the way.

Keywords: generating function; series transformation; gamma function; Hankel contour

MSC: 05A15; 30E20; 31B10; 11B73

---

1. Introduction

1.1. Definitions

Given a sequence $\{f_n\}_{n\geq 0}$, we adopt the notation for the respective ordinary generating function (OGF), $F(z)$, and exponential generating function (EGF), $\widehat{F}(z)$, of the sequence in some formal indeterminate parameter $z \in \mathbb{C}$:

$$F(z) = \sum_{n\geq 0} f_n z^n \tag{1}$$

$$\widehat{F}(z) = \sum_{n\geq 0} \frac{f_n}{n!} z^n.$$

Notice that we can always construct these functions over any sequence $\{f_n\}_{n \in \mathbb{N}}$ and formally perform operations on these functions within the ring of formal power series in z without any considerations on the constraints imposed by the convergence of the underlying series as a complex function of z. If we assume that the respective series for $F(z)$ or $\widehat{F}(z)$ is analytic, or converges absolutely, for all $z \in \mathbb{C}$ with $0 < |z| < \sigma_f$, then we can apply complex function theory to these sequence generating functions and treat them as analytic functions of z on this region.

We can precisely define the form of an integral transformation (in one variable) as [1] (§ 1.4)

$$\mathcal{I}[f(x)](k) := \int_a^b \mathcal{K}(x,k) f(x) dx, \tag{2}$$

for $-\infty \leq b < a \leq +\infty$ and where the function $\mathcal{K} : \mathbb{R} \times \mathbb{C} \to \mathbb{C}$ is called the kernel of the transformation. When the function f which we operate on in the formula given by the last equation corresponds to an OGF or EGF of a sequence with which we are concerned in applications, we consider integrals of the form in (2) to be so-called generating function transformations. Such generating function transformations are employed to transform the ordinary power series of the target generating function for one sequence into the form of a generating function which enumerates another sequence we are interested in studying.

Generating function transformations form a useful combinatorial and analytic method (depending on perspective) which can be combined and employed to study new sequences of many forms. Our focus in this article is to motivate the constructions of generating function transformations as meaningful and indispensable tools in enumerative combinatorics, combinatorial number theory, and in the theory of partitions, among other fields where such applications live. The particular modus operandi within this article shows the evolution of integral transforms for the reciprocal gamma function, and its multi-factorial integer sequence special cases, as a motivating method for enumerating several types of special sequences and series which we will consider in the next sections.

The references [2–4] provide a much broader sense of the applications of generating function techniques in general to those readers who are not familiar with this topic as a means for sequence enumeration. A comprehensive array of analytic and experimental techniques in the theory of integral transformations is also treated in the references [1,5]. We focus on only a comparatively few concrete examples of integral and sequence transformations in the next subsections with hopes to motivate our primary results proved in this article from this perspective. We hope that the discussion of these techniques in this short note provide motivation and useful applications to readers in a broader range of mathematical areas.

1.2. From Hobby To Short Note: OGF-to-EGF Conversion Formulas

A time consuming hobby that the author assumes from time to time is rediscovering old and unusual identities in mathematics textbooks– particularly in the areas of combinatorics and discrete mathematics. Favorite books to search include Comtet's *Advanced Combinatorics* and the exercises and their solutions found in *Concrete Mathematics* by Graham, Knuth and Patashnik. One curious and interesting conversion operation discussed in the exercises to Chapter 7 of the latter book involves a pair of integral formulas for converting an arbitrary sequence OGF into its EGF and vice versa provided the resulting integral is suitably convergent. The exercise listed in Concrete Mathematics suggests the second form of the operation. Namely, that of converting a sequence EGF into its OGF.

In this direction, we have an easy conversion integral for converting from the EGF of a sequence $\{f_n\}_{n \geq 0}$, denoted by $\widehat{F}(z)$, and its corresponding OGF, denoted by $F(z)$, given by the Laplace–Borel transform [6] (§ B.14):

$$\mathcal{L}[\widehat{F}](z) = F(z) = \int_0^\infty \widehat{F}(tz) e^{-t} dt.$$

Other integral formulas for conversions between specified generating function "types" can be constructed similarly as well (see Section 1.3). The key facets in constructing these semi-standard, or at least known, conversion integrals is in applying a termwise series operation which generates a factor, or reciprocal factor, of the gamma function $\Gamma(z+1)$ when $z \in \mathbb{N}$. The corresponding "reversion" operation of converting from a sequence's OGF to its EGF requires a more careful treatment of the properties of the reciprocal gamma function, $1/\Gamma(z+1)$, and the construction of integral formulas which generate it for $z \in \mathbb{N}$ involving the Hankel loop contour described in Section 2.

That being said, Graham, Knuth and Patshnik already suggest a curious "known" integral formula for performing this corresponding OGF-to-EGF conversion operation of the following form [3] (p. 566):

$$\widehat{F}(z) = \frac{1}{2\pi} \int_{-\pi}^{\pi} F\left(ze^{-\imath t}\right) e^{e^{\imath t}} dt. \tag{3}$$

The statement of this result is given without proof in the identity-full appendix section of the textbook. When first (re)-discovered many years back, the author assumed that the motivation for this integral transformation must correspond to the non-zero paths of a complex contour integral for the reciprocal gamma function. For many years the precise formulation of a proof of this termwise integral formula and its generalization to enumerating terms of reciprocal generalized multi-factorial functions, such as $1/(2n-1)!!$, remained a mystery and curiosity of periodic interest to the author. In the summer of 2017, the author finally decided to formally inquire about the proof and possible generalizations in an online mathematics forum. The question went unanswered for over a year until by chance the author stumbled onto a Fourier series identity which finally motivated a rigorous proof of the formula in (3). This note explains this proof and derives another integral formula for this operation of OGF-to-EGF inversion based on the Hankel loop contour. The preparation of this article is intended to be expository in nature in the hope of inspiring the creativity of more researchers towards developing related integral transformations of sequence generating functions.

1.3. Examples: Integral transformations of a Sequence Generating Function

Integral transformations are a powerful and convenient formal and analytic tool which are used to study sequences and their properties. Moreover, they are easy to parse and apply in many contexts with only basic knowledge of infinitesimal calculus making them easy-to-understand operations which we can apply to sequence generating functions. The author is an enthusiast for particularly pretty or interesting integral representations (cf. [5,7]) and has taken a special research interest in finding integral formulas of the ordinary generating function of sequence which transform the series into another generating function enumerating a modified special sequence.

One notable example of such an integral transformation given in [8] (§ 2) allows us to construct generalized polylogarithm-like and Dirichlet-like series over any prescribed sequence in the following forms for integers $r \geq 1$:

$$\sum_{n\geq 0} \frac{f_n}{(n+1)^r} z^n = \frac{(-1)^{r-1}}{(r-1)!} \int_0^1 \log^{r-1}(t) F(tz) dt \qquad (4)$$

$$= \frac{1}{r!} \int_0^\infty t^{r-1} e^{-t} F\left(e^{-t} z\right) dt.$$

Another source of generating function transformation identities correspond to the bilateral series given by Lindelöf in [9] (§ 2) of the form

$$\sum_{n=-\infty}^{\infty} f(n) z^n = -\frac{1}{2\pi i} \oint_\gamma \pi \cot(\pi w) f(w) z^w dw, \qquad (5)$$

where γ is any closed contour in $\mathbb{C}$ which contains all of the singular points of f in its interior. In this note, we will focus on integral formulas for generating function transformations of an arbitrary sequence, $\{f_n\}_{n\geq 0}$.

Additional series transformations involving a sequence generating function into the form of $\sum_{n\geq 0} f_n z^n / g(n)^s$ for $\operatorname{Re}(s) > 1$ and non-zero sequences $\{g(n)\}_{n\geq 0}$ are proved in [10,11]. Note that the harmonic-number-related coefficients implicit to these series transformations satisfy summation formulas which are readily expressed by Nörlund-Rice contour integral formulas as well. The author has proved in [12] so-called square series transformations providing that

$$\sum_{n\geq 0} f_n q^{n^2} z^n = \frac{1}{\sqrt{2\pi}} \int_0^\infty \left[\sum_{b=\pm 1} F\left(e^{bt\sqrt{2\operatorname{Log}(q)}}\right) \right] e^{-t^2/2} dt, \quad |q|, |qz| < 1. \qquad (6)$$

Applications of these square series integral representations include many new integral formulas for theta functions and classical q-series identities such as the Jacobi triple product and the partition function generating function, $(q;q)_\infty^{-1}$, expanded by Euler's pentagonal number theorem.

There are more general Meinardus methods for computing asymptotics of the coefficients of classes of partition number generating functions of the form [13]

$$\sum_{n\geq 0} p_n(b) z^n := \prod_{k\geq 1} \left(1 - z^k\right)^{-b_k}, \tag{7}$$

where $p_n(b)$ denotes the number of weighted partitions of n corresponding to the parameter weights b_k for $k \geq 1$. Generating functions enumerating partition function sequences of this type are related to a known Euler transform of a sequence $\{a_n\}_{n\geq 1}$ given by [14]

$$1 + \sum_{n\geq 1} b_n z^n := \prod_{j\geq 1} \frac{1}{(1-z^j)^{a_j}} \implies \log(1 + B(z)) = \sum_{k\geq 1} \frac{A(z^k)}{k}, \tag{8}$$

where $A(z) := \sum_n a_n z^n$ and $B(z) := \sum_n b_n z^n$ are the respective OGFs of the component sequences. In this case the right-hand-side generating function in the last equation is generated succinctly by a q-integral for the q-beta function of the form [15]

$$\frac{1}{1-q} \int_0^1 f(x) d(z, x) = \sum_{i \geq 0} f(z^i) z^i,$$

where inputting the modified generating function, $\widetilde{A}_z(t) := A(t) \log(z) / (t \log t)$ for fixed z, into this integral formula generates the second to last series result.

1.4. Results Proved in This Note

In this short note we provide proofs of known integral formulas providing an ordinary-to-exponential generating function operation. We prove the following theorem using the Hankel loop contour for the reciprocal gamma function in Section 2.

Theorem 1 (OGF-to-EGF Integral Formula I). *For any real $c > 0$, provided that $F(z)$ is analytic for $0 < |z| \leq c$, we have that*

$$\widehat{F}(z) = \sum_{n\geq 0} f_n z^n \int_{-\infty}^{\infty} \frac{e^{c+\imath t}}{(c + \imath t)^{n+1}} dt = \int_{-\infty}^{\infty} \frac{e^{c+\imath t}}{(c + \imath t)} F\left(\frac{z}{c + \imath t}\right) dt.$$

We also give a rigorous proof of the next integral formula relating $F(z)$ and $\widehat{F}(z)$.

Theorem 2 (OGF-to-EGF Integral Formula II). *If $F(z)$ is analytic for $0 < |z| < \sigma_f$, we have that (3) holds. Namely, we have that*

$$\widehat{F}(z) = \frac{1}{2\pi} \int_{-\pi}^{\pi} F\left(z e^{\imath t}\right) e^{e^{\imath t}} dt.$$

The proof of Theorem 2 is given in Section 3.

2. Integral Representations of the Reciprocal Gamma Function

Since $\Gamma(z)$ is a meromorphic function of z with poles at the non-positive integers, it follows that the reciprocal gamma function, $1/\Gamma(z)$, is an entire function (of order one) with zeros at $z = 0, -1, -2, \ldots$ [16] (§ 5.1). Indeed, as $|z| \to \infty$ at a constant $|\arg(z)| < \pi$, we can expand

$$\log\left[\frac{1}{\Gamma(z)}\right] \sim -z\log z + z + \frac{1}{2}\log\left(\frac{z}{2\pi}\right) - \frac{1}{12z} + \frac{1}{360z^3} - \frac{1}{1260z^5}, \tag{9}$$

which can be computed via the infinite products

$$\frac{1}{\Gamma(z)} = z \prod_{n \geq 1} \frac{\left(1 + \frac{z}{n}\right)}{\left(1 + \frac{1}{n}\right)^z} = ze^{\gamma z} \prod_{n \geq 1} \left(1 + \frac{z}{n}\right) e^{-z/n},$$

where $\gamma \approx 0.577216$ is Euler's gamma constant. Classically, Karl Weierstrass called the function $1/\Gamma(z)$ the "factorielle" function, and used its representation to prove his famous Weierstrass factorization theorem in complex analysis [17] (§ 2).

For $z \in \mathbb{C}$ such that $\mathrm{Re}(z) > 0$ we have a known series expansion for the reciprocal gamma function given by

$$\frac{1}{\Gamma(z)} = \sum_{k=1}^{\infty} a_k z^k = z + \gamma z^2 + \left(\frac{\gamma^2}{2} - \frac{\pi^2}{12}\right) z^3 + \left(\frac{\gamma^3}{6} - \frac{\gamma \pi^2}{12} + \frac{\zeta(3)}{3}\right) z^4 + \cdots. \tag{10}$$

The coefficients a_k in this expansion satisfy many known recurrence relations and expansions by the Riemann zeta function. In [18] an exact integral formula for these coefficients is given by

$$a_n = \frac{(-1)^n}{\pi \cdot n!} \int_0^\infty e^{-t} \mathrm{Im}\left\{(\log t - \imath \pi)^n\right\} dt.$$

This integral formula is obtained in the reference using Euler's reflection formula for the gamma function given by

$$\frac{1}{\Gamma(z)} = \frac{\sin(\pi z)}{\pi} \Gamma(1-z),$$

and then applying a standard known real integral to express the gamma function on the right-hand-side of the previous equation. Equivalently, the reflection formula can be stated as

$$\frac{1}{\Gamma(1+z)\Gamma(1-z)} = \frac{\sin(\pi z)}{\pi z}.$$

2.1. The Hankel Loop Contour for the Reciprocal Gamma Function

We seek an exact integral representation for the reciprocal gamma function, not just an integral formula defining the coefficients of its Taylor series expansion about zero in this case. To find such a formula we must use the Hankel loop contour $H_{\delta,\varepsilon}$ shown in Figure 1 and consider the contributions of each component section of the contour in the limiting cases for increasingly small $\delta, \varepsilon \to 0$. We prove Theorem 1 using the next lemma derived from this contour below.

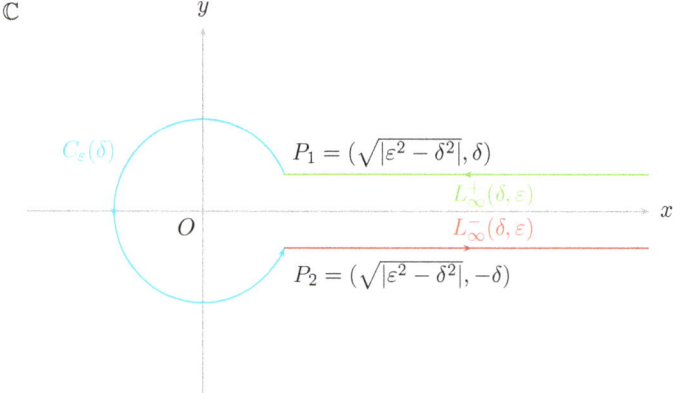

Figure 1. The Hankel loop contour providing an integral representation of the reciprocal gamma function when $\operatorname{Re}(z) > 0$. This contour starts positively from the right, traverses the horizontal line $L_\infty^+(\delta,\varepsilon)$ at distance $+\delta$ from the x-axis from $+\infty \to \sqrt{|\varepsilon^2 - \delta^2|}$, then enters the semi-circular loop about the origin of radius ε denoted by $C_\varepsilon(\delta)$ at the point P_1, and then at the point $P_2 = (\sqrt{|\varepsilon^2 - \delta^2|}, -\delta)$ traverses the last horizontal line $L_\infty^-(\delta,\varepsilon)$ back to infinity parallel to the x-axis.

Lemma 1. *For any real $c > 0$ and $z \in \mathbb{C}$ such that $\operatorname{Re}(z) > 0$,*

$$\frac{1}{\Gamma(z)} = \frac{1}{2\pi} \int_{-\infty}^{\infty} (c + \imath t)^{-z} e^{c + \imath t} dt. \tag{11}$$

Proof. Working from the figure, we have that [16] (§5.9)

$$\frac{1}{\Gamma(z)} = \lim_{d,\varepsilon \to 0} \frac{1}{2\pi\imath} \oint_{H_{H_{\delta,\varepsilon}}} (-t)^{-z} e^{-t} dt \tag{12a}$$

$$= \lim_{d,\varepsilon \to 0} \frac{1}{2\pi\imath} \left[\int_{C_\varepsilon(\delta)} + \int_{L_\infty^+(\delta,\varepsilon)} + \int_{L_\infty^-(\delta,\varepsilon)} \right] \left(e^{-\imath\pi z} t^{-z} e^{-t} \right) dt. \tag{12b}$$

We will first approach the contribution of the section of the contour given by C_ε which is a path enclosing the origin along the circle of radius ε centered at $(0,0)$. This portion of the contour is oriented in the positive direction and begins at the point $P_1 := (\sqrt{|\varepsilon^2 - \delta^2|}, \delta)$ and ends at the point $P_2 := (\sqrt{|\varepsilon^2 - \delta^2|}, -\delta)$. By parameterizing t along this circle, we obtain the real integral giving

$$I_C := \lim_{d,\varepsilon \to 0} \int_{\sin^{-1}(-\frac{\delta}{\varepsilon})}^{\sin^{-1}(\frac{\delta}{\varepsilon})} \imath\varepsilon^2 e^{-\imath\pi z} e^{-2\imath z t} e^{-e^{2\imath t}} dt = 0, \tag{12c}$$

since $\sin^{-1}\left(\frac{\delta}{\varepsilon}\right) = \frac{\delta}{\varepsilon} + \frac{\delta^3}{6\varepsilon^3} + O\left(\frac{\delta^5}{\varepsilon^5}\right) \to 0$ as δ, ε independently tend to zero. Now we can easily parameterize each of the sections of the contour on the horizontal lines each at distance δ from the x-axis. In particular, let's define our integrand in the complex parameters z, w as $f_\Gamma(z, w) := e^{-\imath\pi z} w^{-z} e^{-w}$. Then we consider the limiting cases of the following parameterizations of the two line segments $\{(s, \pm\delta) : s \in [\sqrt{|\varepsilon^2 - \delta^2|}, T]\}$ on $L_\infty^+(\delta, \varepsilon)$ and $L_\infty^-(\delta, \varepsilon)$, respectively, by evaluating the limit of $\delta, \varepsilon \to 0$ and then letting T tend to $+\infty$:

$$z_\pm(\delta, \varepsilon; t) := \sqrt{|\varepsilon^2 - \delta^2|} \pm \imath\delta + t\left(T - \sqrt{|\varepsilon^2 - \delta^2|}\right) \tag{12d}$$

$$z'_\pm(\delta, \varepsilon; t) = T - \sqrt{|\varepsilon^2 - \delta^2|}, \text{ for } t \in [0, 1]. \tag{12e}$$

When we take the first small-order limits we obtain

$$\lim_{\delta,\varepsilon \to 0} \int_0^1 f_\Gamma\left(z_\pm(\delta,\varepsilon;t)\right) \cdot z'_\pm(\delta,\varepsilon;t) dt = \frac{1}{2\pi \imath} \int_0^T e^{-\imath \pi z} s^{-z} e^{-s} ds, \tag{12f}$$

which by substitution provides us with the symmetric bounds of integration given by

$$\lim_{T \to \infty} \frac{1}{2\pi \imath} \int_0^T e^{-\imath \pi z} s^{-z} e^{-s} ds = \pm \int_{\mp \infty}^0 s^{-z} e^s ds. \tag{12g}$$

We then finally arrive at the stated known integral formula for the reciprocal gamma function which holds for any fixed real $c > 0$. □

Proof of Theorem 1. Since we are initially motivated by finding a general conversion integral from a sequence OGF into its EGF, we notice that we require an application of (11) termwise to the Taylor series expansions of our prescribed generating function by setting $z = n+1$. For example, if we assume that our sequence OGF at hand is well enough behaved when its argument satisfies $0 < \operatorname{Re}(z) < c$ for some fixed choice of the real $c > 0$ in the integral formula from above, we can sum the integrand of (11) termwise to obtain

$$\widehat{F}(z) = \sum_{n \geq 0} f_n z^n \int_{-\infty}^\infty \frac{e^{c+\imath t}}{(c+\imath t)^{n+1}} dt = \int_{-\infty}^\infty \frac{e^{c+\imath t}}{(c+\imath t)} F\left(\frac{z}{c+\imath t}\right) dt. \quad □$$

2.2. Examples: Applications of the Integral Formula on the Real Line

We can perform the same "trick" of the generating function trades to sum a "doubly exponential" sequence generating function when we replace the sequence OGF by its EGF in the previous equation:

$$\sum_{n \geq 0} \frac{f_n z^n}{(n!)^2} = \int_{-\infty}^\infty \frac{e^{c+\imath t}}{(c+\imath t)} \widehat{F}\left(\frac{z}{c+\imath t}\right) dt. \tag{13}$$

Perhaps at first glance this iterated integral formula is somewhat unsatisfying since we have really just repeated the procedure for constructing the first integral twice, but in fact there are notable special case applications which we can derive from this method of summation which provide new integral representations for otherwise hard-to-sum hypergeometric series.

For example, if we take the geometric series sequence case where $f_n \equiv 1$ for all $n \geq 0$, then we can arrive at a new integral formula for the doubly exponential series expansion of the incomplete Bessel function, $I_0(2\sqrt{z}) = \sum_{n \geq 0} z^n/(n!)^2$ [3] (§ 5.5). In particular, we easily obtain that

$$I_0(2\sqrt{z}) = \int_{-\infty}^\infty \frac{e^{c+\imath t}}{c+\imath t} \exp\left(\frac{z}{c+\imath t}\right) dt. \tag{14}$$

There is an integral representation for this function which is simpler to evaluate in the general case given in (3). We elaborate more on this identity, its proof, and the corresponding series involving Stirling numbers which it implies in the next section.

3. An Integral Formula from Fourier Analysis

One curious identity that the author has come across relating the OGF of a sequence to its EGF is found in the appendices of the Concrete Mathematics reference [3] (p. 566). It states (3) without proof, again providing that

$$\widehat{F}(z) = \frac{1}{2\pi} \int_{-\pi}^\pi F\left(z e^{-\imath t}\right) e^{e^{\imath t}} dt.$$

Finding a precise method of verifying this unproven identity is the initial motivation for this note. Given the discussion and lead up to an integral for the reciprocal gamma function taken over the real line via the Hankel loop contour in the last section, the author initially assumed—and asked with no replies in online math forums—that this computationally correct integral representation must correspond to the non-zero components of some complex contour integral. It turns out that this formula follows from the basic theory and constructions of Fourier analysis.

Proof of Theorem 2. Given a sequence, $\{f_n\}_{n\geq 0}$, its (mostly convergent) Fourier series is given by $f(x) = \sum_{n\geq 0} f_n e^{\imath \pi n}$. The terms of this sequence are then generated by this Fourier series according to the standard integral formula [1,19]

$$f_m = \frac{1}{2\pi} \int_{-\pi}^{\pi} f(x) e^{-\imath m x} dx,$$

for natural numbers $m \geq 0$. If we can assume that the Fourier series, $f(x)$, or equivalently the OGF, $F(e^{\imath x})$, is absolutely convergent for all $x \in [-\pi, \pi]$ then we can sum over the integral formula in the previous equation to obtain the first key component to this proof:

$$\sum_{m\geq 0} \frac{f_m z^m}{m!} = \frac{1}{2\pi} \int_{-\pi}^{\pi} F(e^{\imath x}) e^{z e^{-\imath x}} dx.$$

The change of variables $e^{\imath x} = z \cdot e^{-\imath t}$ for fixed z shows that this formula is equivalent to the integral formula in (3) directly by a change of variables. Also, by expanding the integrand in powers of $e^{\pm \imath x}$ where

$$\int_{-\pi}^{\pi} e^{\imath (n-k)x} dx = 2\pi \cdot \delta_{n,k},$$

it is apparent that these two formulas in fact generate the same power series representation for $\widehat{F}(z)$. □

Alternate Proof of Theorem 2. Another satisfyingly less analytical and more formally motivated explanation for this behavior can be given by considering known integral formulas for the Hadamard product of two series given in terms of the orthogonal set $\{e^{\imath kx}\}_{k=-\infty}^{\infty}$ for x on the symmetric interval $[-\pi, \pi]$ [2] (§ 1.12(V); Ex. 1.30, p. 85) [4] (cf. §6.3). This perspective on the formulations of these two series allows us to swap the series variables $z e^{\pm \imath x} \mapsto e^{\pm \imath x}$ from the input of one function in the product to another and similarly in the reverse direction. Thus we can effectively pick and choose where we would like to position the generating function parameter z in each component of the integrand—whether it be situated more naturally as an argument to F as in (3), or whether we choose to keep it nested in the corresponding multiplier function as in the previous equation. We shall see other examples of these integral formula variants in the next remark and following examples. □

Remark 1 (Generalizations of series expansions from Fourier series). *This technique of using a convergent Fourier series and the corresponding integral operation for extracting its coefficients can be generalized to generate many other series variants. For example, there are many zeta function and polylogarithm-related series which are summed by modifying a polylogarithmic series of the form expanded in Section 1.2 by the reciprocal of the central binomial coefficients, $\binom{2n}{n}$. In particular, in the exponential-series-based generating function cases we have that*

$$\sum_{n\geq 0} \frac{f_n z^n}{n! \cdot \binom{2n}{n}} = \frac{2}{\pi} \int_{-\pi}^{\pi} F(e^{-\imath x}) \frac{\left[\sqrt{4 - z e^{\imath x}} + \sqrt{z e^{\imath x}} \sin^{-1}\left(\frac{\sqrt{z e^{\imath x}}}{2}\right)\right]}{(4 - z e^{\imath x})^{3/2}} dx \qquad (15)$$

$$= \frac{2}{\pi} \int_{-\pi}^{\pi} F(z e^{-\imath x}) \frac{\left[\sqrt{4 - e^{\imath x}} + \sqrt{e^{\imath x}} \sin^{-1}\left(\frac{\sqrt{e^{\imath x}}}{2}\right)\right]}{(4 - e^{\imath x})^{3/2}} dx,$$

and in the geometric-series-based OGF cases we recover the exponential error function by

$$\sum_{n \geq 0} \frac{f_n z^n}{\binom{2n}{n}} = \frac{1}{4\pi} \int_{-\pi}^{\pi} F(e^{-ix}) \left[2 + e^{\frac{ze^{ix}}{4}} \sqrt{\pi z e^{ix}} \operatorname{erf}\left(\frac{\sqrt{z e^{ix}}}{2}\right) \right] dx \tag{16}$$

$$= \frac{1}{4\pi} \int_{-\pi}^{\pi} F(z e^{-ix}) \left[2 + e^{\frac{e^{ix}}{4}} \sqrt{\pi e^{ix}} \operatorname{erf}\left(\frac{\sqrt{e^{ix}}}{2}\right) \right] dx.$$

There are many other possibilities for constructing integral transformations for modified generating function types. All one needs to do is be creative and consult a detailed reference of compendia such as [5,7].

Examples: Generalizations and Solutions to a Long-Standing Forum Post

The primary goal of the first post [20] mentioned in the introduction was to eventually generalize the integral formula in (3) to enumerate the modified EGF sequences of the form

$$\widehat{F}_{a,b}(z) := \sum_{n \geq 0} \frac{f_n z^n}{\Gamma(an+b+1)},$$

for integers $a \geq 1$ and $b \geq 0$, or over factors of the generalized integer multifactorials defined in [21] as

$$\widetilde{F}_{a,d}(z) := \sum_{n \geq 0} \frac{f_n z^n}{(an+d)!_{(a)}}.$$

In the spirit of our realization that the integral representation in (3) is derived from a Fourier series coefficient formula, we may similarly complete our initial goal to sum the second forms of these series in the special cases where $(a,b) = (2,0), (2,1)$. In particular, we can sum these cases of the modified EGFs defined above in closed-form as explicit integral formulas in the forms

$$\widetilde{F}_{2,0}(z) = \frac{1}{2\pi} \int_{-\pi}^{\pi} F\left(ze^{-it}\right) e^{\frac{1}{2}e^{it}} dt \tag{17}$$

$$\widetilde{F}_{2,1}(z) = \frac{1}{2\pi} \int_{-\pi}^{\pi} F\left(ze^{-it}\right) e^{\frac{1}{2}[e^{it} - it]} \operatorname{erf}\left(\sqrt{\frac{e^{it}}{2}}\right) dt.$$

The modified exponential series of the first type identified above are primarily summed in closed-form using expansions of the Mittag–Leffler functions, $E_{a,b}(z) := \sum_{n \geq 0} z^n / \Gamma(an+b)$, and powers of primitive a^{th} roots of unity [16] (§ 10.46). For example, let's take $(a,b) := (3,0)$ and observe that

$$E_{3,0}(t) = \sum_{m \geq 0} \frac{t^m}{\Gamma(3m+1)} = \frac{e^{t^{1/3}}}{3} + \frac{2e^{-t^{1/3}/2}}{3} \cos\left(\frac{\sqrt{3} t^{1/3}}{2}\right). \tag{18}$$

Then we arrive at a corresponding explicit integral representation for the modified EGF of any sequence of the form

$$\widehat{F}_{3,0}(z) = \frac{1}{2\pi} \int_{-\pi}^{\pi} F(ze^{-it}) E_{3,0}\left(e^{it}\right) dt.$$

4. Concluding Remarks

We have proved two key new forms of integral representations for the reciprocal gamma function on the real line. By composition and the uniform convergence of power series for functions defined on some disc $|z| < \sigma_f$, these results effectively provide us with OGF-to-EGF conversion formulas between the generating functions for some $F(z)$. These integral formulas for OGF-to-EGF conversion can be applied termwise, or in analytic estimates of the asymptotic growth of the coefficients in the power series expansions of the functions defined by the corresponding integral transformation.

We have provided several examples of motivating cases of our so-termed generating function transformations by integral-based methods in Section 1.3. The broader applications of these transformation methods to other fields and phrasings of problems is certainly possible given a suitable context waiting for a new method from which to be approached. We hope that readers come away from this article with a new understanding of how useful and sometimes indispensable integral transformation methods are in sequence analysis.

Funding: This research received no external funding.

Conflicts of Interest: The author declare no conflict of interest.

References

1. Debnath, L.; Bhatta, D. *Integral Transforms And Their Applications*, 3rd ed.; CRC Press: Boca Raton, FL, USA, 2015.
2. Comtet, L. *Advanced Combinatorics: The Art of Finite and Infinite Expansions*; D. Reidel Publishing Company: Dordrecht, The Netherlands, 1974.
3. Graham, R.L.; Knuth, D.E.; Patashnik, O. *Concrete Mathematics: A Foundation for Computer Science*; Addison-Wesley: Boston, MA, USA, 1994.
4. Stanley, R.P. *Enumerative Combinatorics, Volume 2*; Cambridge University Press: Cambridge, UK, 2005.
5. Boros, G.; Moll, V.H. *Irresistible Integrals: Symbolics, Analysis and Experiments in the Evaluation of Integrals*; Cambridge University Press: Cambridge, UK, 2006.
6. Flajolet, P.; Sedgewick, R. *Analytic Combinatorics*; Cambridge University Press: Cambridge, UK, 2009.
7. Gradshetyn, I.S.; Rhyzhik, I.M. *Table of Integrals, Series, and Products*; Academic Press: Cambridge, MA, USA, 1994.
8. Borwein, D.; Borwein, J.M.; Girgensohn, R. Explicit evaluation of Euler sums. *Proc. Edinb. Math. Soc.* **1995**, *38*, 277–294. [CrossRef]
9. Milgram, M.S. Integral and Series Representations of Riemann's Zeta Function, Dirichlet's eta Function and A Medley of Related Results. *arXiv* **2012**, arXiv:1208.3429.
10. Schmidt, M.D. Zeta Series Generating Function Transformations Related to Generalized Stirling Numbers and Partial Sums of the Hurwitz Zeta Function. *Online J. Anal. Comb.* **2018**, *13*, 1–31.
11. Schmidt, M.D. Zeta series generating function transformztions related to polylogarithm functions and the *k*–order harmonic numbers. *Online J. Anal. Comb.* **2017**, *12*, 1–22.
12. Schmidt, M.D. Square series generating function transformations. *Inequalities Spec. Funct.* **2017**, arxiv:1609.02803.
13. Granovskya, B.L.; Starkb, D.; Erlihson, M. Meinardus' theorem on weighted partitions: Extensions and a probabilistic proof. *Adv. Appl. Math.* **2008**, *41*, 307–328. [CrossRef]
14. Bernstein, M.; Sloane, N.J.A. Some Cannonical Sequences of Integers. *arXiv* **2002**, arxiv:math/0205301.
15. Andrews, G.E. *q-Series: Their Development and Application in Analysis, Number Theory, Combinatorics, Physics, and Computer Algebra*; American Mathematical Society: Providence, RI, USA, 1986.
16. Olver, F.W.J.; Lozier, D.W.; Boisvert, R.F.; Clark, C.W. (Eds.) *NIST Handbook of Mathematical Functions*; Cambridge University Press: Cambridge, UK, 2010.
17. Remmert, R. *Classical Topics in Complex Function Theory*; Springer: Berlin/Heidelberg, Germany, 1991.
18. Fekih-Ahmed, L. On the power series expansions of the reciprocal gamma function. *arXiv* **2017**, arxiv:abs/1407.5983.
19. Tolstov, G.P. *Fourier Series*; Dover Publications: Mineola, NY, USA, 2014.
20. An Integral Formula For The Reciprocal Gamma Function. Available online: https://math.stackexchange.com/questions/2274972/an-integral-formula-for-the-reciprocal-gamma-function (accessed on 22 May 2018).
21. Schmidt, M.D. Generalized *j*–factorial functions, polynomials, and applications. *J. Integer Seq.* **2010**, *13*, 1–54.

© 2019 by the author. Licensee MDPI, Basel, Switzerland. This article is an open access article distributed under the terms and conditions of the Creative Commons Attribution (CC BY) license (http://creativecommons.org/licenses/by/4.0/).

Article

Fixed Point Theorems through Modified ω-Distance and Application to Nontrivial Equations

Tariq Qawasmeh [1], Abdalla Tallafha [1] and Wasfi Shatanawi [2,3,*]

[1] Department of mathematics, School of Science, University of Jordan, Amman 11942, Jordan; jorqaw@yahoo.com or tar9160251@fgs.ju.edu.jo (T.Q.); a.tallafha@ju.edu.jo (A.T.)
[2] Department of Mathematics and general courses, Prince Sultan University, Riyadh 11586, Saudi Arabia
[3] Department of Medical Research, China Medical University Hospital, China Medical University, Taichung 40402, Taiwan
* Correspondence: wshatanawi@psu.edu.sa or wshatanawi@yahoo.com

Received: 27 April 2019; Accepted: 5 May 2019; Published: 8 May 2019

Abstract: In this manuscript, we utilize the concept of modified ω-distance mapping, which was introduced by Alegre and Marin [Alegre, C.; Marin, J. Modified ω-distance on quasi metric spaces and fixed point theorems on complete quasi metric spaces. *Topol. Appl.* **2016**, *203*, 120–129] in 2016 to introduce the notions of (ω, φ)-Suzuki contraction and generalized (ω, φ)-Suzuki contraction. We employ these notions to prove some fixed point results. Moreover, we introduce an example to show the novelty of our results. Furthermore, we introduce some applications for our results.

Keywords: quasi metric space; Suzuki contractions; fixed point theorems; modified ω-distance; almost perfect functions

1. Introduction and Preliminaries

Constructing new contractions and formulating new fixed point theorems are very important subjects in mathematics since active researchers employ the existence and uniqueness of the fixed point to solve some integral equations, differential equations, etc.

Banach was the first pioneer mathematician who constructed and formulated the first fixed point theorem, which was called after him as the Banach contraction principle [1].

Suzuki [2] introduced a new contraction and generalized the Banach contraction principle.

In the rest of this paper, the letter d refers to a metric on a set B and f_1 refers to self-mappings on B.

One of the important contractions is the Kannan contraction [3]:

$$d(f_1 l_1, f_1 l_2) \leq \alpha [d(l_1, f_1 l_1) + d(l_2, f_1 l_2)] \text{ for all } l_1, l_2 \in B,$$

where $\alpha \in [0, \frac{1}{2})$.

Moreover, Kannan proved that if f_1 satisfies Kannan contraction, then f_1 has a unique fixed point.

In 1931, Wilson [4] generalized the notion of metric spaces to a new notion called quasi metric spaces.

Definition 1. *We call $q : B \times B \to [0, \infty)$ a quasi metric if q satisfies:*

(i) $q(l_1, l_2) = 0 \iff l_1 = l_2$
and:
(ii) $q(l_1, l_2) \leq q(l_1, l_3) + q(l_3, l_2)$ for all $l_1, l_2, l_3 \in B.$

(B, q) *is called a quasi metric space.*

From now on, by (B, q), we mean a quasi metric space.
Defining $q_m : B \times B \to [0, +\infty)$ via

$$q_m(l_1, l_2) = \max\{q(l_1, l_2), q(l_2, l_1)\},$$

we generate a metric on B.

Recall the following definitions.

Definition 2. *[5,6] The sequence (l_t) converges to $l \in B$ if $\lim_{t \to \infty} q(l_t, l) = \lim_{n \to \infty} q(l, l_t) = 0$.*

Definition 3. *[6] Let (l_t) be a sequence in (B, q). Then, we say that:*

(i) (l_t) is left-Cauchy if for any $\varepsilon > 0$, there exists $n_0 \in \mathbb{N}$ such that $q(l_t, l_m) < \varepsilon \; \forall \; t \geq m > n_0$.
(ii) (l_t) is right-Cauchy if for any $\varepsilon > 0$, there exists $n_0 \in \mathbb{N}$ such that $q(l_t, l_m) < \varepsilon \; \forall \; m \geq t > n_0$.

Definition 4. *[5,6] We say that (l_t) is Cauchy if for any $\varepsilon > 0$, there exists $n_0 \in \mathbb{N}$ such that $q(l_t, l_m) \leq \varepsilon \; \forall \; t, m > n_0$.*

We note that (l_t) in (B, q) is Cauchy if and only if (l_t) is right and left Cauchy.

Definition 5. *[5,6] We say that (B, q) is complete if every Cauchy sequence in B is convergent.*

For some theorems in quasi-metric space, see [5–9].

Alegre and Marin [10] introduced the concept of modified ω-distance mappings on (B, d).

Definition 6. *[10] A modified ω-distance (shortened as $m\omega$-distance) on (B, q) is a function $p : B \times B \to [0, \infty)$, which satisfies:*

(W1) $p(l_1, l_2) \leq p(l_1, l_3) + p(l_3, l_2)$ for all $l_1, l_2, l_3 \in B$;
(W2) $p(l, .) : B \to [0, \infty)$ is lower semi-continuous for all $l \in B$; and
(mW3) for each $\varepsilon > 0$, there exist $\nu > 0$ such that if $p(l_1, l_2) \leq \nu$ and $p(l_2, l_3) \leq \nu$, then $q(l_1, l_3) \leq \varepsilon$ for all $l_1, l_2, l_3 \in B$.

Definition 7. *[10] We call an $m\omega$-distance function a p strong $m\omega$-distance if p is lower semi-continuous on its second coordinate.*

Remark 1. *[10] If q is a quasi metric on B, then q is $m\omega$-distance.*

Lemma 1. *[11] Let (α_t), (β_t) be two sequences of nonnegative real numbers converging to zero. Assume that p is $m\omega$-distance. Then, we have the following:*

(i) *If $p(l_t, l_m) \leq \alpha_t$ for any $t, m \in \mathbb{N}$ with $m \geq t$, then (l_t) is right Cauchy in (B, q).*
(ii) *If $p(l_t, l_m) \leq \beta_m$ for any $t, m \in \mathbb{N}$ with $t \geq m$, then (l_t) is left Cauchy in (B, q).*

Remark 2. *[11] The above lemma implies that if $\lim_{m, t \to \infty} p(l_t, l_m) = 0$, then (l_t) is Cauchy in (B, q).*

For some works on ω-distance, we ask the readers to see [11–13].

Abodayeh et al. [14] generalized the definition of altering the distance function [15] to the concept of the almost perfect function.

Definition 8. *We call a non-decreasing function $\varphi : [0, \infty) \to [0, \infty)$ almost perfect if φ satisfies:*

(i) $\varphi(l) = 0$ if and only if $l = 0$.
(ii) *If (l_t) is a sequence in $[0, \infty)$ such that $\lim_{t \to \infty} \varphi(l_t) = 0$, then $\lim_{t \to \infty} l_t = 0$.*

2. Main Results

We begin our work with the following definition:

Definition 9. Let $\varphi : \mathbb{R}^+ \to \mathbb{R}^+$ be an almost perfect function and p be modified ω-distance on B. We say that p is bounded with respect to φ if there exists an integer $A > 0$ such that:

$$\varphi p(l, e) \leq A \text{ for all } l, e \in B.$$

Definition 10. Equip (B, q) with an $m\omega$-distance mapping p. Then, we call that $f_1 : B \to B$ an (ω, φ)-Suzuki contraction if there are an almost perfect function φ and a constant $k \in [0, 1)$ such that for all $l, e \in X$ and $t \in \mathbb{N}$, we have:

$$(1-k)p\left(l, f_1^t l\right) \leq p(l,e) \implies \varphi p(f_1 l, f_1 e) \leq k\varphi p(l,e),$$

and:

$$(1-k)p\left(f_1^t l, l\right) \leq p(e,l) \implies \varphi p(f_1 e, f_1 l) \leq k\varphi p(e,l).$$

Now, we introduce and prove our first result.

Theorem 1. Equip (B, q) with an $m\omega$-distance mapping p. Let p be bounded with respect to the almost perfect function φ and f_1 be an (ω, φ)-Suzuki contraction mapping. Suppose that:

(i) f_1 is continuous,

or

(ii) if $u^* \in B$ and $u^* \neq f_1 u^*$, then:

$$\inf\{p(e, u^*) + p(f_1 e, u^*) : e \in B\} > 0. \tag{1}$$

Then, f_1 has a unique fixed point in B.

Proof. By starting with $l_0 \in B$, we produce a sequence (l_t) in B inductively by putting $l_{t+1} = f_1 l_t$ for all $t \in \mathbb{N} \cup \{0\}$. Given $m, t \in \mathbb{N} \cup \{0\}$ with $m > t$, then $m = t + s$ for some $s \in \mathbb{N}$. From the definition, we have:

$$\begin{aligned}(1-k)p(l_{t-1}, l_{m-1}) &= (1-k)p(l_{t-1}, l_{t+s-1}) \\ &\leq p(l_{t-1}, l_{t+s-1}).\end{aligned}$$

Therefore, we get that:

$$\begin{aligned}\varphi p(l_t, l_m) &= \varphi p(f_1 l_{t-1}, f_1^s l_{t-1}) \\ &= \varphi p(f_1 l_{t-1}, f_1 l_{t+s-1}) \\ &\leq k\varphi p(l_{t-1}, l_{t+s-1}).\end{aligned} \tag{2}$$

Repeating (2) t-times, we get that:

$$\varphi p(l_t, l_m) \leq k^t \varphi p(l_0, l_s). \tag{3}$$

Since (B, p) is bounded with respect to φ, then we have:

$$\varphi p(l_t, l_m) \leq k^t A \text{ for some integer } A > 0. \tag{4}$$

By letting $t, m \to \infty$, we get that:

$$\lim_{t,m \to \infty} \varphi p(l_t, l_m) = 0. \tag{5}$$

By the definition of φ, we get that:
$$\lim_{t,m\to\infty} p(l_t, l_m) = 0. \tag{6}$$

Since $m > t$, Lemma 1 implies that (l_t) is right Cauchy. Now, suppose that $t, m \in \mathbb{N} \cup \{0\}$ with $t > m$. Then, $t = m + q$ for some $q \in \mathbb{N}$. We note that:
$$(1-k)p(l_{t-1}, l_{m-1}) \leq p(l_{m+q-1}, l_{m-1}).$$

Therefore, we get that:
$$\varphi p(l_t, l_m) = \varphi p(f_1 l_{t-1}, f_1 x_{m-1})$$
$$\leq \cdots \leq k^m \varphi p(l_q, l_0) \tag{7}$$

$$\varphi p(l_t, l_m) \leq k^m \varphi p(l_q, l_0). \tag{8}$$

Since (B, p) is bounded with respect to φ, we get that:
$$\varphi p(l_n, l_m) \leq k^m A \text{ for some integer } A > 0. \tag{9}$$

By letting $t, m \to \infty$, we have:
$$\lim_{t,m\to\infty} \varphi p(l_n, l_m) = 0. \tag{10}$$

Therefore,
$$\lim_{t,m\to\infty} p(l_t, l_m) = 0. \tag{11}$$

Since $t > m$, Lemma 1 implies that (l_t) is left Cauchy. Therefore, we deduce that (l_t) is Cauchy. The completeness of (B, q) implies that there exists an element $l^* \in B$ such that $l_t \to l^*$. If f_1 is continuous, then $l_{t+1} = f_1 l_t$ converges to $f_1 l^*$. The uniqueness of the limit ensures that $f_1 l^* = l^*$. Let $\epsilon > 0$. Since $\lim_{t,m\to\infty} p(l_t, l_m) = 0$, we choose $k_0 \in \mathbb{N}$ such that $p(l_t, l_m) \leq \frac{\epsilon}{2}$ for all $l, m \geq k_0$. The lower semi continuity of p implies that:
$$p(l_t, l^*) \leq \liminf_{j\to\infty} p(l_t, l_j) \leq \frac{\epsilon}{2} \text{ for all } n \geq k_0.$$

Assume that $l^* \neq f_1 l^*$. Then, by (1), we have:
$$\inf\{p(e, l^*) + p(f_1 e, l^*) : e \in B\} \leq \inf\{p(l_t, l^*) + p(f_1 l_t, l^*) : t \in \mathbb{N}\}$$
$$= \inf\{p(l_t, l^*) + p(l_{t+1}, l^*) : t \in \mathbb{N}\} \leq \epsilon,$$

a contradiction. Therefore, $l^* = f_1 l^*$. Now, assume that $z^* \in B$ is a fixed point of f_1. Therefore:
$$(1-k)p(z^*, f_1^t z^*) = (1-k)p(z^*, z^*) \leq p(z^*, z^*).$$

Thus,
$$\varphi p(z^*, z^*) = \varphi p(f_1 z^*, f_1 z^*) \leq k \varphi p(z^*, z^*).$$

Since $k < 1$ and φ is an almost perfect function, we conclude that $p(z^*, z^*) = 0$. Assume that there exists $v^* \in B$ such that $v^* = f_1 v^*$. Since $p(z^*, z^*) = 0$, we have:
$$(1-k)p(z^*, f_1^t z^*) = (1-k)p(z^*, z^*) \leq p(z^*, v^*).$$

Therefore,
$$\varphi p(z^*, v^*) = \varphi p(f_1 z^*, f_1 v^*) \leq k \varphi p(z^*, v^*).$$

Thus, we have $\varphi p(z^*, v^*) = 0$, and so, $p(z^*, v^*) = 0$. Hence, by (mW3), we have $q(z^*, v^*) = 0$. Thus, $v^* = z^*$. Therefore, the fixed point of f_1 is unique. □

Corollary 1. *Equip (B, q) with an mω-distance mapping p. Assume p is bounded with respect to φ. Assume for all $e, l \in B$, we have:*

$$\varphi p(f_1 e, f_1 l) \leq k \varphi(p(e, l)), \text{ where } k \in [0, 1). \tag{12}$$

Furthermore, assume that:

(i) f_1 *is continuous,*
   or
(ii) *if $u^* \in B$ and $u^* \neq f_1 u^*$, then:*

$$\inf \{p(e, u^*) + p(f_1 e, u^*) : e \in B\} > 0.$$

Then, f_1 has a unique fixed point in B.

By taking the almost perfect function φ in Corollary 1 as follows:
$\varphi(e) = e$, we get the following result:

Corollary 2. *Equip (B, q) with an mω-distance mapping p. Assume there exists $A > 0$ such that $p(e, l) \leq A$ for all $e, l \in B$. Furthermore, assume that there exists $k \in [0, 1)$ such that for all $e, l \in B$, we have:*

$$p(f_1 e, f_1 l) \leq k p(e, l), \text{ where } k \in [0, 1).$$

Furthermore, assume that:

(i) f_1 *is continuous,*
   or
(ii) *if $u^* \in B$ and $u^* \neq f_1 u^*$, then:*

$$\inf \{p(e, u^*) + p(f_1 e, u^*) : e \in B\} > 0.$$

Then, f_1 has a unique fixed point in B.

Example 1. *Let $B = \{0, 1, 2, \cdots, n\}$, where $n \in \mathbb{N}$. Define $p, q : B \times B \to [0, +\infty)$ as follows:*

$$q(e, l) = \begin{cases} 0 & \text{if } e = l; \\ 3e + l & \text{if } e \neq l, \end{cases}$$

and:

$$p(e, l) = \begin{cases} 0 & \text{if } e = l; \\ \frac{1}{2}(3e + l) & \text{if } e \neq l. \end{cases}$$

Furthermore, define $f_1 : B \to B$ by:

$$f_1 e = \begin{cases} 0 & \text{if } e = 0, 1; \\ 1 & \text{if } e = 2, 3, \cdots, n, \end{cases}$$

and $\varphi : \mathbb{R}^+ \to \mathbb{R}^+$ by:

$$\varphi(l) = \begin{cases} 3^l - 1 & \text{if } l \in [0, n]; \\ 3^l & \text{if } l > n. \end{cases}$$

Then,

1. φ is an almost perfect function.
2. p is an mω-distance function on q.
3. q is a quasi metric on B.
4. (B, q) is complete.
5. f_1 satisfies (ω, φ)-Suzuki contraction with $k = \frac{1}{\sqrt{3}}$, i.e., $\forall e, l \in B, j \in \mathbb{N}$, we have:

$$\left(1 - \frac{1}{\sqrt{3}}\right) p\left(e, f_1^j e\right) \leq p(e, l) \implies \varphi p(f_1 e, f_1 l) \leq k \varphi p(e, l),$$

and:

$$\left(1 - \frac{1}{\sqrt{3}}\right) p\left(f_1^j e, e\right) \leq p(l, e) \implies \varphi p(f_1 l, f_1 e) \leq k \varphi p(l, e).$$

Proof. The proofs of (1), (2), and (3) are obvious. To show that q is complete, let (l_t) be a Cauchy sequence in B. Then, for each $t, m \in \mathbb{N}$, we have:

$$\lim_{m, t \to \infty} q(l_t, l_m) = 0.$$

Therefore, we deduce that $l_t = l_m$ for all $t, m \in \{0, 1, 2, \cdots\}$, but possible for finitely many. Thus, (l_t) converges in B. Hence, (B, q) is complete. To prove (5), given $e, l \in B$, we divide our proof into the following cases: Case (1): $e = 0$. Here, we have:

$$\left(1 - \frac{1}{\sqrt{3}}\right) p(0, 0) = \left(1 - \frac{1}{\sqrt{3}}\right) p(e, f_1^j e) \leq p(0, l) \text{ where } l = 0, 1, \cdots, n.$$

If $l \in \{0, 1\}$, then:

$$\varphi p(f_1 0, f_1 l) = \varphi p(0, 0) = 0 \leq \left(\frac{1}{\sqrt{3}}\right) \varphi p(0, l).$$

If $l \in \{2, 3, \cdots, n\}$, then:

$$\varphi p(f_1 0, f_1 l) = \varphi p(0, 1) = \varphi\left(\frac{1}{2}\right) = 3^{\frac{1}{2}} - 1.$$

Therefore,

$$\varphi p(0, l) = \varphi\left(\frac{l}{2}\right) = 3^{\frac{l}{2}} - 1.$$

$$\varphi p(f_1 0, f_1 l) = 3^{\frac{1}{2}} - 1 \leq \left(\frac{1}{\sqrt{3}}\right)\left(3^{\frac{l}{2}} - 1\right).$$

Case (2): $e = 1$. Here:

$$\left(1 - \frac{1}{\sqrt{3}}\right) p(e, 0) = \left(1 - \frac{1}{\sqrt{3}}\right) p(1, f_1 1) \leq p(1, l) \text{ where } l = 0, 2, 3, \cdots, n.$$

If $l = 0$, then we have $\varphi p(f1, fl) = 0$. Therefore,

$$\varphi p(f1, fl) = 0 \leq \left(\frac{1}{\sqrt{3}}\right)\left(3^{\frac{3}{2}} - 1\right).$$

If $l = 2, 3, \cdots, n$, then:

$$\varphi p(f_1 1, f_1 l) = \varphi p(0, 1) = \varphi\left(\frac{1}{2}\right) = 3^{\frac{1}{2}} - 1.$$

Now,
$$\varphi p(1,l) = \varphi\left(\frac{3+l}{2}\right) = 3^{\frac{3+l}{2}} - 1.$$

Thus,
$$\varphi p(f_1 1, f_1 l) = 3^{\frac{1}{2}} - 1 \leq \left(\frac{1}{\sqrt{3}}\right) \varphi\left(\frac{3+l}{2}\right) = \left(\frac{1}{\sqrt{3}}\right)\left(3^{\frac{3+l}{2}} - 1\right).$$

Case (3): $e \in \{2, 3, \cdots, n\}$. Here,
$$\left(1 - \frac{1}{\sqrt{3}}\right) p(e, 1) = \left(1 - \frac{1}{\sqrt{3}}\right) p(e, f_1 e) \leq p(e, l) \text{ where } l = 1, 2, \cdots, n.$$

If $l = 1$, then:
$$\varphi p(fe, f1) = \varphi p(1, 0) = \varphi\left(\frac{3}{2}\right) = 3^{\frac{3}{2}} - 1.$$

$$\varphi p(e, 1) = \varphi\left(\frac{3e+1}{2}\right) = \begin{cases} 3^{\frac{7}{2}} - 1 & \text{if } e = 2 \\ 3^{\frac{3e+1}{2}} & \text{if } 3 \leq e \leq n. \end{cases}$$

$$\varphi p(fe, f1) = 3^{\frac{3}{2}} - 1 \leq \left(\frac{1}{\sqrt{3}}\right) \varphi p(e, 1).$$

If $l \in \{2, 3, \cdots, n\}, e \in \{2, 3, \cdots, n\}$ and $e \neq l$, then:
$$\varphi p(f_1 e, f_1 l) = \varphi p(1, 1) = \varphi(0) = 0.$$

$$\varphi p(e, l) = \varphi\left(\frac{3e+l}{2}\right) = \begin{cases} 3^{\frac{3e+l}{2}} - 1 & \text{if } 3e + l \leq 2n \\ 3^{\frac{3e+l}{2}} & \text{if } 3e + l > 2n. \end{cases}$$

Similarly, we can show that:
$$(1 - \frac{1}{\sqrt{3}}) p\left(f^t e, e\right) \leq p(l, e) \implies \varphi p(fl, fe) \leq k \varphi p(l, e).$$

Hence, f_1 satisfies (ω, φ)-Suzuki contraction. Therefore, f_1 has a unique fixed point. □

Next, we introduce the definition of a generalized (ω, φ)-Suzuki contraction.

Definition 11. *Equip (B, q) with an mω-distance mapping p. We call $f_1 : B \to B$ a generalized (ω, φ)-Suzuki contraction if there exists an ultra distance function φ and a constant $k \in [0, 1)$ such that for all $e, l \in B, j \in \mathbb{N}$, we have:*
$$(1-k) p\left(e, f_1^j e\right) \leq p(e, l) \implies \varphi p(f_1 e, f_1 l) \leq k \max\{\varphi p(e, f_1 e), \varphi p(l, f_1 l)\},$$

and:
$$(1-k) p\left(f_1^j e, e\right) \leq p(l, e) \implies \varphi p(f_1 l, f_1 e) \leq k \max\{\varphi p(e, f_1 e), \varphi p(l, f_1 l)\}.$$

We introduce and prove the second result:

Theorem 2. *Equip (X, q) with an mω-distance mapping p. Assume that p is bounded with respect to the almost perfect function φ. Assume that f_1 is a generalized (ω, φ)-Suzuki contraction mapping. Furthermore, suppose that:*

(i) *f_1 is continuous,*

or

(ii) *if $u^* \in B$ and $u^* \neq f_1 u^*$, then:*
$$\inf \{p(e, u^*) + p(f_1 e, u^*) : e \in B\} > 0. \tag{13}$$

Then, f_1 has a unique fixed point in B.

Proof. Start with $l_0 \in B$ to construct (l_n) in B inductively by putting $l_{t+1} = f_1 l_t$ for all $t \in \mathbb{N} \cup \{0\}$. Given $t, m \in \mathbb{N} \cup \{0\}$ with $t < m$, let $m = t + j$ with $j \in \mathbb{N}$. We note that:

$$(1-k)p(l_{t-1}, l_{m-1}) = (1-k)p(l_{t-1}, f_1^j l_{t-1})$$
$$\leq p(l_{t-1}, l_{m-1}).$$

Since f_1 is a generalized (ω, φ)-Suzuki contraction, we have:

$$\varphi p(l_t, l_m) = \varphi p(f_1 l_{t-1}, f_1 l_{m-1})$$
$$\leq k \max\{\varphi p(l_{t-1}, f_1 l_{t-1}), \varphi p(l_{m-1}, f_1 l_{m-1})\} \qquad (14)$$
$$= k \max\{\varphi p(l_{t-1}, l_t)), \varphi p(l_{m-1}, l_m)\}.$$

Now,
$$(1-k)p(l_{t-2}, l_{t-1}) = (1-k)p(l_{t-2}, f_1 l_{t-2})$$
$$\leq p(l_{t-2}, l_{t-1}).$$

Therefore, we get that:

$$\varphi p(l_{t-1}, l_t) = \varphi p(f_1 l_{t-2}, f_1 l_{t-1})$$
$$\leq k \max\{\varphi p(l_{t-2}, f_1 l_{t-2}), \varphi p(l_{t-1}, f_1 l_{t-1})\} \qquad (15)$$
$$= k \max\{\varphi p(l_{t-2}, l_{t-1}), \varphi p(l_{t-1}, l_t)\}.$$

Since $k < 1$, we get that:
$$\varphi p(l_{t-1}, l_t) \leq k \varphi p(l_{t-2}, l_{t-1}). \qquad (16)$$

Repeating (16) t-times, we get that:
$$\varphi p(l_{t-1}, l_t) \leq k^{t-1} \varphi p(l_0, l_1). \qquad (17)$$

Similarly, we get that that:
$$\varphi p(l_{m-1}, l_m) \leq k^{m-1} \varphi p(l_0, l_1). \qquad (18)$$

Using Equations (14), (17), and (18), we get:
$$\varphi p(l_t, l_m) \leq k \max\{k^{t-1} \varphi p(l_0, l_1), k^{m-1} \varphi p(l_0, l_1)\}. \qquad (19)$$

Since $t < m$, we get that:
$$\varphi p(l_t, l_m) \leq k^t \varphi p(l_0, l_1). \qquad (20)$$

The boundedness property of p with respect to φ implies that:
$$\varphi p(l_t, l_m) \leq k^t A \text{ for some integer } A \geq 0. \qquad (21)$$

By letting $t, m \to \infty$, we get that:
$$\lim_{t,m \to \infty} \varphi p(l_t, l_m) = 0. \qquad (22)$$

Thus,
$$\lim_{t,m \to \infty} p(l_t, l_m) = 0. \qquad (23)$$

Since $t < m$, Lemma 1 implies that (l_t) is right Cauchy. In a similar manner, we can show that (l_t) is left Cauchy. Hence, (l_t) is Cauchy. The completeness of q ensures that there exists $l^* \in B$ such that

(l_t) converges to l^*. If f_1 is continuous, then $(l_{t+1}) = (f_1 l_t)$ converges to $f_1 l^*$. The uniqueness of the limit implies that $f_1 l^* = l^*$. Given $\varepsilon > 0$. Since $\lim_{t,m\to\infty} p(l_t, l_m)) = 0$, there exists $n_0 \in \mathbb{N}$ such that $p(l_t, l_m) \leq \frac{\varepsilon}{2}$ for all $t, m \geq n_0$. The lower semi continuity of p implies that:

$$p(l_t, l^*) \leq \liminf_{i\to\infty} p(l_t, l_i) \leq \frac{\varepsilon}{2} \text{ for all } t \geq n_0.$$

Assume that $l^* \neq f_1 l^*$, then by (13), we have:

$$\inf\{p(e, l^*) + p(f_1 e, l^*) : e \in B\}$$
$$\leq \inf\{p(l_m, l^*) + p(f_1 l_t, l^*) : t \in \mathbb{N}\}$$
$$= \inf\{p(l_t, l^*) + p(l_{t+1}, l^*) : n \in \mathbb{N}\} \leq \varepsilon,$$

a contradiction. Therefore, $l^* = f_1 l^*$. Assume $z^* \in B$ such that $f_1 z^* = z^*$. First, we prove that $p(z^*, z^*) = 0$. Since:

$$(1-k)p(z^*, f_1^j z^*) = (1-k)p(z^*, z^*) \leq p(z^*, z^*),$$

then:

$$\varphi p(z^*, z^*) = \varphi p(f_1 z^*, f_1 z^*) \leq k\varphi p(z^*, z^*).$$

Since $k < 1$ and φ is an almost perfect function, then $p(z^*, z^*) = 0$. Therefore,

$$(1-k)p(z^*, f_1^t z^*) = (1-k)p(z^*, z^*) \leq p(z^*, l^*).$$

Therefore,

$$\varphi p(z^*, l^*) = \varphi p(f_1 z^*, f_1 l^*)$$
$$\leq k \max\{\varphi p(z^*, f_1 z^*), \varphi p(l^*, f_1 l^*)\}$$
$$= k \max\{\varphi(p(z^*, z^*)), \varphi(p(v^*, v^*))\}$$
$$= 0.$$

The definition of φ informs us that $p(z^*, l^*) = 0$. The definition of p implies that $q(z^*, l^*) = 0$. Hence: $z^* = l^*$. □

Corollary 3. *Equip (B, q) with an mw-distance mapping p. Assume p is bounded with respect to the almost perfect function φ. Suppose that for all $e, l \in B$, we have:*

$$\varphi p(f_1 e, f_1 l) \leq k \max\{\varphi p(e, f_1 e), \varphi p(l, f_1 l)\}, \text{ where } k \in [0, 1). \tag{24}$$

Furthermore, assume that:

(i) *f is continuous;*

or

(ii) *if $u^* \in B$ and $u^* \neq f_1 u^*$, then:*

$$\inf\{p(e, u^*) + p(f_1 e, u^*) : e \in B\} > 0.$$

Then, f_1 has a unique fixed point in B.

Corollary 4. *Equip (B, q) with an mw-distance mapping p. Assume that there exists $A > 0$ such that $p(e, l) \leq A$ for all $e, l \in B$. Furthermore, assume that for all $e, l \in B$, we have:*

$$p(f_1 e, f_1 l) \leq \alpha(p(e, f_1 e) + p(l, f_1 l)), \text{ where } 0 \leq \alpha < \frac{1}{2}.$$

Assume that:

(i) f_1 is continuous;

or

(ii) if $u^* \in B$ and $u^* \neq f_1 u^*$, then:

$$\inf \{p(e, u^*) + p(f_1 e, u^*) : e \in B\} > 0.$$

Then, f_1 has a unique fixed point in B.

Proof. Define the almost perfect function φ via $\varphi(e) = e$ in Corollary 3. Then:

$$\varphi(p(f_1 e, f_1 l)) = p(f_1 e, f_1 l)$$
$$\leq \lambda(p(e, f_1 e) + p(l, f_1 l))$$
$$\leq 2\lambda \max \{p(e, f_1 e), p(l, f_1)\}$$
$$= 2\lambda \max \{\varphi(p(e, f_1 e)), \varphi(p(l, f_1 l))\}.$$

□

3. Application

In this section, we utilize Corollaries 1 and 4 to give some applications of our work.

Theorem 3. *For any positive integer n, the equation:*

$$nx^n - x^{n-1} + 4nx - 2 = 0$$

has a unique solution in [0, 1].

Proof. Let $B = [0, 1]$. Define $q : B \times B \to \mathbb{R}^+$ by $q(x, y) = |x - y|$. Then, (B, q) is a complete quasi metric space. Furthermore, define $p : B \times B \to [0, \infty)$ by $p(x, y) = |x - y|$. Then, p is an $m\omega$-distance mapping. Now, equip (B, q) with p.
Define $f_1 : B \to B$ by:

$$f_1(x) = \frac{x^{n-1} + 2}{n(x^{n-1} + 4)}.$$

Furthermore, define $\varphi : [0, \infty) \to [0, \infty)$ by:

$$\varphi(a) = \begin{cases} a^2 & \text{if } a \in [0, 1]; \\ a^2 + \frac{1}{2} & \text{if } a > 1. \end{cases}$$

Note that φ is an almost perfect function and p is bounded with respect to φ. For $x, y \in B$, we have:

$$\varphi p(f_1 x, f_1 y) = \frac{1}{n^2} \left| \frac{x^{n-1} + 2}{x^{n-1} + 4} - \frac{y^{n-1} + 2}{y^{n-1} + 4} \right|^2$$

$$= \frac{1}{n^2} \left| \frac{2x^{n-1} - 2y^{n-1}}{(x^{n-1} + 4)(y^{n-1} + 4)} \right|^2$$

$$\leq \frac{4(n-1)^2}{n^2} \left(\frac{1}{(x^2 + 4)^2 (y^2 + 4)^2} \right) |x - y|^2$$

$$\leq \frac{(n-1)^2}{64 n^2} |x - y|^2$$

$$= \frac{(n-1)^2}{64 n^2} \varphi p(x, y).$$

By taking $k = \frac{(n-1)^2}{64n^2}$ and noting that f_1 is continuous, we conclude that f_1 satisfies all conditions of Corollary 1. Thus, f_1 has a unique fixed point. Note that the unique fixed point of f_1 is the unique solution of:
$$nx^n - x^{n-1} + 4nx - 2 = 0.$$

□

Example 2. *The equation:*
$$1000x^{1000} - x^{999} + 4000x - 2 = 0$$
has a unique solution in [0, 1].

Proof. It follows from Theorem 3 by taking $n = 1000$. □

Let Y be the set of non-decreasing functions $\tau : \mathbb{R}^+ \to \mathbb{R}^+$ such that τ is Lebesgue integrable for all compact sets in $\mathbb{R}^+$ and:
$$\int_0^\mu \tau(v)dv > 0 \text{ where } \mu > 0.$$

Theorem 4. *Equip (B, q) with an mw-distance mapping p. Assume that there exists $A > 0$ such that $p(e, l) \leq A$ for all $e, l \in B$. Furthermore, suppose the following condition:*

(i) f_1 *is continuous.*
(ii) *There exists $\tau \in Y$ and $\alpha \in [0, 1/2)$ such that for all $e, l \in B$, we have:*

$$\int_0^{p(f_1e, f_1l)} \tau(v)dv \leq \alpha \left(\int_0^{p(e, f_1e)} \tau(v)dv + \int_0^{p(l, f_1l)} \tau(v)dv \right).$$

Then, f_1 has a unique fixed point in B.

Proof. Let $\varphi = \int_0^t \tau(v)dv$. Then, φ is an almost perfect function. Corollary 4 ensures that f_1 has a unique fixed point in B. □

4. Conclusions

The notions of (ω, φ)-Suzuki contraction and generalized (ω, φ)-Suzuki contraction are introduced. According to these nations many fixed point results are investigated. Some applications are introduced on the obtained results.

Author Contributions: Supervision: A.T. and W.S. Writing, original draft: T.Q., A.T., and W.S. Writing, review and editing: W.S.

Funding: This research is funded by Prince Sultan University through research group NAMAM, Group Number RG-DES-2017-01-17.

Acknowledgments: The authors would like to thank the anonymous reviewers and Editor for their valuable remarks on our paper. Furthermore, the third author thanks Prince Sultan University for funding this work through research group NAMAM, Group Number RG-DES-2017-01-17.

Conflicts of Interest: The authors declare no conflict of interest.

References

1. Banach, S. Sur les Opération dans les ensembles abstraits et leur application aux equations integral. *Fundam. Math.* **1922**, *3*, 133–181. [CrossRef]
2. Suzuki, T. A generalized Banach contraction principle that characterizes metric completeness. *Proc. Am. Math. Soc.* **2008**, *136*, 1861–1869. [CrossRef]
3. Kannan, R. Some results on fixed point. *Bull. Calcutta Math. Soi.* **1968**, *60*, 71–76.

4. Wilson, W.A. On quasi-metric spaces. *Am. J. Math.* **1931**, *53*, 675–684. [CrossRef]
5. Aydi, H.; Jellali, M.; Karapınar, E. On fixed point results for α-implicit contractions in quasi metric spaces and consequences. *Nonlinear Anal. Model. Control* **2016**, *21*, 40–56. [CrossRef]
6. Jleli, M.; Samet, B. Remarks on G-metric spaces and fixed point theorems. *Fixed Point Theory Appl.* **2012**, *2012*, 210. [CrossRef]
7. Bilgili, N.; Karapınar, E.; Samet, B. Eneralized α-ψ contractive mapping in quasi metric spaces and related fixed point theorems. *J. Inequal. Appl.* **2014**, *2014*, 36. [CrossRef]
8. Shatanawi, W.; Noorani, M.S.; Alsamir, H.; Bataihah, A. Fixed and common fixed point theorems in partially ordered quasi metric spaces. *Am. J. Math.* **2012**, *16*, 516–528 [CrossRef]
9. Shatanawi, W.; Pitea, A. Some coupled fixed point theorems in quasi-partial metric spaces. *Fixed Point Theory Appl.* **2013**, *2013*, 153. [CrossRef]
10. Alegre, C.; Marin, J. Modified ω-distance on quasi metric spaces and fixed point theorems on complete quasi metric spaces. *Topol. Appl.* **2016**, *203*, 120–129. [CrossRef]
11. Nuseir, I.; Shatanawi, W.; Abu-Irwaq, I.; Bataihah, A. Nonlinear contraction and fixed point theorems with modified ω-distance mappings in complete quasi metric spaces. *J. Nonlinear Sci. Appl.* **2017**, *10*, 5342–5350. [CrossRef]
12. Abodayeh, K.; Bataiahah, A.; Shatanawi, W. Generalized Ω-distance mappings and some fixed point theorems. *Politehn. Univ. Bucharest Sci. Bull. Ser. A Appl. Math. Phys.* **2017**, *79*, 223–232.
13. Mongkolkeha, C.; Cho, Y.J. Some coincidence point theorems in ordered metric spaces via ω-distance. *Carapthan J.* **2018**, *34*, 207–214.
14. Khan, M.S.; Swaleh, M.; Sessa, S. Some fixed point and common fixed point results through Ω-distance under nonlinear contraction. *Gazi Univ. J. Sci.* **2017**, *30*, 293–302.
15. Abodayeh, K.; Shatanawi, W.; Bataihah, A.; Ansari, A.H. Fixed point theorems by altering distances between the points. *Bull. Aust. Math. Soc.* **1984**, *30*, 1–9.

© 2019 by the authors. Licensee MDPI, Basel, Switzerland. This article is an open access article distributed under the terms and conditions of the Creative Commons Attribution (CC BY) license (http://creativecommons.org/licenses/by/4.0/).

Article

A Note on the Displacement Problem of Elastostatics with Singular Boundary Values

Alfonsina Tartaglione

Dipartimento di Matematica e Fisica, Università degli Studi della Campania "Luigi Vanvitelli", 81100 Caserta, Italy; alfonsina.tartaglione@unicampania.it

Received: 18 March 2019; Accepted: 16 April 2019; Published: 19 April 2019

Abstract: The displacement problem of linear elastostatics in bounded and exterior domains with a non-regular boundary datum a is considered. Precisely, if the elastic body is represented by a domain of class C^k ($k \geq 2$) of $\mathbb{R}^3$ and $a \in W^{2-k-1/q,q}(\partial\Omega)$, $q \in (1,+\infty)$, then it is proved that there exists a solution which is of class C^∞ in the interior and takes the boundary value in a well-defined sense. Moreover, it is unique in a natural function class.

Keywords: linear elastostatics; simple layer potentials; displacement problem; existence and uniqueness theorems; Fredholm alternative; singular data

MSC: 74B05; 35Q74; 45B05

1. Introduction

The *displacement problem* (classically known as the *Dirichlet problem*) in linear elastostatics consists of finding solutions to the differential system [1]

$$\begin{aligned} \operatorname{div} \mathbb{C}[\nabla u] &= 0 \quad \text{in } \Omega, \\ u &= a \quad \text{on } \partial\Omega. \end{aligned} \quad (1)$$

In (1) Ω is a bounded domain of $\mathbb{R}^3$, standing for the reference configuration of a linearly elastic body whose unknown displacement field $u = u(x)$ ($x \in \Omega$) we are looking for, supposing it is assigned on the boundary $\partial\Omega$ through condition $(1)_2$. Concrete examples of displacement problems can be found, for example, in [2], Chapter XIV. Using the components, (1) can be written as

$$\partial_j C_{ijhk} \partial_k u_h = 0,$$

where ∂_i is the derivative with respect to x_i and, hereafter, the summation over repeated indexes is understood. We suppose that the elasticity tensor $\mathbb{C} = (C_{ijhk})$, representing the material properties of the body, is independent of the point (or, in other words, that the body is homogeneous). Recall that $\mathbb{C}$ is a fourth-order tensor, that is, it is a linear map from Lin to Sym, where Lin is the linear space of all second–order tensors and Sym is its subspace of symmetric tensors, such that $\mathbb{C}[W] = 0$ for all skew tensors W. We require that $\mathbb{C}$ is *symmetric* (or, in other words, that the body is *hyperelastic*), that is,

$$E \cdot \mathbb{C}[L] = L \cdot \mathbb{C}[E], \quad \forall\, E, L \in \text{Lin}. \quad (2)$$

Furthermore, we require that it is *strongly elliptic*, that is,

$$(b \otimes c) \cdot \mathbb{C}[b \otimes c] = b_i c_j C_{ijhk} b_h c_k > 0, \quad \forall b, c \neq 0. \tag{3}$$

Hereafter, we say that Ω is of class C^k ($k \geq 2$) if for every $\xi \in \partial\Omega$ there is a neighborhood of ξ (on $\partial\Omega$) which is the graph of a function of class C^k. Moreover, $W^{k,q}(\Omega), q \in (1, +\infty)$, is the Sobolev space of all $\varphi \in L^1_{\text{loc}}(\Omega)$ such that $\|\varphi\|_{W^{k,q}(\Omega)} = \|\varphi\|_{L^q(\Omega)} + \|\nabla_k \varphi\|_{L^q(\Omega)} < +\infty$; $W^{k,q}_0(\Omega)$ is the completion of $C_0^\infty(\Omega)$ with respect to $\|\varphi\|_{W^{k,q}(\Omega)}$ and $W^{-k,q'}(\Omega), 1/q + 1/q' = 1$, is its dual space; $W^{k-1/q,q}(\partial\Omega)$ is the trace space of $W^{k,q}(\Omega)$ and $W^{1-k-1/q',q'}(\partial\Omega)$ is its dual space.

If Ω is of class C^k ($k \geq 2$) and $a \in W^{k-1/q,q}(\partial\Omega), q \in (1, +\infty)$, then (1) has a unique solution $u \in W^{k,q}(\Omega)$ and natural estimates hold (see [3–7]). This result also holds when the elastic body is subjected to a body force, that is, if in place of $(1)_1$ we consider the system

$$\text{div } \mathbb{C}[\nabla u] = f \quad \text{in } \Omega \tag{4}$$

with $f \in C_0^\infty(\Omega)$.

As, in applications, the boundary data are often represented by singular fields, it is undoubtly interesting to investigate problem (1) when a satisfies weaker regularity hypotheses.

Using the theory of layer integral equations (see [8], Chapters 2/3 and [2], Chapters IV/V) and the Fredholm alternative (see Section 2), we prove (in Theorem 1) that if $a \in W^{2-k-1/q,q}(\partial\Omega)$, then (1) has a solution, u, expressed by a simple layer potential and, thus, taking the boundary value in a well-defined sense. Moreover, it is unique in a reasonable function class. The result also holds for exterior domains (see Theorem 2).

To obtain these results, we recall some established facts about simple layer potentials associated to the system $(1)_1$.

2. The Simple Layer Potentials

For every $\psi \in L^1(\partial\Omega)$, the field

$$v[\psi](x) = \int_{\partial\Omega} U(x - \zeta) \psi(\zeta) d\sigma_\zeta, \tag{5}$$

where $U(x - y)$ is the fundamental solution to $(1)_1$ (see, e.g., [9], Chapter III), defines the *simple layer potential* with density ψ. Recall that (see, e.g., [2,8]) $v[\psi]$ is an analytical solution of $(1)_1$ in $\mathbb{R}^3 \setminus \partial\Omega$ and inherits from U the following asymptotic behavior

- $\nabla_k v[\psi](x) = O(|x|^{-1-k})$;
- $\int_{\partial\Omega} \psi = 0 \Rightarrow \nabla_k v[\psi](x) = O(|x|^{-2-k})$.

If $\psi \in W^{k-1-1/q,q}(\partial\Omega)$, then

$$\|v[\psi]\|_{W^{k,q}(\Omega)} \leq c \|\psi\|_{W^{k-1-1/q,q}(\partial\Omega)} \tag{6}$$

with c independent of ψ, and the following limit exists

$$\lim_{\epsilon \to 0^\pm} v[\psi](\xi - \epsilon l(\xi)) = S[\psi](\xi) \tag{7}$$

for almost all $\xi \in \partial\Omega$ and axis l in a ball tangent to $\partial\Omega$ at ξ.

The map
$$S: W^{k-1-1/q,q}(\partial\Omega) \to W^{k-1/q,q}(\partial\Omega) \tag{8}$$
defined by (7) and representing the trace of the simple layer potential with density ψ, is continuous, so that
$$\|S[\psi]\|_{W^{k-1/q,q}(\partial\Omega)} \leq c\|\psi\|_{W^{k-1-1/q,q}(\partial\Omega)}, \tag{9}$$
for some constant c depending only on k, q, and Ω. Moreover, S can be extended to a linear and continuous operator
$$S': W^{1-k-1/q',q'}(\partial\Omega) \to W^{2-k-1/q',q'}(\partial\Omega),$$
which coincides with the adjoint of S and defines the trace of the simple layer with density $\psi \in W^{1-k-1/q',q'}(\partial\Omega)$:
$$v[\psi](x) = \int_{\partial\Omega}^* U(x-\zeta)\psi(\zeta)d\sigma_\zeta. \tag{10}$$

In (10) and hereafter, we use the notation $\int_{\partial\Omega}^* f\varphi$ to denote the duality pairing $\langle,\rangle$ between f and φ; that is, the value of the functional f belonging to (for instance) $W^{-k,q'}(\partial\Omega)$ at $\varphi \in W_0^{k,q}(\partial\Omega)$.

By (6), one obtains
$$\|v[\psi]\|_{W^{2-k,q'}(\Omega)} \leq c\|\psi\|_{W^{1-k-1/q',q'}(\partial\Omega)}. \tag{11}$$

In the next section, we will prove the existence of a solution to (1) with singular boundary values by making use of the Fredholm alternative—we recall for the sake of completeness—applied to a suitable functional equation translating the boundary value problem (1).

If $\mathcal{B}$ and $\mathcal{D}$ are two Banach spaces and $\mathcal{B}'$, $\mathcal{D}'$ are their dual spaces, a linear and continuous map $\mathcal{T}: \mathcal{B} \to \mathcal{D}$ is said to be *Fredholmian* if its range is closed and $\dim \text{Kern}\,\mathcal{T} = \dim \text{Kern}\,\mathcal{T}' \in \mathbb{N}_0$, where $\mathcal{T}': \mathcal{D}' \to \mathcal{B}'$ is the adjoint of $\mathcal{T}$. The classical *Fredholm alternative* (see [10], Chapter 5) assures us that the equation
$$a = \mathcal{T}[u]$$
has a solution if and only if
$$\langle \phi', a \rangle = 0, \quad \forall \phi' \in \text{Kern}\,\mathcal{T}'.$$

Moreover, the equation
$$a' = \mathcal{T}'[u']$$
has a solution if and only if
$$\langle a', \phi \rangle = 0, \quad \forall \phi \in \text{Kern}\,\mathcal{T}.$$

3. Existence and Uniqueness of Solutions to (1) with Singular Data

We are in a position to prove the following existence and uniqueness theorem for the displacement problem (1) with non-regular boundary data. To this end, we need the following result (Theorem 1 in [11]).

Lemma 1. *Let Ω be a bounded domain of class C^k ($k \geq 2$). The operator S is Fredholmian and $\text{Kern}\,S = \text{Kern}\,S' = 0$.*

Theorem 1. *Let Ω be a bounded domain of class C^k ($k \geq 2$). If $a \in W^{2-k-1/q,q}(\partial\Omega)$, $q \in (1,+\infty)$, then, (1) has a solution u expressed by a simple layer potential with density $\psi \in W^{1-k-1/q,q}(\partial\Omega)$. It satisfies the estimate*
$$\|u\|_{W^{2-k,q}(\Omega)} \leq c\|a\|_{W^{2-k-1/q,q}(\partial\Omega)}, \tag{12}$$

and is unique in the class of all $u \in W^{2-k,q}(\Omega)$ such that

$$\int_\Omega^* u \cdot \phi = \int_{\partial\Omega}^* a \cdot \mathbb{C}[\nabla z]n, \tag{13}$$

for all $\phi \in C_0^\infty(\Omega)$, where n denotes the unit normal to $\partial\Omega$ (exterior with respect to Ω) and z is the solution of

$$\begin{aligned} \operatorname{div} \mathbb{C}[\nabla z] &= \phi \quad \text{in } \Omega, \\ z &= 0 \quad \text{on } \partial\Omega. \end{aligned} \tag{14}$$

Proof. In order to prove the existence of a solution to (1) in the form of a simple layer potential $u = v[\psi]$, we have to require that the boundary condition (1)$_2$ is met. Thus, in terms of the operator $\mathcal{S}'$, we have to analyse the functional equation

$$\mathcal{S}'[\psi] = a. \tag{15}$$

By virtue of Lemma 1, (15) has a solution $\psi \in W^{1-k-1/q,q}(\partial\Omega)$ and the field $u = v[\psi]$ is a solution to (1) which is C^∞ in Ω and satisfies (1)$_2$ in the sense of (15). Let a_j be a regular sequence on $\partial\Omega$ which converges to a strongly in $W^{2-k-1/q,q}(\partial\Omega)$. Let $v[\psi_j]$ be the solution of (1) with datum a_j:

$$\begin{aligned} \operatorname{div} \mathbb{C}[\nabla v[\psi_j]] &= 0 \quad \text{in } \Omega, \\ v[\psi_j] &= a_j \quad \text{on } \partial\Omega. \end{aligned} \tag{16}$$

By (11) $v[\psi_j]$ converges to $v[\psi]$ strongly in $W^{2-k,q}(\Omega)$. Let consider the scalar product of (14)$_1$ and $v[\psi_j]$ and the scalar product of (16)$_1$ and z. Taking into account the boundary conditions (14)$_2$ and (16)$_2$, then integrating by parts twice gives

$$\int_\Omega v[\psi_j] \cdot \phi = \int_\Omega v[\psi_j] \cdot \operatorname{div} \mathbb{C}[\nabla z] = \int_{\partial\Omega} a_j \cdot \mathbb{C}[\nabla z]n - \int_\Omega \nabla v[\psi_j] \cdot \mathbb{C}[\nabla z] \tag{17}$$

and

$$0 = \int_\Omega z \cdot \operatorname{div} \mathbb{C}[\nabla v[\psi_j]] = -\int_\Omega \nabla z \cdot \mathbb{C}[\nabla v[\psi_j]]. \tag{18}$$

As $\mathbb{C}$ is symmetric, from (17) and (18), we obtain

$$\int_\Omega v[\psi_j] \cdot \phi = \int_{\partial\Omega} a_j \cdot \mathbb{C}[\nabla z]n. \tag{19}$$

By the trace theorem and well-known estimates for the solutions of system (14), we obtain

$$\begin{aligned} \left| \int_{\partial\Omega} a_j \cdot \mathbb{C}[\nabla z]n \right| &\leq \|a_j\|_{W^{2-k-1/q,q}(\partial\Omega)} \|\mathbb{C}[\nabla z]n\|_{W^{k-1-1/q',q'}(\partial\Omega)} \\ &\leq \|a_j\|_{W^{2-k-1/q,q}(\partial\Omega)} \|\phi\|_{W^{k-2,q'}(\Omega)} \end{aligned} \tag{20}$$

Hence, by letting $j \to +\infty$ in (19) we obtain (13) and (12) by a duality argument. □

We can also consider the problem

$$\begin{aligned} \operatorname{div} \mathbb{C}[\nabla u] &= 0 \quad \text{in } \Omega, \\ u &= a \quad \text{on } \partial\Omega, \\ \lim_{|x| \to +\infty} u(x) &= 0, \end{aligned} \tag{21}$$

where Ω in now an exterior domain of $\mathbb{R}^3$, that is, $\Omega = \mathbb{R}^3 \setminus \overline{\Omega'}$, with Ω' a bounded domain (see, e.g., [12–14]). This problem is very intriguing in applications, where one has to consider, for example, the deformations of an elastic body with some holes (defects).

With a proof analogous to the above one for bounded domains, we obtain the following result.

Theorem 2. *Let Ω be an exterior domain of class C^k ($k \geq 2$). If $a \in W^{2-k-1/q,q}(\partial\Omega)$, with $q \in (1,+\infty)$, then (21) has a solution u expressed by a simple layer potential with density $\psi \in W^{1-k-1/q,q}(\partial\Omega)$. It satisfies the estimate*

$$\|u\|_{W^{2-k,q}(\Omega)} \leq c \|a\|_{W^{2-k-1/q,q}(\partial\Omega)}, \tag{22}$$

and is unique in the class of all $u \in W^{2-k,q}_{loc}(\Omega)$ such that

$$\int_\Omega^* u \cdot \phi = -\int_{\partial\Omega}^* a \cdot \mathbb{C}[\nabla z] n, \tag{23}$$

for all $\phi \in C_0^\infty(\Omega)$, where n denotes the unit normal to $\partial\Omega$ (exterior with respect to Ω') and z is the solution of

$$\begin{aligned} \operatorname{div} \mathbb{C}[\nabla z] &= \phi \quad \text{in } \Omega, \\ z &= 0 \quad \text{on } \partial\Omega, \\ \lim_{|x| \to +\infty} z(x) &= 0. \end{aligned} \tag{24}$$

Proof. First of all, we observe that Lemma 1 also holds for exterior domains (Theorem 1 in [11]). Thus, we can apply the Fredholm alternative again, obtaining a solution ψ to (15) and the corresponding solution $u = v[\psi]$ to (21). Then, with the analogous meaning of a_j and $v[\psi_j]$, in place of (17) and (18), we get

$$\int_{\Omega \cap B_R} v[\psi_j] \cdot \phi = -\int_{\partial\Omega} a_j \cdot \mathbb{C}[\nabla z] n + \int_{\partial B_R} v[\psi_j] \cdot \mathbb{C}[\nabla z] e_R - \int_{\Omega \cap B_R} \nabla v[\psi_j] \cdot \mathbb{C}[\nabla z] \tag{25}$$

and

$$0 = \int_{\partial B_R} z \cdot \mathbb{C}[\nabla v[\psi_j]] e_R - \int_{\Omega \cap B_R} \nabla z \cdot \mathbb{C}[\nabla v[\psi_j]], \tag{26}$$

where B_R is a ball of sufficiently large radius R containing $\partial\Omega$ and e_R is the unit normal to its boundary ∂B_R. By virtue of (2), we obtain

$$\int_{\Omega \cap B_R} v[\psi_j] \cdot \phi = -\int_{\partial\Omega} a_j \cdot \mathbb{C}[\nabla z] n + \int_{\partial B_R} v[\psi_j] \cdot \mathbb{C}[\nabla z] e_R - \int_{\partial B_R} z \cdot \mathbb{C}[\nabla v[\psi_j]] e_R. \tag{27}$$

Taking into account the asymptotic behavior of $v[\psi]$ and z, we obtain the thesis by first letting $R \to +\infty$, and then $j \to +\infty$. □

4. Conclusions

In this paper, existence and uniqueness theorems for the displacement problem of linear elastostatics with singular data are proved for three-dimensional bounded and exterior domains of class C^k ($k \geq 2$). The difficulty of the problem lies in defining the attainability of the boundary datum, which belongs to a space of non-regular fields (namely, $W^{2-k-1/q,q}(\partial\Omega)$, $q \in (1,+\infty)$). The proofs of the theorems make use of the theory of layer integral equations, of the existence and uniqueness results for regular data and of the analysis of the trace operator associated to the simple layer potentials.

As far as the two-dimensional case is concerned, the situation is more involved (also for regular data) because of the behavior of the fundamental solution ($U(x-y) = O(\ln(|x-y|))$). As pointed out in [15] (see also [16]), in this case, the search for a solution in the form of a simple layer potential $v[\psi]$ could not lead to existence and uniqueness for degenerate-scale problems. To overcome this difficulty, one may search for the solution in the form of a sum $v[\psi] + c$, with c constant and $\int_{\partial\Omega} \psi = 0$ [15]. This could be the starting point for further research into existence and uniqueness with singular data in two-dimensional domains.

Funding: This research was supported by Programma VALERE plus—Università degli Studi della Campania "Luigi Vanvitelli".

Conflicts of Interest: The author declares no conflict of interest.

References

1. Gurtin, M.E. The linear theory of elasticity. In *Handbuch der Physik*; Truesedell, C., Ed.; Springer: Berlin/Heidelberg, Germany, 1972; Volume VIa/2.
2. Kupradze, V.D.; Gegelia, T.G.; Basheleishvili, M.O.; Burchuladze, T.V. *Three Dimensional Problems of the Mathematical Theory of Elasticity and Thermoelasticity*; North–Holland: Amsterdam, The Netherlands, 1979.
3. Duvant, G.; Lions, J.L. *Inequalities in Mechanics and Physics*; Springer: Berlin/Heidelberg, Germany, 1976.
4. Fichera, G. Sull'esistenza e sul calcolo delle soluzioni dei problemi al contorno, relativi all'equilibrio di un corpo elastico. *Ann. Sc. Norm. Super. Pisa Cl. Sci. (III)* **1950**, *4*, 35–99.
5. Fichera, G. Existence theorems in elasticity. In *Handbuch der Physik*; Truesedell, C., Ed.; Springer: Berlin/Heidelberg, Germany, 1972; Volume VIa/2.
6. Lions, J.L.; Magenes, E. *Non–Homogeneous Boundary–Value Problems and Applications*; Springer: Berlin/Heidelberg, Germany, 1972; Volume I.
7. Nečas, J. *Les Méthodes Directes en Théorie des Équations Élliptiques*; Masson: Paris, France; Academie: Prague, Czech Republic, 1967.
8. Miranda, C. *Partial Differential Equations of Elliptic Type*; Springer: Berlin/Heidelberg, Germany, 1970.
9. John, F. *Plane Waves and Spherical Means Applied to Partial Differential Equations*; Interscience: New York, NY, USA, 1955.
10. Schechter, M. *Principles of Functional Analysis*; Graduate Studies in Mathematics; American Mathematical Society: Providence, RI, USA, 2002; Volume 36.
11. Starita, G.; Tartaglione, A. On the Fredholm Property of the Trace Operators Associated with the Elastic Layer Potentials. *Mathematics* **2019**, *7*, 134. [CrossRef]
12. Russo, A.; Tartaglione, A. On the contact problem of classical elasticity. *J. Elast.* **2010**, *99*, 19–38. [CrossRef]
13. Russo, A.; Tartaglione, A. Strong uniqueness theorems and the Phragmen-Lindelof principle in nonhomogeneous elastostatics. *J. Elast.* **2011**, *102*, 133–149. [CrossRef]
14. Tartaglione, A. On existence, uniqueness and the maximum modulus theorem in plane linear elastostatics for exterior domains. *Ann. Univ. Ferrara Sez. VII Sci. Mat.* **2001**, *47*, 89–106.
15. Chen, J.T.; Huang, W.S.; Lee, Y.T.; Kao, S.K. A necessary and sufficient BEM/BIEM for two-dimensional elasticity problems. *Eng. Anal. Bound. Elem.* **2016**, *67*, 108–114. [CrossRef]
16. Hong, H.K.; Chen, J.T. Derivations of Integral Equations of Elasticity. *J. Eng. Mech.* **1988**, *114*, 1028–1044. [CrossRef]

© 2019 by the author. Licensee MDPI, Basel, Switzerland. This article is an open access article distributed under the terms and conditions of the Creative Commons Attribution (CC BY) license (http://creativecommons.org/licenses/by/4.0/).

Article

Fixed Point Results in Partial Symmetric Spaces with an Application

Mohammad Asim [1,*], A. Rauf Khan [2] and Mohammad Imdad [1]

1. Department of Mathematics, Aligarh Muslim University, Aligarh-202002, India; mhimdad@gmail.com
2. Department of Applied sciences, University Polytechnic, Aligarh Muslim University, Aligarh-202002, India; abdurraufamu@gmail.com
* Correspondence: mohdasim.rs@amu.ac.in

Received: 8 December 2018; Accepted: 15 January 2019; Published: 22 January 2019

Abstract: In this paper, we first introduce the class of partial symmetric spaces and then prove some fixed point theorems in such spaces. We use one of the our main results to examine the existence and uniqueness of a solution for a system of Fredholm integral equations. Furthermore, we introduce an analogue of the Hausdorff metric in the context of partial symmetric spaces and utilize the same to prove an analogue of the Nadler contraction principle in such spaces. Our results extend and improve many results in the existing literature. We also give some examples exhibiting the utility of our newly established results.

Keywords: partial symmetric; fixed point; contraction and weak contraction; Nadler's theorem

MSC: 47H10; 54H25

1. Introduction

The classical Banach contraction principle is one of the most powerful and effective results in analysis established by Banach [1], which guarantees the existence and uniqueness of fixed points in complete metric spaces. This principle has been extended and generalized in many different directions. One of these ways is to enlarge the class of spaces, such as partial metric spaces [2], metric-like spaces [3], b-metric spaces [4], rectangular metric spaces [5], cone metric spaces [6], and several others. Sometimes, one may come across situations wherein all the metric conditions are not needed (see [7–11]). Motivated by this reality, several authors established fixed point and common fixed point results in symmetric spaces (or semi-metric spaces).

A symmetric d on a non-empty set X is a function $d : X \times X \to \mathbb{R}_+$ which satisfies $d(x,y) = d(y,x)$ and $d(x,y) = 0$ if and only if $x = y$, for all $x, y \in X$. Unlike the metric, the symmetric is not generally continuous. Due to the absence of a triangular inequality, the uniqueness of the limit of a sequence is no longer ensured. To have a workable setting, Wilson [12] suggested several related weaker conditions to overcome the earlier mentioned difficulties, which we will adopt to our setting. Such weaker conditions will be stated in the preliminaries.

In 1969, Nadler [13] initiated the study of fixed points for multi-valued contractions using the Hausdorff metric, and extended the Banach fixed point theorem to set-valued contractive maps. The theory of multi-valued maps has applications in control theory, convex optimization, differential equations, economics, and so on.

On the other hand, Matthews [2] introduced the concept of partial metric spaces as a part of the study of denotational semantics of dataflow networks, and proved an analogue of the Banach contraction theorem, and Kannan-Ćirić and Ćirić quasi-type fixed point results.

Combining the ideas involved in the concepts of partial metric spaces and symmetric spaces, we introduce the class of partial symmetric spaces, wherein we prove existence and uniqueness fixed point results for certain types of contractions in partial symmetric spaces. Furthermore, with a view to prove a multivalued analogue of Nadler's fixed point theorem, we adopt the idea of the Hausdorff metric in the setting of partial symmetric spaces. Finally, we use one of the our main results to examine the existence and uniqueness of a solution for a system of Fredholm integral equations.

2. Preliminaries

In this section, we collect some relevant definitions and examples which are needed in our subsequent discussions.

Now, we introduce the partial symmetric space as follows:

Definition 1. *Let X be a non-empty set. A mapping $\mathcal{P} : X \times X \to \mathbb{R}_+$ is said to be a partial symmetric if, for all $x, y, z \in X$:*

(1$\mathcal{P}$) $x = y$ if and only if $\mathcal{P}(x,x) = \mathcal{P}(y,y) = \mathcal{P}(x,y)$;
(2$\mathcal{P}$) $\mathcal{P}(x,x) \leq \mathcal{P}(x,y)$;
(3$\mathcal{P}$) $\mathcal{P}(x,y) = \mathcal{P}(y,x)$.

Then, the pair $(X, \mathcal{P})$ is said to be a partial symmetric space.

A partial symmetric space $(X, \mathcal{P})$ reduces to a symmetric space if $\mathcal{P}(x,x) = 0$, for all $x \in X$. Obviously, every symmetric space is a partial symmetric space, but not conversely.

Example 1. *Let $X = \mathbb{R}$ and define a mapping $\mathcal{P} : X \times X \to \mathbb{R}_+$ for all $x, y \in X$ and $p, q > 1$, as follows:*

$$\mathcal{P}(x,y) = |x-y|^p + |x-y|^q.$$

Then, the pair $(X, \mathcal{P})$ is a partial symmetric space.

Example 2. *Let $X = \mathbb{R}_+$ and define a mapping $\mathcal{P} : X \times X \to \mathbb{R}_+$ for all $x, y \in X$ and $p, q > 1$, as follows:*

$$\mathcal{P}(x,y) = (\max\{x,y\})^p + (\max\{x,y\})^q.$$

Then, the pair $(X, \mathcal{P})$ is a partial symmetric space.

Example 3. *Let $X = [0, \pi)$ and define a mapping $\mathcal{P} : X \times X \to \mathbb{R}_+$ for all $x, y \in X$ and $\alpha > 0$, as follows:*

$$\mathcal{P}(x,y) = \sin|x-y| + \alpha.$$

Then, the pair $(X, \mathcal{P})$ is a partial symmetric space.

Let $(X, \mathcal{P})$ be a partial symmetric space. Then, the $\mathcal{P}$-open ball, with center $x \in X$ and radius $\epsilon > 0$, is defined by:
$$B_{\mathcal{P}}(x, \epsilon) = \{y \in X : \mathcal{P}(x,y) < \mathcal{P}(x,x) + \epsilon\}.$$

Similarly, the $\mathcal{P}$-closed ball, with center $x \in X$ and radius $\epsilon > 0$, is defined by:
$$B_{\mathcal{P}}[x, \epsilon] = \{y \in X : \mathcal{P}(x,y) \leq \mathcal{P}(x,x) + \epsilon\}.$$

The family of $\mathcal{P}$-open balls for all $x \in X$ and $\epsilon > 0$,
$$\mathcal{U}_{\mathcal{P}} = \{B_{\mathcal{P}}(x, \epsilon) : x \in X, \epsilon > 0\},$$

forms a basis of some topology $\tau_{\mathcal{P}}$ on X.

Lemma 1. *Let $(X, \tau_\mathcal{P})$ be a topological space and $f : X \to X$. If f is continuous then, for every convergent sequence $x_n \to x$ in X, the sequence fx_n converges to fx. The converse holds if X is metrizable.*

In subsequent future discussions, we need some more basic definitions, namely: Convergent sequences, Cauchy sequences, and complete partial symmetric spaces, which are outlined in the following:

Definition 2. *A sequence $\{x_n\}$ in $(X, \mathcal{P})$ is said to be $\mathcal{P}$-convergent to $x \in X$, with respect to $\tau_\mathcal{P}$, if*

$$\mathcal{P}(x,x) = \lim_{n \to \infty} \mathcal{P}(x_n, x).$$

Definition 3. *A sequence $\{x_n\}$ in $(X, \mathcal{P})$ is said to be $\mathcal{P}$-Cauchy if and only if $\lim_{n,m \to \infty} \mathcal{P}(x_n, x_m)$ exists and is finite.*

Definition 4. *A partial symmetric space $(X, \mathcal{P})$ is said to be $\mathcal{P}$-complete if every $\mathcal{P}$-Cauchy sequence $\{x_n\}$ in X is $\mathcal{P}$-convergent, with respect to $\tau_\mathcal{P}$ to a point in $x \in X$, such that*

$$\mathcal{P}(x,x) = \lim_{n \to \infty} \mathcal{P}(x_n, x) = \lim_{n,m \to \infty} \mathcal{P}(x_n, x_m).$$

Now, we adopt some definitions from symmetric spaces in the setting of partial symmetric spaces:

Definition 5. *Let $(X, \mathcal{P})$ be a partial symmetric. Then*

(A1) $\lim_{n \to \infty} \mathcal{P}(x_n, x) = \mathcal{P}(x, x)$ and $\lim_{n \to \infty} \mathcal{P}(x_n, y) = \mathcal{P}(x, y)$ imply that $x = y$, for a sequence $\{x_n\}$, x, and y in X.

(A2) *A partial symmetric $\mathcal{P}$ is said to be 1-continuous if* $\lim_{n \to \infty} \mathcal{P}(x_n, x) = \mathcal{P}(x, x)$ *implies that* $\lim_{n \to \infty} \mathcal{P}(x_n, y) = \mathcal{P}(x, y)$, *where $\{x_n\}$ is a sequence in X and $x, y \in X$.*

(A3) *A partial symmetric $\mathcal{P}$ is said to be continuous if* $\lim_{n \to \infty} \mathcal{P}(x_n, x) = \mathcal{P}(x, x)$ *and* $\lim_{n \to \infty} \mathcal{P}(x_n, y) = \mathcal{P}(x, y)$ *imply that* $\lim_{n \to \infty} \mathcal{P}(x_n, y_n) = \mathcal{P}(x, y)$ *where $\{x_n\}$ and $\{y_n\}$ are sequences in X and $x, y \in X$.*

(A4) $\lim_{n \to \infty} \mathcal{P}(x_n, x) = \mathcal{P}(x, x)$ and $\lim_{n \to \infty} \mathcal{P}(x_n, y_n) = \mathcal{P}(x, x)$ imply $\lim_{n \to \infty} \mathcal{P}(y_n, x) = \mathcal{P}(x, x)$, for sequences $\{x_n\}, \{y_n\}$, and x in X.

(A5) $\lim_{n \to \infty} \mathcal{P}(x_n, y_n) = \mathcal{P}(x, x)$ and $\lim_{n \to \infty} \mathcal{P}(y_n, z_n) = \mathcal{P}(x, x)$ imply $\lim_{n \to \infty} \mathcal{P}(x_n, z_n) = \mathcal{P}(x, x)$, for sequences $\{x_n\}, \{y_n\}, \{z_n\}$, and x in X.

Remark 1. *From the Definition 5, it is observed that $(A3) \Rightarrow (A2)$, $(A4) \Rightarrow (A1)$, and $(A2) \Rightarrow (A1)$ but, in general, the converse implications are not true.*

3. Fixed Point Results

Let $(X, \mathcal{P})$ be a partial symmetric space and $f : X \to X$. Then, for every $x \in X$ and for all $i, j \in \mathbb{N}$, we define

$$\mathfrak{S}(\mathcal{P}, f, x) = \sup\{\mathcal{P}(f^i x, f^j x) : i, j \in \mathbb{N}\}. \tag{1}$$

Definition 6. *Let $(X, \mathcal{P})$ be a partial symmetric space. A mapping $f : X \to X$ is said to be a κ-contraction if*

$$\mathcal{P}(fx, fy) \leq \kappa \mathcal{P}(x, y), \quad \forall \, x, y \in X, \tag{2}$$

where $\kappa \in (0, 1)$.

Now, we prove an analogue of the Banach contraction principle in the setting of partial symmetric spaces:

Theorem 1. Let $(X, \mathcal{P})$ be a complete partial symmetric space and $f : X \to X$. Assume that the following conditions are satisfied:

(i) f is a κ-contraction, for some $\kappa \in [0, 1)$;
(ii) there exists $x_0 \in X$ such that $\mathfrak{S}(\mathcal{P}, f, x_0) < \infty$; and
(iii) either
  (a) f is continuous, or
  (b) $(X, \mathcal{P})$ satisfies the $(A1)$ property.

Then, f has a unique fixed point $x \in X$ such that $\mathcal{P}(x, x) = 0$.

Proof. Choose $x_0 \in X$ and construct an iterative sequence $\{x_n\}$ by:

$$x_1 = fx_0, \; x_2 = f^2 x_0, \; x_3 = f^3 x_0, \ldots, x_n = f^n x_0, \ldots$$

Now, from (2) (for all $i, j \in \mathbb{N}$), we have

$$\mathcal{P}(f^{n+i} x_0, f^{n+j} x_0) \leq \kappa \mathcal{P}(f^{n-1+i} x_0, f^{n-1+j} x_0).$$

The above inequality holds for all $i, j \in \mathbb{N}$; therefore, by conditions (ii) and (1), we have

$$\mathfrak{S}(\mathcal{P}, f, f^n x_0) \leq \kappa \mathfrak{S}(\mathcal{P}, f, f^{n-1} x_0).$$

Repeating this procedure indefinitely, we have (for every $n \in \mathbb{N}$)

$$\mathfrak{S}(\mathcal{P}, f, f^n x_0) \leq \kappa^n \mathfrak{S}(\mathcal{P}, f, x_0). \tag{3}$$

Let $n, m \in \mathbb{N}$, such that $m = n + p$ for some $p \in \mathbb{N}$. Using (3), we have

$$\mathcal{P}(f^n x_0, f^{n+p} x_0) \leq \mathfrak{S}(\mathcal{P}, f, f^n x_0) \leq \kappa^n \mathfrak{S}(\mathcal{P}, f, x_0).$$

As $\mathfrak{S}(\mathcal{P}, f, x_0) < \infty$ and $\kappa \in (0, 1)$, we have

$$\lim_{n,m \to \infty} \mathcal{P}(x_n, x_m) = 0,$$

so that $\{x_n\}$ is a $\mathcal{P}$-Cauchy sequence in X. In light of the $\mathcal{P}$-completeness of X, there exists $x \in X$ such that $\{x_n\}$ $\mathcal{P}$-converges to x. Now, we show that $x \in X$ is a fixed point of f.

Assume that f is continuous. Then,

$$x = \lim_{n \to \infty} x_{n+1} = f(\lim_{n \to \infty} x_n) = fx.$$

Alternately, assume that $(X, \mathcal{P})$ satisfies the $(A1)$ property. Now, we have

$$\mathcal{P}(fx_n, fx) \leq \mathcal{P}(x_n, x),$$

which, on taking $n \to \infty$, implies that $\lim_{n \to \infty} \mathcal{P}(x_{n+1}, fx) = 0$. Thus, from the $(A1)$ property, $fx = x$. Therefore, x is a fixed point of f. To prove the uniqueness of the fixed point, let on contrary that there exist $x, y \in X$ such that $fx = x$ and $fy = y$. Then, by the definition of κ-contraction, we have

$$\mathcal{P}(x, y) = \mathcal{P}(fx, fy) \leq \kappa \mathcal{P}(x, y),$$

a contradiction. Hence, $x = y$; that is, x is a unique fixed point of f. Finally, we show that $\mathcal{P}(x,x) = 0$. Since, f is κ-contraction mapping, we have

$$\mathcal{P}(x,x) = \mathcal{P}(fx, fx) \leq \kappa \mathcal{P}(x,x).$$

This implies that $\mathcal{P}(x,x) < 0$, implying thereby that $\mathcal{P}(x,x) = 0$. This completes the proof. □

Now, we recall the definition of the Kannan-Ćirić contraction condition [14]:

Definition 7. *Let $(X, \mathcal{P})$ be a partial symmetric space. A mapping $f : X \to X$ is said to be a Kannan-Ćirić type κ-contraction if, for all $x, y \in X$,*

$$\mathcal{P}(fx, fy) \leq \kappa \max\{\mathcal{P}(x, fx), \mathcal{P}(y, fy)\}, \tag{4}$$

where $\kappa \in [0, 1)$.

Next, we prove a fixed point result via Kannan-Ćirić type κ-contractions in the setting of partial symmetric spaces:

Theorem 2. *Let $(X, \mathcal{P})$ be a complete partial symmetric space and $f : X \to X$. Assume that the following conditions are satisfied:*

(i) f is a Kannan-Ćirić type κ-contraction mapping,
(ii) f is continuous.

Then, f has a unique fixed point $x \in X$ such that $\mathcal{P}(x,x) = 0$.

Proof. Take $x_0 \in X$, and construct an iterative sequence $\{x_n\}$ by:

$$x_1 = fx_0, \ x_2 = f^2 x_0, \ x_3 = f^3 x_0, \ldots, x_n = f^n x_0, \ldots$$

Now, we assert that $\lim_{n \to \infty} \mathcal{P}(x_n, x_{n+1}) = 0$. On setting $x = x_n$ and $y = x_{n+1}$ in (4), we get

$$\begin{aligned}
\mathcal{P}(x_n, x_{n+1}) &= \mathcal{P}(fx_{n-1}, fx_n) \\
&\leq \kappa \max\{\mathcal{P}(x_{n-1}, fx_{n-1}), \mathcal{P}(x_n, fx_n)\} \\
&\leq \kappa \max\{\mathcal{P}(x_{n-1}, x_n), \mathcal{P}(x_n, x_{n+1})\}.
\end{aligned} \tag{5}$$

Assume that $\max\{\mathcal{P}(x_{n-1}, x_n), \mathcal{P}(x_n, x_{n+1})\} = \mathcal{P}(x_n, x_{n+1})$, then from (5), we have

$$\mathcal{P}(x_n, x_{n+1}) \leq \kappa \mathcal{P}(x_n, x_{n+1}),$$

a contradiction (since $\kappa \in (0, 1)$). Thus, $\max\{\mathcal{P}(x_{n-1}, x_n), \mathcal{P}(x_n, x_{n+1})\} = \mathcal{P}(x_{n-1}, x_n)$. Therefore, (5) gives rise

$$\mathcal{P}(x_n, x_{n+1}) = \kappa \mathcal{P}(x_{n-1}, x_n) \text{ for all } n \in \mathbb{N}.$$

Thus, inductively, we have

$$\mathcal{P}(x_n, x_{n+1}) = \kappa^n \mathcal{P}(x_0, x_1) \text{ for all } n \in \mathbb{N}.$$

On taking the limit as $n \to \infty$, we get

$$\lim_{n \to \infty} \mathcal{P}(x_n, x_{n+1}) = 0. \tag{6}$$

Now, we assert that $\{x_n\}$ is a $\mathcal{P}$-Cauchy sequence. From (4), we have, for $n, m \in \mathbb{N}$,

$$\begin{aligned}\mathcal{P}(x_n, x_m) &= \mathcal{P}(fx_{n-1}, fx_{m-1}) \\ &\leq \kappa \max\{\mathcal{P}(x_{n-1}, fx_{n-1}), \mathcal{P}(x_{m-1}, fx_{m-1})\} \\ &\leq \kappa \max\{\mathcal{P}(x_{n-1}, x_n), \mathcal{P}(x_{m-1}, x_m)\}.\end{aligned}$$

By taking the limit as $n, m \to \infty$ and using (6), we have

$$\lim_{n,m \to \infty} \mathcal{P}(x_n, x_m) = 0. \tag{7}$$

Hence, $\{x_n\}$ is a $\mathcal{P}$-Cauchy sequence. Since $(X, \mathcal{P})$ is $\mathcal{P}$-complete, there exists $x \in X$ such that $\lim_{n \to \infty} \mathcal{P}(x_n, x) = 0$. Now, we show that $x \in X$ is a fixed point of f. By the continuity of f, we have

$$x = \lim_{n \to \infty} x_{n+1} = f(\lim_{n \to \infty} x_n) = fx.$$

Therefore, x is a fixed point of f. For the uniqueness part, let on contrary that there exist $x, y \in X$ such that $fx = x$ and $fy = y$. Then, from (4), we have

$$\begin{aligned}\mathcal{P}(x, y) &= \mathcal{P}(fx, fy) \\ &\leq \kappa \max\{\mathcal{P}(x, fx), \mathcal{P}(y, fy)\}, \\ &\leq \kappa \max\{\mathcal{P}(x, x), \mathcal{P}(y, y)\}.\end{aligned}$$

So, either $\mathcal{P}(x, y) \leq \kappa \mathcal{P}(x, x)$ or $\mathcal{P}(x, y) \leq \kappa \mathcal{P}(y, y)$, which is a contradiction. Therefore, x is a unique fixed point of f. Finally, we show that $\mathcal{P}(x, x) = 0$. From (4), we have

$$\begin{aligned}\mathcal{P}(x, x) &= \kappa \mathcal{P}(fx, fx) \\ &\leq \kappa \max\{\mathcal{P}(x, fx), \mathcal{P}(x, fx)\}, \\ &\leq \kappa \max\{\mathcal{P}(x, x), \mathcal{P}(x, x)\}, \\ &\leq \kappa \mathcal{P}(x, x),\end{aligned}$$

this implies that $\mathcal{P}(x, x) < 0$, implying thereby that $\mathcal{P}(x, x) = 0$. This completes the proof. □

Now, we present some fixed point results for Ćirić quasi contractions in the setting of partial symmetric spaces. We start with the following definition.

Definition 8. *Let $(X, \mathcal{P})$ be a partial symmetric space and $f : X \to X$. Then f is said to be κ-weak contraction if, for all $x, y \in X$, and $\kappa \in (0, 1)$*

$$\mathcal{P}(fx, fy) \leq \kappa \max\{\mathcal{P}(x, y), \mathcal{P}(x, fx), \mathcal{P}(y, fy), \mathcal{P}(x, fy), \mathcal{P}(y, fx)\}. \tag{8}$$

Proposition 1. *Let f be a κ-weak contraction for any $\kappa \in (0, 1)$. If x is a fixed point of f, then $\mathcal{P}(x, x) = 0$.*

Proof. Suppose $x \in X$ is a fixed point of f. Since f is a κ-weak contraction, we have that

$$\begin{aligned}\mathcal{P}(x, x) &= \mathcal{P}(fx, fx) \\ &\leq \kappa \max\{\mathcal{P}(x, x), \mathcal{P}(x, fx), \mathcal{P}(x, fx), \mathcal{P}(x, fx), \mathcal{P}(x, fx)\} \\ &= \kappa \max\{\mathcal{P}(x, x), \mathcal{P}(x, x), \mathcal{P}(x, x), \mathcal{P}(x, x), \mathcal{P}(x, x)\} \\ &= \kappa \mathcal{P}(x, x),\end{aligned}$$

this implies that $\mathcal{P}(x, x) < 0$, implying thereby $\mathcal{P}(x, x) = 0$. □

Theorem 3. Let $(X, \mathcal{P})$ be a $\mathcal{P}$-complete partial symmetric space and $f : X \to X$. Suppose that the following conditions hold:

(i) f is a κ-weak contraction for some $\kappa \in [0, 1)$;
(ii) there exists $x_0 \in X$ such that $\mathfrak{S}(\mathcal{P}, f, x) < \infty$; and
(iii) f is continuous.

Then, f has a unique fixed point.

Proof. Assume $x_0 \in X$, and construct an iterative sequence $\{x_n\}$ by:

$$x_1 = fx_0, \; x_2 = f^2 x_0, \; x_3 = f^3 x_0, \ldots, x_n = f^n x_0, \ldots$$

Let n be an arbitrary positive integer. Since f is a κ-weak contraction, for all $i, j \in \mathbb{N}$, we have

$$\mathcal{P}(f^{n+i} x_0, f^{n+j} x_0) \leq \kappa \max \{ \mathcal{P}(f^{n-1+i} x_0, f^{n-1+j} x_0), \mathcal{P}(f^{n-1+i} x_0, f^{n+i} x_0),$$
$$\mathcal{P}(f^{n-1+j} x_0, f^{n+j} x_0), \mathcal{P}(f^{n-1+i} x_0, f^{n+j} x_0), \mathcal{P}(f^{n-1+j} x_0, f^{n+i} x_0) \}.$$

Since the above inequality is true for all $i, j \in \mathbb{N}$, therefore by conditions (ii) and (1), we have

$$\mathfrak{S}(\mathcal{P}, f, f^n x_0) \leq \kappa \mathfrak{S}(\mathcal{P}, f, f^{n-1} x_0).$$

Continuing this process indefinitely, we have, for all $n \geq 1$,

$$\mathfrak{S}(\mathcal{P}, f, f^n x_0) \leq \kappa^n \mathfrak{S}(\mathcal{P}, f, x_0). \tag{9}$$

Now, for each $n, m \in \mathbb{N}$, such that $m = n + p$ for some $p \in \mathbb{N}$, we have, due to (9), that

$$\mathcal{P}(f^n x_0, f^{n+p} x_0) \leq \mathfrak{S}(\mathcal{P}, f, f^n x_0) \leq \kappa^n \mathfrak{S}(\mathcal{P}, f, x_0). \tag{10}$$

Since $\mathfrak{S}(\mathcal{P}, f, x_0) < \infty$ and $\kappa \in (0, 1)$, we have

$$\lim_{n, m \to \infty} \mathcal{P}(x_n, x_m) = 0,$$

so $\{x_n\}$ is a $\mathcal{P}$-Cauchy sequence in X. In view of the $\mathcal{P}$-completeness of X, there exists $x \in X$ such that $\{x_n\}$ $\mathcal{P}$-converges to x. Now, we show that x is a fixed point of f. By the continuity of f, we have

$$x = \lim_{n \to \infty} x_{n+1} = f(\lim_{n \to \infty} x_n) = fx.$$

Therefore, x is a fixed point of f. For the uniqueness part, let on contrary that there exist $x, y \in X$ such that $fx = x$ and $fy = y$. Thus, by using the condition (8), we have

$$\mathcal{P}(x, y) = \mathcal{P}(fx, fy)$$
$$\leq \kappa \max \{ \mathcal{P}(x, y), \mathcal{P}(x, fx), \mathcal{P}(y, fy), \mathcal{P}(x, fy), \mathcal{P}(y, fx) \}$$
$$= \kappa \max \{ \mathcal{P}(x, y), \mathcal{P}(x, x), \mathcal{P}(y, y), \mathcal{P}(x, y), \mathcal{P}(y, x) \}.$$

By using the property $(2\mathcal{P})$, we have

$$\mathcal{P}(x, y) \leq \kappa \mathcal{P}(x, y) < \mathcal{P}(x, y),$$

a contradiction, and so $\mathcal{P}(x, y) = 0$; which implies that $x = y$. Thus, f has a unique fixed point. This completes the proof. □

Now, we furnish the following example, which illustrates Theorem 3.

Example 4. Consider $X = [0,1]$ and a partial symmetric $\mathcal{P} : X \times X \to \mathbb{R}_+$ defined by $\mathcal{P}(x,y) = \max\{x,y\}$, for all $x, y \in X$. Define a self-mapping f on X by

$$fx = \frac{2x^2}{5}, \text{ for all } x \in X.$$

Observe that

$$\begin{aligned}
\mathcal{P}(fx, fy) &= \max\{fx, fy\} \\
&= \max\left\{\frac{2x^2}{5}, \frac{2y^2}{5}\right\} \\
&\leq \frac{2}{5}\max\{x,y\} = \frac{2}{5}\mathcal{P}(x,y) \\
&\leq \frac{2}{5}\max\{\mathcal{P}(x,y), \mathcal{P}(x,fx), \mathcal{P}(y,fy), \mathcal{P}(x,fy), \mathcal{P}(y,fx)\},
\end{aligned}$$

for all $x, y \in X$. Observe that f is continuous and condition (ii) holds. Thus, all the conditions of Theorem 3 are satisfied and so f has a unique fixed point (i.e., $x = 0$).

Notice that this example can not be covered by metrical fixed point theorems.

Corollary 1. *The conclusions of Theorem 3 remain true, if the contractive condition (8) is replaced by any one of the following:*

(i) $\mathcal{P}(fx, fy) \leq \frac{\kappa}{2}[\mathcal{P}(x, fy) + \mathcal{P}(y, fx)]$;
(ii) $\mathcal{P}(fx, fy) \leq \kappa \max\{\mathcal{P}(x,y), \mathcal{P}(x, fx), \mathcal{P}(y, fy)\}$;
(iii) $\mathcal{P}(fx, fy) \leq \kappa \max\{\mathcal{P}(x,y), \mathcal{P}(x, fy), \mathcal{P}(y, fx)\}$;
(iv) $\mathcal{P}(fx, fy) \leq \kappa \max\left\{\mathcal{P}(x,y), \frac{\mathcal{P}(x,fx)+\mathcal{P}(y,fy)}{2}, \frac{\mathcal{P}(x,fy)+\mathcal{P}(y,fx)}{2}\right\}$;
(v) $\mathcal{P}(fx, fy) \leq \kappa \max\left\{\mathcal{P}(x,y), \frac{\mathcal{P}(x,fx)+\mathcal{P}(y,fy)}{2}, \mathcal{P}(x,fy), \mathcal{P}(y,fx)\right\}$; or
(vi) $\mathcal{P}(fx, fy) \leq \kappa \max\left\{\mathcal{P}(x,y), \mathcal{P}(x,fx), \mathcal{P}(y,fy), \frac{\mathcal{P}(x,fy)+\mathcal{P}(y,fx)}{2}\right\}$.

4. Application

In this section, we endeavor to apply Theorem 1 to prove the existence and uniqueness of a solution of the following integral equation of Fredholm type:

$$x(t) = \int_a^b G(t, s, x(s))ds + h(t) \text{ for all } t, s \in [a,b], \tag{11}$$

where $G, h \in C([a,b], \mathbb{R})$ (say, $X = C([a,b], \mathbb{R})$. Define a partial symmetric space $\mathcal{P}$ on X:

$$\mathcal{P}(x,y) = \sup_{t \in [a,b]} |x(t) - y(t)|^p + \sup_{t \in [a,b]} |x(t) - y(t)|^q, \text{ for all } x, y \in X, \text{ and } p, q > 1.$$

Then, $(X, \mathcal{P})$ is a complete partial symmetric space.

Now we are equipped to state and prove our result, as follows:

Theorem 4. *Assume that, for all $x, y \in C([a,b], \mathbb{R})$,*

$$|G(t, s, x(s)) - G(t, s, y(s))| \leq \frac{1}{2(b-a)}|x(s) - y(s)|, \tag{12}$$

for all $t, s \in [a, b]$. Then, Equation (11) has a unique solution.

Proof. Define $f : X \to X$ by

$$fx(t) = \int_a^b G(t, s, x(s))ds + h(t) \text{ for all } t, s \in [a, b].$$

It is clear that x is a fixed point of the operator f if and only if it is a solution of Equation (11). Now, for all $x, y \in X$, we have

$$\begin{aligned}
|fx(t) - fy(t)|^p + |fx(t) - fy(t)|^q &\leq \left(\int_a^b |G(t, s, x(s)) - G(t, s, y(s))|ds\right)^p \\
&+ \left(\int_a^b |G(t, s, x(s)) - G(t, s, y(s))|ds\right)^q \\
&\leq \left(\int_a^b \frac{1}{2(b-a)}|x(s) - y(s)|ds\right)^p \\
&+ \left(\int_a^b \frac{1}{2(b-a)}|x(s) - y(s)|ds\right)^q \\
&\leq \frac{1}{2^p(b-a)^p} \sup_{t \in [a,b]} |x(t) - y(t)|^p \left(\int_a^b ds\right)^p \\
&+ \frac{1}{2^q(b-a)^q} \sup_{t \in [a,b]} |x(t) - y(t)|^q \left(\int_a^b ds\right)^q \\
&\leq \lambda \left(\sup_{t \in [a,b]} |x(t) - y(t)|^p + \sup_{t \in [a,b]} |x(t) - y(t)|^q\right).
\end{aligned}$$

Thus, condition (12) is satisfied, with $\lambda = \max\{\frac{1}{2^p}, \frac{1}{2^q}\} \in [0, 1)$. Hence, the operator f has a unique fixed point; that is, the Fredholm integral Equation (11) has a unique solution. $\square$

5. Results Involving Set-Valued Map

In this section, first we extend the idea of Hausdorff distance to partial symmetric spaces. Let $(X, \mathcal{P})$ be a partial symmetric space and $\mathcal{CB}^{\mathcal{P}}(X)$ be the family of all nonempty, $\tau_{\mathcal{P}}$-closed, and bounded subsets of $(X, \mathcal{P})$. Observe that A will be bounded if there exist $x_0 \in X$ and $M \geq 0$ such that, for all $a \in A$, $\mathcal{P}(x_0, a) \leq \mathcal{P}(a, a) + M$.

Moreover, for $A, B \in \mathcal{CB}^{\mathcal{P}}(X)$ and $x \in X$, we define:

$$dist_{\mathcal{P}}(x, A) = \inf\{\mathcal{P}(x, a) : a \in A\};$$

$$\delta_{\mathcal{P}}(A, B) = \sup\{dist_{\mathcal{P}}(a, B) : a \in A\}; \text{ and}$$

$$\delta_{\mathcal{P}}(B, A) = \sup\{dist_{\mathcal{P}}(b, A) : b \in B\}.$$

Remark 2. Let $(X, \mathcal{P})$ be a partial symmetric space and A a non-empty subset of X, then

$$a \in \overline{A} \text{ if and only if } dist_{\mathcal{P}}(a, A) = \mathcal{P}(a, a),$$

where $\overline{A}$ denotes the closure of A, with respect to the partial symmetric $\mathcal{P}$. Also, A is $\mathcal{P}$-closed in $(X, \mathcal{P})$ if and only if $A = \overline{A}$.

Proposition 2. Let $(X, \mathcal{P})$ be a partial symmetric space. For $A, B, C \in \mathcal{CB}^{\mathcal{P}} X$, we have the following:

(i) $\delta_{\mathcal{P}}(A, A) = \sup\{\mathcal{P}(a, a) : a \in A\};$
(ii) $\delta_{\mathcal{P}}(A, A) \leq \delta_{\mathcal{P}}(A, B);$
(iii) $B \subset C \Rightarrow \delta_{\mathcal{P}}(A, C) \leq \delta_{\mathcal{P}}(A, B);$

(iv) $\delta_\mathcal{P}(A, B) = 0 \Rightarrow A \subseteq B$; and
(v) $\delta_\mathcal{P}(A \cup B, C) = \max\{\delta_\mathcal{P}(A, C), \delta_\mathcal{P}(B, C)\}$.

Proof. (i) Suppose $A \in \mathcal{CB}^\mathcal{P}(X)$. Since $a \in \overline{A}$ if and only if $\mathcal{P}(a, A) = \mathcal{P}(a, a)$,

$$\delta_\mathcal{P}(A, A) = \sup\{dist_\mathcal{P}(a, A) : a \in A\} = \sup\{\mathcal{P}(a, a) : a \in A\}.$$

(ii) Suppose $a \in A$. By definition of the partial symmetric space, we know that $\mathcal{P}(a, a) \leq \mathcal{P}(a, b)$, which implies that

$$\mathcal{P}(a, a) \leq dist_\mathcal{P}(a, B) \leq \delta_\mathcal{P}(A, B).$$

Hence, condition (i) gives rise to

$$\delta_\mathcal{P}(A, A) \leq \delta_\mathcal{P}(A, B).$$

(iii) Suppose $A, B, C \in \mathcal{CB}^\mathcal{P}(X)$, such that $B \subseteq C$. Then,

$$dist_\mathcal{P}(x, B) \leq dist_\mathcal{P}(x, C) \text{ for all } a \in X.$$

Thus,

$$B \subset C \Rightarrow \delta_\mathcal{P}(A, C) \leq \delta_\mathcal{P}(A, B).$$

(iv) Suppose $A, B \in \mathcal{CB}^\mathcal{P}(X)$, such that $\delta_\mathcal{P}(A, B) = 0$. Then,

$$\sup\{dist_\mathcal{P}(a, A) : a \in A\} = 0 \Rightarrow dist_\mathcal{P}(a, B) = 0 \text{ for all } a \in A.$$

In view of the above conditions (i) and (ii), we have

$$\mathcal{P}(a, a) \leq \delta_\mathcal{P}(A, B) = 0 \Rightarrow \mathcal{P}(a, a) = 0, \text{ for all } a \in A.$$

Therefore, $dist_\mathcal{P}(a, B) = \mathcal{P}(a, a)$ for all $a \in A$ implies that 'a' is in the closure of B for all $a \in A$. Since B is $\mathcal{P}$-closed, we have $A \subseteq B$.

(v) Suppose $A, B, C \in \mathcal{CB}^\mathcal{P}(X)$. Then,

$$\begin{aligned}
\delta_\mathcal{P}(A \cup B, C) &= \sup\{dist_\mathcal{P}(x, C) : x \in A \cup B\} \\
&= \max\{\sup\{dist_\mathcal{P}(x, C) : x \in A\}, \sup\{dist_\mathcal{P}(x, C) : x \in B\}\} \\
&= \max\{\delta_\mathcal{P}(A, C), \delta_\mathcal{P}(B, C)\}.
\end{aligned}$$

□

Next, let $(X, \mathcal{P})$ be a partial symmetric space. Define

$$H_\mathcal{P}(A, B) = \max\{\delta_\mathcal{P}(A, B), \delta_\mathcal{P}(B, A)\}.$$

Proposition 3. *Let $(X, \mathcal{P})$ be a partial symmetric space. For $A, B, C, D \in \mathcal{CB}^\mathcal{P}(X)$, we have the following:*

(1H) $H_\mathcal{P}(A, A) \leq H_\mathcal{P}(A, B)$;
(2H) $H_\mathcal{P}(A, B) = H_\mathcal{P}(B, A)$; and
(3H) $H_\mathcal{P}(A \cup B, C \cup D) = \max\{H_\mathcal{P}(A, C), H_\mathcal{P}(B, D)\}$.

Proof. (1H) By condition (ii) of Proposition 2, we have

$$H_\mathcal{P}(A, A) = \delta_\mathcal{P}(A, A) \leq \delta_\mathcal{P}(A, B) \leq H_\mathcal{P}(A, B).$$

(2H) By the definition of $H_\mathcal{P}$, we have

$$H_\mathcal{P}(A,B) = \max\{\delta_\mathcal{P}(A,B), \delta_\mathcal{P}(B,A)\} = \max\{\delta_\mathcal{P}(B,A), \delta_\mathcal{P}(A,B)\} = H_\mathcal{P}(B,A).$$

(3H) By condition (v) of Proposition 2, we have

$$\begin{aligned}\delta_\mathcal{P}(A\cup B, C\cup D) &= \max\{\delta_\mathcal{P}(A, C\cup D), \delta_\mathcal{P}(B, C\cup D)\} \\ &\leq \max\{\delta_\mathcal{P}(A,C), \delta_\mathcal{P}(B,D)\} \\ &\leq \max\{H_\mathcal{P}(A,C), H_\mathcal{P}(B,D)\}.\end{aligned}$$

Similarly, we obtain

$$\delta_\mathcal{P}(C\cup D, A\cup B) \leq \max\{H_\mathcal{P}(A,C), H_\mathcal{P}(B,D)\}.$$

Hence, by the definition of $H_\mathcal{P}$, we have, for all $A, B, C, D \in \mathcal{CB}^\mathcal{P}(X)$, that

$$H_\mathcal{P}(A\cup B, C\cup D) = \max\{H_\mathcal{P}(A,C), H_\mathcal{P}(B,D)\}.$$

□

Proposition 4. Let $(X, \mathcal{P})$ be a partial symmetric space. For $A, B \in \mathcal{CB}^\mathcal{P}(X)$, we have

$$H_\mathcal{P}(A,B) = 0 \Rightarrow A = B.$$

Proof. Let $H_\mathcal{P}(A,B) = 0$. Then, by the definition of $H_\mathcal{P}$, we have

$$\delta_\mathcal{P}(A,B) = \delta_\mathcal{P}(B,A) = 0.$$

Thus, by condition (iii) of Proposition 2, we get $A \subseteq B$ and $B \subseteq A$, which implies $A = B$. □

Now, we prove the following lemma which is needed in the sequel:

Lemma 2. Let $(X, \mathcal{P})$ be partial symmetric space and $A, B \in \mathcal{CB}^\mathcal{P}(X)$. Then, for any $h > 1$ and $a \in A$, there exists $b \in B$ such that

$$\mathcal{P}(a,b) \leq h H_\mathcal{P}(A,B). \tag{13}$$

Proof. First, we consider $A = B$. From (i) of Proposition 2,

$$H_\mathcal{P}(A,B) = H_\mathcal{P}(A,A) = \delta_\mathcal{P}(A,A) = \sup_{a\in A} \mathcal{P}(a,a).$$

Observe that, for any $a \in A$ and any $h > 1$, we have

$$\mathcal{P}(a,a) \leq \sup_{a\in A} \mathcal{P}(a,a) = H_\mathcal{P}(A,B) \leq h H_\mathcal{P}(A,B).$$

Consequently, $b = a$ satisfies the inequality (13). Now, let $A \neq B$. Then, there exists $a \in A$ such that

$$\mathcal{P}(a,b) > h H_\mathcal{P}(A,B) \text{ for all } b \in B.$$

Then,

$$\inf\{\mathcal{P}(a,b) : b \in B\} \geq h H_\mathcal{P}(A,B),$$

so that

$$\text{dist}_\mathcal{P}(a,B) \geq h H_\mathcal{P}(A,B).$$

Hence,
$$H_{\mathcal{P}}(A,B) \geq \delta_{\mathcal{P}}(A,B) = \sup_{a\in A} dist_{\mathcal{P}}(a,B) \geq dist_{\mathcal{P}}(a,B) \geq hH_{\mathcal{P}}(A,B),$$

a contradiction, since $h > 1$. □

Recall that, if $f : X \to \mathcal{CB}^{\mathcal{P}}(X)$ is a mapping, then an element $x \in X$ is said to be a fixed point of f if $x \in fx$.

Now, we state and prove our main result in this section:

Theorem 5. *Let $(X, \mathcal{P})$ be a complete partial symmetric space and $f : X \to \mathcal{CB}^{\mathcal{P}}(X)$. Assume that the following conditions are satisfied:*

(i) *there exists $\kappa \in [0, 1)$ such that*
$$H_{\mathcal{P}}(fx, fy) \leq \kappa \mathcal{P}(x,y) \text{ for all } x,y \in X;$$

(ii) *there exists $x_0 \in X$ such that $\mathfrak{S}(\mathcal{P}, f, x_0) < \infty$; and*

(iii) *f is continuous.*

Then, f has a unique fixed point $x \in X$, such that $\mathcal{P}(x,x) = 0$.

Proof. Suppose $x_0 \in X$ and $x_1 \in fx_0$. From Lemma 2 with $h = \frac{1}{\sqrt{\kappa}}$, there exists $x_2 \in fx_1$ such that $\mathcal{P}(x_1, x_2) \leq \frac{1}{\sqrt{\kappa}} H_{\mathcal{P}}(fx_0, fx_1)$. Since $H_{\mathcal{P}}(fx_0, fx_1) \leq \kappa \mathcal{P}(x_0, x_1)$, then $\mathcal{P}(x_1, x_2) \leq \sqrt{\kappa}\mathcal{P}(x_0, x_1)$. Similarly, for $x_2 \in fx_1$ there exists $x_3 \in fx_2$ such that

$$\mathcal{P}(x_2, x_3) \leq \frac{1}{\sqrt{\kappa}} H_{\mathcal{P}}(fx_1, fx_2) \leq \sqrt{\kappa}\mathcal{P}(x_1, x_2).$$

Inductively, we obtain a sequence $\{x_n\}$ in X, such that

$$x_{n+1} \in fx_n \text{ and } \mathcal{P}(x_{n+1}, x_n) \leq \sqrt{\kappa}\mathcal{P}(x_n, x_{n-1}), \text{ for all } n \in \mathbb{N}.$$

By condition (i), for all $i, j \in \mathbb{N}$ we have

$$\mathcal{P}(f^{n+i}x_0, f^{n+j}x_0) \leq \sqrt{\kappa}\mathcal{P}(f^{n-1+i}x_0, f^{n-1+j}x_0).$$

Therefore, by condition (ii) and (1), we have

$$\mathfrak{S}(\mathcal{P}, f, f^n x_0) \leq \sqrt{\kappa}\mathfrak{S}(\mathcal{P}, f, f^{n-1}x_0).$$

Continuing this process, we have, for every $n \in \mathbb{N}$,

$$\mathfrak{S}(\mathcal{P}, f, f^n x_0) \leq (\sqrt{\kappa})^n \mathfrak{S}(\mathcal{P}, f, x_0). \tag{14}$$

By using (3), we have, for $n, m, p \in \mathbb{N}$ such that $m = n + p$,

$$\mathcal{P}(f^n x_0, f^{n+p}x_0) \leq \mathfrak{S}(\mathcal{P}, f, f^n x_0) \leq (\sqrt{\kappa})^n \mathfrak{S}(\mathcal{P}, f, x_0).$$

Since $\mathfrak{S}(\mathcal{P}, f, x_0) < \infty$ and $\kappa \in (0, 1)$, then

$$\lim_{n,m\to\infty} \mathcal{P}(x_n, x_m) = 0,$$

so that $\{x_n\}$ is a $\mathcal{P}$-Cauchy sequence in X. In view of the $\mathcal{P}$-completeness of X, there exists $x \in X$ such that $\{x_n\}$ $\mathcal{P}$-converges to x. Therefore,

$$\mathcal{P}(x,x) = \lim_{n \to \infty} \mathcal{P}(x_n, x) = \lim_{n,m \to \infty} \mathcal{P}(x_n, x_m) = 0.$$

As $H_\mathcal{P}(fx_n, fx) \leq \kappa \mathcal{P}(x_n, x)$, implies that

$$\lim_{n \to \infty} H_\mathcal{P}(fx_n, fx) = 0.$$

Hence, $x_{n+1} \in fx_n$. Therefore,

$$dist_\mathcal{P}(x_{n+1}, fx) \leq \delta_\mathcal{P}(fx_n, fx) \leq H_\mathcal{P}(fx_n, fx).$$

Hence,

$$\lim_{n \to \infty} dist_\mathcal{P}(x_{n+1}, fx) = 0.$$

By the continuity of f, we obtain

$$dist_\mathcal{P}(x, fx) = \lim_{n \to \infty} dist_\mathcal{P}(x_{n+1}, fx) = 0.$$

Thus, we have $\mathcal{P}(x,x) = dist_\mathcal{P}(x, fx) = 0$. As fx is $\mathcal{P}$-closed, then we have $x \in fx$. Hence, x is a fixed point of f in X. This completes the proof. □

Next, we adopt the following example to demonstrate Theorem 5.

Example 5. Consider $X = \{0, 1, 2\}$ equipped with the partial symmetric $\mathcal{P} : X \times X \to \mathbb{R}_+$ defined by

$$\mathcal{P}(x,y) = \frac{1}{2}|x-y|^2 + \frac{1}{4}(\max\{x,y\})^2, \text{ for all } x, y \in X.$$

Then $(X, \mathcal{P})$ is a $\mathcal{P}$-complete symmetric space. Note that $\{0\}$ and $\{0, 1\}$ are bounded sets in $(X, \mathcal{P})$. In fact, if $x \in \{0, 1, 2\}$ then

$$x \in \overline{\{0\}} \Leftrightarrow dist_\mathcal{P}(x, \{0\}) = \mathcal{P}(x,x)$$
$$\Leftrightarrow \frac{3x^2}{4} = \frac{x^2}{4}$$
$$\Leftrightarrow x = 0 \Leftrightarrow x \in \{0\}.$$

Hence, $\{0\}$ is closed with respect to the partial symmetric $\mathcal{P}$. Next,

$$x \in \overline{\{0,1\}} \Leftrightarrow dist_\mathcal{P}(x, \{0,1\}) = \mathcal{P}(x,x)$$
$$\Leftrightarrow \min\left\{\frac{3x^2}{2}, \frac{1}{2}|x-1|^2 + \frac{1}{4}(\max\{x,1\})^2\right\} = \frac{x^2}{4}$$
$$\Leftrightarrow x \in \{0,1\}.$$

Hence, $\{0, 1\}$ is also closed with respect to the partial symmetric $\mathcal{P}$.
Now, define $f : X \to C\mathcal{B}^\mathcal{P}(X)$ by:

$$f0 = f1 = \{0\} \text{ and } f2 = \{0, 1\}.$$

Notice that f is continuous under the partial symmetric $\mathcal{P}$. Now, to show that the contractive condition (i) of Theorem 5 is satisfied, we distinguish the following cases:
Case 1. Let $x, y \in \{0, 1\}$. Then,

$$H_\mathcal{P}(fx, fy) = H_\mathcal{P}(0,0) = 0,$$

so that the contractive condition (i) satisfied.
Case 2. Let $x \in \{0,1\}$ and $y = 2$. Then, with $k = \frac{1}{2}$, we have

$$\begin{aligned}
H_\mathcal{P}(fx, f2) &= H_\mathcal{P}(\{0\}, \{0,1\}) \\
&= \max\{\delta_\mathcal{P}(\{0\}, \{0,1\}), \delta_\mathcal{P}(\{0,1\}, \{0\})\} \\
&= \max\{dist_\mathcal{P}(0, \{0,1\}), \max\{\mathcal{P}(0,0), \mathcal{P}(1,0)\}\} \\
&= \frac{3}{4} \leq kH_\mathcal{P}(x, 2).
\end{aligned}$$

Case 3. Let $x = y = 2$. Then, with $k = \frac{1}{4}$, we have

$$\begin{aligned}
H_\mathcal{P}(f2, f2) &= H_\mathcal{P}(\{0,1\}, \{0,1\}) \\
&= \delta_\mathcal{P}(\{0,1\}, \{0,1\}) \\
&= \max\{dist_\mathcal{P}(0, \{0,1\}), dist_\mathcal{P}(1, \{0,1\})\} \\
&= \max\{0, \min\{\mathcal{P}(1,0), \mathcal{P}(1,1)\}\} \\
&= \frac{1}{4} \leq kH_\mathcal{P}(2, 2).
\end{aligned}$$

Hence, the contractive condition (i) of Theorem 5 is satisfied for $k = \frac{1}{2}$.

By routine calculation, one can verify the other conditions of Theorem 5. Observe that f has a unique fixed point (namely, $x = 0$).

6. Conclusions

First, we enlarged the class of symmetric spaces to the class of partial symmetric spaces, wherein we proved several results which included analogues of the Banach contraction principle, the Kannan-Ćirić fixed theorem, and the Ćirić quasi-fixed point theorem, in such spaces. We also furnished some examples, exhibiting the utility of our newly established results. Furthermore, we used one of the our main results to examine the existence and uniqueness of a solution for a system of Fredholm integral equations. Moreover, we extended the idea of Hausdorff distance to partial symmetric spaces, and proved an analogue of Nadler's fixed point theorem and some related results.

Author Contributions: All authors contributed equally in writing this article. All authors read and approved the final manuscript.

Funding: This research is not funded by any external agency.

Conflicts of Interest: The authors declare no conflict of interest.

References

1. Banach, S. Sur les operations dans les ensembles abstraits et leur application aux equations integrals. *Fund. Math.* **1922**, *3*, 133–181. [CrossRef]
2. Matthews, S.G. Partial metric topology. *Ann. N. Y. Acad. Sci.* **1994**, *728*, 183–197. [CrossRef]
3. Amini-Harandi, A. Metric-like spaces, partial metric spaces and fixed points. *Fixed Point Theory Appl.* **2012**, *2012*, 204. [CrossRef]
4. Czerwik, S. Contraction mappings in b-metric spaces. *Acta Math. Inf. Univ. Ostrav.* **1993**, *1*, 5–11.
5. Branciari, A. A fixed point theorem of Banach–Caccioppoli type on a class of generalized metric spaces. *Publ. Math.* **2000**, *57*, 31–37.
6. Huang, L.G.; Zhang, X. Cone metric spaces and fixed point theorems of contractive mappings. *J. Math. Anal. Appl.* **2007**, *332*, 1468–1476. [CrossRef]
7. Imdad, M.; Asim, M.; Gubran, R. Common fixed point theorems for g-Generalized contractive mappings in b-metric spaces. *Indian J. Math.* **2018**, *60*, 85–105.

8. Ciric, L.B. A generalization of Banach's contraction principle. *Proc. Am. Math. Soc.* **1974**, *45*, 267–273. [CrossRef]
9. Mustafa, Z.; Roshan, J.R.; Parvaneh, V.; Kadelburg, Z. Some common fixed point results in ordered partial b-metric spaces. *J. Inequal. Appl.* **2013**, *2013*, 562. [CrossRef]
10. Imdad, M.; Ali, J. Common fixed point theorems in symmetric spaces employing a new implicit function and common property (EA). *Bull. Belg. Math. Soc. Simon Stev.* **2009**, *16*, 421–433.
11. Jleli, M.; Samet, B. A generalized metric space and related fixed point theorems. *Fixed Point Theory Appl.* **2015**, *2015*, 61. [CrossRef]
12. Wilson, W.A. On semi-metric spaces. *Am. J. Math.* **1931**, *53*, 361–373. [CrossRef]
13. Nadler, S.B. Multivalued contraction mappings. *Pac. J. Math.* **1969**, *30*, 475–488. [CrossRef]
14. Villa-Morales, J. Subordinate Semimetric Spaces and Fixed Point Theorems. *J. Math.* **2018**, *2018*, 7856594. [CrossRef]

© 2019 by the authors. Licensee MDPI, Basel, Switzerland. This article is an open access article distributed under the terms and conditions of the Creative Commons Attribution (CC BY) license (http://creativecommons.org/licenses/by/4.0/).

Article

A New Identity for Generalized Hypergeometric Functions and Applications

Mohammad Masjed-Jamei [1] and Wolfram Koepf [2],*

[1] Department of Mathematics, K.N.Toosi University of Technology, P.O.Box 16315-1618, 11369 Tehran, Iran; mmjamei@kntu.ac.ir or mmjamei@yahoo.com
[2] Institute of Mathematics, University of Kassel, Heinrich-Plett Str. 40, 34132 Kassel, Germany
* Correspondence: koepf@mathematik.uni-kassel.de

Received: 19 November 2018; Accepted: 14 January 2019; Published: 18 January 2019

Abstract: We establish a new identity for generalized hypergeometric functions and apply it for first- and second-kind Gauss summation formulas to obtain some new summation formulas. The presented identity indeed extends some results of the recent published paper (*Some summation theorems for generalized hypergeometric functions*, Axioms, 7 (2018), Article 38).

Keywords: generalized hypergeometric functions; Gauss and confluent hypergeometric functions; summation theorems of hypergeometric functions

MSC: 33C20, 33C05, 65B10

---

1. Introduction

Let $\mathbb{R}$ and $\mathbb{C}$ denote the sets of real and complex numbers and z be a complex variable. For real or complex parameters a and b, the generalized binomial coefficient

$$\binom{a}{b} = \frac{\Gamma(a+1)}{\Gamma(b+1)\Gamma(a-b+1)} = \binom{a}{a-b} \quad (a, b \in \mathbb{C}),$$

in which

$$\Gamma(z) = \int_0^\infty x^{z-1} e^{-x} dx,$$

denotes the well-known gamma function for $\mathrm{Re}(z) > 0$, can be reduced to the particular case

$$\binom{a}{n} = \frac{(-1)^n (-a)_n}{n!},$$

where $(a)_b$ denotes the Pochhammer symbol [1] given by

$$(a)_b = \frac{\Gamma(a+b)}{\Gamma(a)} = \begin{cases} 1 & (b = 0,\ a \in \mathbb{C}\setminus\{0\}), \\ a(a+1)\ldots(a+b-1) & (b \in \mathbb{C},\ a \in \mathbb{C}). \end{cases} \quad (1)$$

By referring to the symbol (1), the generalized hypergeometric functions [2]

$$_pF_q\left(\begin{array}{c} a_1, \ldots, a_p \\ b_1, \ldots, b_q \end{array} \bigg| z\right) = \sum_{k=0}^\infty \frac{(a_1)_k \ldots (a_p)_k}{(b_1)_k \ldots (b_q)_k} \frac{z^k}{k!}, \quad (2)$$

are indeed a Taylor series expansion for a function, say f, as $\sum_{k=0}^{\infty} c_k^* z^k$ with $c_k^* = f^{(k)}(0)/k!$ for which the ratio of successive terms can be written as

$$\frac{c_{k+1}^*}{c_k^*} = \frac{(k+a_1)(k+a_2)...(k+a_p)}{(k+b_1)(k+b_2)...(k+b_q)(k+1)}.$$

According to the ratio test [3,4], the series (2) is convergent for any $p \leq q+1$. In fact, it converges in $|z| < 1$ for $p = q+1$, converges everywhere for $p < q+1$ and converges nowhere ($z \neq 0$) for $p > q+1$. Moreover, for $p = q+1$ it absolutely converges for $|z| = 1$ if the condition

$$A^* = \text{Re}\left(\sum_{j=1}^{q} b_j - \sum_{j=1}^{q+1} a_j\right) > 0,$$

holds and is conditionally convergent for $|z| = 1$ and $z \neq 1$ if $-1 < A^* \leq 0$ and is finally divergent for $|z| = 1$ and $z \neq 1$ if $A^* \leq -1$.

There are two important cases of the series (2) arising in many physics problems [5,6]. The first case (convergent in $|z| \leq 1$) is the Gauss hypergeometric function

$$y = {}_2F_1\left(\begin{array}{c} a, b \\ c \end{array} \bigg| z\right) = \sum_{k=0}^{\infty} \frac{(a)_k (b)_k}{(c)_k} \frac{z^k}{k!},$$

with the integral representation

$${}_2F_1\left(\begin{array}{c} a, b \\ c \end{array} \bigg| z\right) = \frac{\Gamma(c)}{\Gamma(b)\Gamma(c-b)} \int_0^1 t^{b-1}(1-t)^{c-b-1}(1-tz)^{-a} dt,$$

$$(\text{Re } c > \text{Re } b > 0;\ |\arg(1-z)| < \pi), \quad (3)$$

Replacing $z = 1$ in (3) directly leads to the well-known Gauss identity

$${}_2F_1\left(\begin{array}{c} a, b \\ c \end{array} \bigg| 1\right) = \frac{\Gamma(c)\Gamma(c-a-b)}{\Gamma(c-a)\Gamma(c-b)} \qquad \text{Re}(c-a-b) > 0. \qquad (4)$$

The second case, which converges everywhere, is the Kummer confluent hypergeometric function

$$y = {}_1F_1\left(\begin{array}{c} b \\ c \end{array} \bigg| z\right) = \sum_{k=0}^{\infty} \frac{(b)_k}{(c)_k} \frac{z^k}{k!},$$

with the integral representation

$${}_1F_1\left(\begin{array}{c} b \\ c \end{array} \bigg| z\right) = \frac{\Gamma(c)}{\Gamma(b)\Gamma(c-b)} \int_0^1 t^{b-1}(1-t)^{c-b-1} e^{zt} dt,$$

$$(\text{Re } c > \text{Re } b > 0;\ |\arg(1-z)| < \pi).$$

In this paper, we explicitly obtain the simplified form of the hypergeometric series

$${}_pF_q\left(\begin{array}{c} a_1, ..., a_{p-1}, m+1 \\ b_1, ..., b_{q-1}, n+1 \end{array} \bigg| z\right),$$

when m, n are two natural numbers and $m < n$.

2. A New Identity for Generalized Hypergeometric Functions

Let m, n be two natural numbers so that $m < n$. By noting (1), since

$$\frac{(m+1)_k}{(n+1)_k} = \frac{\Gamma(k+m+1)\Gamma(n+1)}{\Gamma(k+n+1)\Gamma(m+1)} = \frac{n!}{m!} \frac{1}{(k+m+1)(k+m+2)...(k+n)},$$

so, we have

$$\frac{(m+1)_k}{k!(n+1)_k} = \frac{\Gamma(k+m+1)\Gamma(n+1)}{k!\Gamma(k+n+1)\Gamma(m+1)} = \frac{n!}{m!} \frac{(k+1)_m}{(k+n)!}. \tag{5}$$

Hence, substituting (5) into a special case of (2) yields

$$_pF_q\left(\begin{array}{c} a_1, \ldots, a_{p-1}, m+1 \\ b_1, \ldots, b_{q-1}, n+1 \end{array} \middle| z\right) = \frac{n!}{m!} \sum_{k=0}^{\infty} \frac{(a_1)_k...(a_{p-1})_k}{(b_1)_k...(b_{q-1})_k} z^k \frac{(k+1)_m}{(k+n)!}$$

$$= \frac{n!}{m!} \sum_{j=n}^{\infty} \frac{(a_1)_{j-n}...(a_{p-1})_{j-n}}{(b_1)_{j-n}...(b_{q-1})_{j-n}} z^{j-n} \frac{(j+1-n)_m}{j!}. \tag{6}$$

In [7], two particular cases of (6) for $m = 0$ and $m = 1$ were considered and other cases have been left as open problems. In this section, we wish to consider those open problems and solve them for any arbitrary value of m. For this purpose, since

$$(a)_{j-n} = \frac{\Gamma(a-n)}{\Gamma(a)}(a-n)_j = (-1)^n \frac{(a-n)_j}{(1-a)_n},$$

relation (6) is simplified as

$$_pF_q\left(\begin{array}{c} a_1, \ldots, a_{p-1}, m+1 \\ b_1, \ldots, b_{q-1}, n+1 \end{array} \middle| z\right) = \frac{n!}{m!} \frac{(-1)^{n(p-q)}}{z^n} \frac{(1-b_1)_n...(1-b_{q-1})_n}{(1-a_1)_n...(1-a_{p-1})_n}$$

$$\times \sum_{j=n}^{\infty} \frac{(a_1-n)_j...(a_{p-1}-n)_j}{(b_1-n)_j...(b_{q-1}-n)_j} \frac{z^j}{j!} (j+1-n)_m. \tag{7}$$

It is clear in (7) that

$$\sum_{j=n}^{\infty} \frac{(a_1-n)_j...(a_{p-1}-n)_j}{(b_1-n)_j...(b_{q-1}-n)_j} \frac{z^j}{j!} (j+1-n)_m = \sum_{j=0}^{\infty} (.) - \sum_{j=0}^{n-1}(.) = S_1^* - S_2^*. \tag{8}$$

To evaluate $S_1^* = \sum_{j=0}^{\infty}(.)$, we can directly use Chu-Vandermonde identity, which is a special case of Gauss identity (4), i.e.,

$$_2F_1\left(\begin{array}{c} -m, q-p \\ q \end{array} \middle| 1\right) = \frac{(p)_m}{(q)_m}. \tag{9}$$

Now if in (9), $p = j - n + 1$ and $q = -n + 1$, we have

$$(j-n+1)_m = (1-n)_m \, _2F_1\left(\begin{array}{c} -m, -j \\ 1-n \end{array} \middle| 1\right) = (1-n)_m \sum_{k=0}^{m} \frac{(-m)_k(-j)_k}{(1-n)_k k!}. \tag{10}$$

Hence, replacing (10) in S_1^* gives

$$S_1^* = \sum_{j=0}^{\infty} \frac{(a_1-n)_j\cdots(a_{p-1}-n)_j}{(b_1-n)_j\cdots(b_{q-1}-n)_j} \frac{z^j}{j!} (1-n)_m \sum_{k=0}^{m} \frac{(-m)_k(-j)_k}{(1-n)_k k!}$$

$$= (1-n)_m \sum_{k=0}^{m} \frac{(-m)_k}{(1-n)_k k!} \left(\sum_{j=k}^{\infty} \frac{(a_1-n)_j\cdots(a_{p-1}-n)_j}{(b_1-n)_j\cdots(b_{q-1}-n)_j} z^j \frac{(-j)_k}{j!} \right). \qquad (11)$$

It is important to note in the second equality of (11) that $(-j)_k = 0$ for any $j = 0, 1, 2, \ldots, k-1$. Therefore, the lower index is starting from $j = k$ instead of $j = 0$. Now since

$$\frac{(-j)_k}{j!} = \frac{(-1)^k}{(j-k)!},$$

relation (11) is simplified as

$$S_1^* = (1-n)_m \sum_{k=0}^{m} \frac{(-m)_k}{(1-n)_k k!} \left(\sum_{j=k}^{\infty} \frac{(a_1-n)_j\cdots(a_{p-1}-n)_j}{(b_1-n)_j\cdots(b_{q-1}-n)_j} z^j \frac{(-1)^k}{(j-k)!} \right)$$

$$= (1-n)_m \sum_{k=0}^{m} \frac{(-m)_k(-z)^k}{(1-n)_k k!} \left(\sum_{r=0}^{\infty} \frac{(a_1-n)_{r+k}\cdots(a_{p-1}-n)_{r+k}}{(b_1-n)_{r+k}\cdots(b_{q-1}-n)_{r+k}} \frac{z^r}{r!} \right). \qquad (12)$$

On the other hand, the well-known identity

$$(a)_{r+k} = (a)_k (a+k)_r,$$

simplifies (12) as

$$S_1^* = (1-n)_m \sum_{k=0}^{m} \frac{(-m)_k(a_1-n)_k\cdots(a_{p-1}-n)_k}{(1-n)_k(b_1-n)_k\cdots(b_{q-1}-n)_k} \frac{(-z)^k}{k!}$$

$$\times \left(\sum_{r=0}^{\infty} \frac{(a_1-n+k)_r\cdots(a_{p-1}-n+k)_r}{(b_1-n+k)_r\cdots(b_{q-1}-n+k)_r} \frac{z^r}{r!} \right)$$

$$= (1-n)_m \sum_{k=0}^{m} \frac{(-m)_k(a_1-n)_k\cdots(a_{p-1}-n)_k}{(1-n)_k(b_1-n)_k\cdots(b_{q-1}-n)_k} \frac{(-z)^k}{k!}$$

$$\times {}_{p-1}F_{q-1}\left(\begin{array}{ccc} a_1-n+k, & \ldots & a_{p-1}-n+k \\ b_1-n+k, & \ldots & b_{q-1}-n+k \end{array} \Big| z \right).$$

To compute the finite sum $S_2^* = \sum_{j=0}^{n-1}(.)$ in (8), we can directly use the identity

$$(j-n+1)_m = \frac{(-n+1)_m(-n+1+m)_j}{(-n+1)_j},$$

to get

$$S_2^* = \sum_{j=0}^{n-1} \frac{(a_1-n)_j\cdots(a_{p-1}-n)_j}{(b_1-n)_j\cdots(b_{q-1}-n)_j} \frac{z^j}{j!}(j+1-n)_m$$

$$= (1-n)_m \sum_{j=0}^{n-1} \frac{(a_1-n)_j\cdots(a_{p-1}-n)_j}{(b_1-n)_j\cdots(b_{q-1}-n)_j} \frac{z^j}{j!} \frac{(-n+1+m)_j}{(-n+1)_j}$$

$$= (1-n)_m \, {}_pF_q\left(\begin{array}{cccc} a_1-n, & \ldots & a_{p-1}-n, & -(n-1-m) \\ b_1-n, & \ldots & b_{q-1}-n, & -(n-1) \end{array} \bigg| z\right). \qquad (13)$$

Finally, by noting the identity

$$\frac{(-n+1)_m}{m!} = (-1)^m \binom{n-1}{m},$$

the main result of this paper is obtained as follows.

Main Theorem. *If m, n are two natural numbers so that $m < n$, then*

$$
{}_pF_q\left(\begin{array}{cccc} a_1, & \ldots a_{p-1}, & m+1 \\ b_1, & \ldots b_{q-1}, & n+1 \end{array} \bigg| z\right) = n! \binom{n-1}{m} \frac{(-1)^{n(p-q)+m}}{z^n} \frac{(1-b_1)_n\cdots(1-b_{q-1})_n}{(1-a_1)_n\cdots(1-a_{p-1})_n}
$$

$$
\times \left\{\sum_{k=0}^{m} \frac{(-m)_k(a_1-n)_k\cdots(a_{p-1}-n)_k}{(1-n)_k(b_1-n)_k\cdots(b_{q-1}-n)_k} {}_{p-1}F_{q-1}\left(\begin{array}{ccc} a_1-n+k, & \ldots & a_{p-1}-n+k \\ b_1-n+k, & \ldots & b_{q-1}-n+k \end{array} \bigg| z\right)\frac{(-z)^k}{k!} \right.
$$

$$
\left. - {}_pF_q\left(\begin{array}{cccc} a_1-n, & \ldots a_{p-1}-n, & -(n-1-m) \\ b_1-n, & \ldots b_{q-1}-n, & -(n-1) \end{array} \bigg| z\right)\right\}, \qquad (14)
$$

where $\{a_k\}_{k=1}^{p-1} \notin \{1,2,\ldots,n\}$ and $\{b_k\}_{k=1}^{q-1} \notin \{n, n-1,\ldots, n-m+1\}$.

Note that the case $m > n$ in (14) leads to a particular case of Karlsson-Minton identity, see e.g., [8,9].

3. Some Special Cases of the Main Theorem

Essentially whenever a generalized hypergeometric series can be summed in terms of gamma functions, the result will be important as only a few such summation theorems are available in the literature. In this sense, the classical summation theorems such as Kummer and Gauss for ${}_2F_1$, Dixon, Watson, Whipple and Pfaff-Saalschutz for ${}_3F_2$, Whipple for ${}_4F_3$, Dougall for ${}_5F_4$ and Dougall for ${}_7F_6$ are well known [1,10]. In this section, we consider some special cases of the above main theorem to obtain new hypergeometric summation formulas.

Special case 1. Note that if $m = 0$, the first equality of (13) reads as

$$S_2^* = \sum_{j=0}^{n-1} \frac{(a_1-n)_j\cdots(a_{p-1}-n)_j}{(b_1-n)_j\cdots(b_{q-1}-n)_j} \frac{z^j}{j!}.$$

Hence, the main theorem is simplified as

$$_pF_q\left(\begin{array}{c} a_1, \ldots a_{p-1}, 1 \\ b_1, \ldots b_{q-1}, n+1 \end{array}\bigg| z\right) = n! \frac{(-1)^{n(p-q)}}{z^n} \frac{(1-b_1)_n \cdots (1-b_{q-1})_n}{(1-a_1)_n \cdots (1-a_{p-1})_n}$$

$$\times \left(_{p-1}F_{q-1}\left(\begin{array}{c} a_1-n, \ldots, a_{p-1}-n \\ b_1-n, \ldots, b_{q-1}-n \end{array}\bigg| z\right) - \sum_{j=0}^{n-1} \frac{(a_1-n)_j \cdots (a_{p-1}-n)_j}{(b_1-n)_j \cdots (b_{q-1}-n)_j} \frac{z^j}{j!}\right),$$

which is a known result in the literature [10] (p. 439).

Special case 2. For $n = m+1$, relation (13) gives $S_2^* = (-1)^m m!$ and the main theorem therefore reads (for $m+1 \to m$) as

$$_pF_q\left(\begin{array}{c} a_1, \ldots a_{p-1}, m \\ b_1, \ldots b_{q-1}, m+1 \end{array}\bigg| z\right) = (-1)^{m(p-q+1)} \frac{m!}{z^m} \frac{(1-b_1)_m \cdots (1-b_{q-1})_m}{(1-a_1)_m \cdots (1-a_{p-1})_m} \times$$

$$\left\{1 - \sum_{k=0}^{m-1} \frac{(a_1-m)_k \cdots (a_{p-1}-m)_k}{(b_1-m)_k \cdots (b_{q-1}-m)_k} {}_{p-1}F_{q-1}\left(\begin{array}{c} a_1-m+k, \ldots, a_{p-1}-m+k \\ b_1-m+k, \ldots, b_{q-1}-m+k \end{array}\bigg| z\right) \frac{(-z)^k}{k!}\right\}.$$

For instance, we have [7]

$$_pF_q\left(\begin{array}{c} a_1, \ldots a_{p-1}, 2 \\ b_1, \ldots b_{q-1}, 3 \end{array}\bigg| z\right) = \frac{2}{z^2} \frac{(1-b_1)_2 \cdots (1-b_{q-1})_2}{(1-a_1)_2 \cdots (1-a_{p-1})_2}$$

$$\times \left(\frac{(a_1-2) \cdots (a_{p-1}-2)}{(b_1-2) \cdots (b_{q-1}-2)} z \, {}_{p-1}F_{q-1}\left(\begin{array}{c} a_1-1, \ldots, a_{p-1}-1 \\ b_1-1, \ldots, b_{q-1}-1 \end{array}\bigg| z\right)\right.$$

$$\left. - {}_{p-1}F_{q-1}\left(\begin{array}{c} a_1-2, \ldots, a_{p-1}-2 \\ b_1-2, \ldots, b_{q-1}-2 \end{array}\bigg| z\right) + 1\right).$$

As a very particular case, replacing $p=3$ and $q=2$ in the above relation yields

$$_3F_2\left(\begin{array}{c} a, b, 2 \\ c, 3 \end{array}\bigg| 1\right)$$

$$= \frac{2}{(a-2)_2(b-2)_2}\left((c-2)_2 + \frac{\Gamma(c)\Gamma(c-a-b+1)}{\Gamma(c-a)\Gamma(c-b)}(ab-a-b-c+3)\right).$$

Special case 3. For $p = q = 1$, the main theorem is simplified as

$$_1F_1\left(\begin{array}{c} m+1 \\ n+1 \end{array}\bigg| z\right)$$

$$= n!\binom{n-1}{m}\frac{(-1)^m}{z^n}\left(e^z {}_1F_1\left(\begin{array}{c} -m \\ -(n-1) \end{array}\bigg| -z\right) - {}_1F_1\left(\begin{array}{c} -(n-1-m) \\ -(n-1) \end{array}\bigg| z\right)\right).$$

For instance, by referring to the special case 1, we have [7,10]

$$_1F_1\left(\begin{array}{c} 1 \\ m \end{array}\bigg| z\right) = \frac{(m-1)!}{z^{m-1}}\left(e^z - \sum_{j=0}^{m-2} \frac{z^j}{j!}\right).$$

Special case 4. For $p = 2$ and $q = 1$, the main theorem is simplified as

$$_2F_1\left(\begin{array}{c}a,\ m+1\\n+1\end{array}\bigg|\ z\right) = n!\binom{n-1}{m}\frac{(-1)^{n+m}}{z^n}\frac{1}{(1-a)_n}$$

$$\times\left\{(1-z)^{n-a}{}_2F_1\left(\begin{array}{c}a-n,\ -m\\-(n-1)\end{array}\bigg|\ \frac{z}{z-1}\right) - {}_2F_1\left(\begin{array}{c}a-n,\ -(n-1-m)\\-(n-1)\end{array}\bigg|\ z\right)\right\},$$

in which we have used the relation $_1F_0\left(\begin{array}{c}a\\-\end{array}\bigg|\ z\right) = (1-z)^{-a}$. For instance, by referring to the special case 1, we have [7,10]

$$_2F_1\left(\begin{array}{c}a,\ 1\\m\end{array}\bigg|\ z\right) = \frac{(m-1)!}{z^{m-1}}\frac{\Gamma(1-a)}{\Gamma(m-a)}\left((1-z)^{m-a-1} - \sum_{j=0}^{m-2}(a-m+1)_j\frac{z^j}{j!}\right).$$

Special case 5. For $p = 3$ and $q = 2$, the main theorem is simplified as

$$_3F_2\left(\begin{array}{c}a_1,\ a_2,\ m+1\\b_1,\ n+1\end{array}\bigg|\ z\right) = n!\binom{n-1}{m}\frac{(-1)^{n+m}}{z^n}\frac{(1-b_1)_n}{(1-a_1)_n(1-a_2)_n}$$

$$\times\left\{\sum_{k=0}^{m}\frac{(-m)_k(a_1-n)_k(a_2-n)_k}{(1-n)_k(b_1-n)_k}{}_2F_1\left(\begin{array}{c}a_1-n+k,\ a_2-n+k\\b_1-n+k\end{array}\bigg|\ z\right)\frac{(-z)^k}{k!}\right.$$

$$\left. -{}_3F_2\left(\begin{array}{c}a_1-n,\ a_2-n,\ -(n-1-m)\\b_1-n,\ -(n-1)\end{array}\bigg|\ z\right)\right\}.\quad(15)$$

As a particular case and by noting the first kind of Gauss formula (4), if $z = 1$ is replaced in (15) then we get

$$_3F_2\left(\begin{array}{c}a_1,\ a_2,\ m+1\\b_1,\ n+1\end{array}\bigg|\ 1\right) = (-1)^{n+m}n!\binom{n-1}{m}\frac{(1-b_1)_n}{(1-a_1)_n(1-a_2)_n}$$

$$\times\left\{\sum_{k=0}^{m}\frac{(-m)_k(a_1-n)_k(a_2-n)_k}{(1-n)_k(b_1-n)_k}\frac{\Gamma(b_1-n+k)\Gamma(b_1-a_1-a_2+n-k)}{\Gamma(b_1-a_1)\Gamma(b_1-a_2)}\frac{(-1)^k}{k!}\right.$$

$$\left. -{}_3F_2\left(\begin{array}{c}a_1-n,\ a_2-n,\ -(n-1-m)\\b_1-n,\ -(n-1)\end{array}\bigg|\ 1\right)\right\}.$$

Therefore, we get

$$_3F_2\left(\begin{array}{c}a_1,\ a_2,\ m+1\\b_1,\ n+1\end{array}\bigg|\ 1\right) = \binom{n-1}{m}\frac{(-1)^m n!}{(1-a_1)_n(1-a_2)_n}$$

$$\times\left\{(b_1-a_1-a_2)_n{}_2F_1\left(\begin{array}{c}a_1,\ a_2\\b_1\end{array}\bigg|\ 1\right){}_3F_2\left(\begin{array}{c}a_1-n,\ a_2-n,\ -m\\1-n+a_1+a_2-b_1,\ 1-n\end{array}\bigg|\ 1\right)\right.$$

$$\left. -(-1)^n(1-b_1)_n{}_3F_2\left(\begin{array}{c}a_1-n,\ a_2-n,\ -(n-1-m)\\b_1-n,\ 1-n\end{array}\bigg|\ 1\right)\right\}.\quad(16)$$

As a numerical example for the result (16), we have

$$_3F_2\left(\begin{array}{c} 1/5,\ 3/10,\ 2 \\ 4/5,\ 5 \end{array}\bigg|\ 1\right) = \frac{72}{(4/5)_4(7/10)_4} \times$$

$$\left((1/5)_4 \sum_{k=0}^{2} \frac{(-2)_k(-19/5)_k(-37/10)_k}{(-3)_k(-16/5)_k k!}\right.$$

$$\left. - (3/10)_4 \frac{\Gamma(4/5)\Gamma(3/10)}{\Gamma(3/5)\Gamma(1/2)} \sum_{k=0}^{1} \frac{(-1)_k(-19/5)_k(-37/10)_k}{(-3)_k(-33/10)_k k!}\right).$$

It is clear that the right-hand side of this equality can be easily computed and therefore the infinite series in the left-hand side has been evaluated.

Similarly, by noting the second kind of Gauss formula [1]

$$_2F_1\left(\begin{array}{c} a,\ b \\ (a+b+1)/2 \end{array}\bigg|\ \frac{1}{2}\right) = \frac{\sqrt{\pi}\,\Gamma((a+b+1)/2)}{\Gamma((a+1)/2)\Gamma((b+1)/2)},$$

relation (15) takes the form

$$_3F_2\left(\begin{array}{c} a_1,\ a_2,\ m+1 \\ b_1,\ n+1 \end{array}\bigg|\ \frac{1}{2}\right) = (-1)^{n+m} 2^n n! \binom{n-1}{m} \frac{(1-b_1)_n}{(1-a_1)_n(1-a_2)_n}$$

$$\times \left\{ \sqrt{\pi} \sum_{k=0}^{m} \frac{(-m)_k(a_1-n)_k(a_2-n)_k}{(1-n)_k(b_1-n)_k} \frac{\Gamma(-n+k+b_1)}{\Gamma((a_1-n+k+1)/2)\Gamma((a_2-n+k+1)/2)} \frac{(-1)^k}{2^k k!} \right.$$

$$\left. -\,_3F_2\left(\begin{array}{c} a_1-n,\ a_2-n,\ -(n-1-m) \\ b_1-n,\ -(n-1) \end{array}\bigg|\ \frac{1}{2}\right) \right\},$$

where $b_1 = (a_1 + a_2 + 1)/2$.

Author Contributions: Both authors have contributed the same amount in all sections.

Funding: The work of the first author has been supported by the Alexander von Humboldt Foundation under the grant number: Ref 3.4-IRN-1128637-GF-E.

Conflicts of Interest: The authors declare no conflict of interest.

References

1. Koepf, W. *Hypergeometric Summation: An Algorithmic Approach to Summation and Special Function Identities*, 2nd ed.; Springer Universitext; Springer: London, UK, 2014.
2. Slater, L.J. *Generalized Hypergeometric Functions*; Cambridge University Press: Cambridge, UK, 1966.
3. Andrews, G.E.; Askey, R.; Roy, R. *Special Functions, Encyclopedia of Mathematics and Its Applications*; Cambridge University Press: Cambridge, UK, 1999; Voulme 71.
4. Arfken, G. *MathematicaL Methods for Physicists*; Academic Press Inc.: New York, NY, USA, 1985.
5. Mathai, A.M.; Saxena, R.K. Generalized hypergeometric functions with applications in statistics and physical sciences. In *Lecture Notes in Mathematics*; Springer: Berlin/Heidelberg, Germany; New York, NY, USA, 1973; Volume 348.
6. Nikiforov, A.F.; Uvarov, V.B. *Special Functions of Mathematical Physics*; Birkhäuser: Basel, Switzerland, 1988.
7. Masjed-Jamei, M.; Koepf, W. Some summation theorems for generalized hypergeometric functions. *Axioms* **2018**, *7*, 38. [CrossRef]
8. Karlsson, P.W. Hypergeometric functions with integral parameter differences. *J. Math. Phys.* **1971**, *12*, 270–271. [CrossRef]

9. Minton, B. Generalized hypergeometric function of unit argument. *J. Math. Phys.* **1970**, *11*, 1375–1376. [CrossRef]
10. Prudnikov, A.P.; Brychkov, Y.A.; Marichev, O.I. *Integrals and Series. Vol. 3. More Special Functions*; Gordon and Breach Science Publishers: Amsterdam, The Netherlands, 1990.

 © 2019 by the authors. Licensee MDPI, Basel, Switzerland. This article is an open access article distributed under the terms and conditions of the Creative Commons Attribution (CC BY) license (http://creativecommons.org/licenses/by/4.0/).

Article
Extended Partial S_b-Metric Spaces

Aiman Mukheimer

Department of Mathematics and General Sciences, Prince Sultan University, 11586 Riyadh, Saudi Arabia; mukheimer@psu.edu.sa

Received: 18 October 2018; Accepted: 18 November 2018; Published: 21 November 2018

Abstract: In this paper, we introduce the concept of extended partial S_b-metric spaces, which is a generalization of the extended S_b-metric spaces. Basically, in the triangle inequality, we add a control function with some very interesting properties. These new metric spaces generalize many results in the literature. Moreover, we prove some fixed point theorems under some different contractions, and some examples are given to illustrate our results.

Keywords: extended partial S_b-metric spaces; S_b-metric spaces; fixed point

1. Introduction

Fixed point theory has become the focus of many researchers lately due its applications in many fields see [1–12]). The concept of b-metric space was introduced by Bakhtin [13], which is a generalization of metric spaces.

Definition 1 ([13]). *A b-metric on a non empty set X is a function $d : X^2 \to [0, \infty)$ such that, for all $x, y, z \in X$ and $k \geq 1$, the following three conditions are satisfied:*
(i) $d(x, y) = 0$ if and only if $x = y$,
(ii) $d(x, y) = d(y, x)$,
(iii) $d(x, y) \leq k[d(x, z) + d(z, y)]$.
As usual, the pair (X, d) is called a b-metric spaces.

A three-dimensional metric space was introduced by Sedghi et al. [14], and it is called S-metric spaces. Later, and as a generalization of the S-metric spaces, S_b-metric spaces were introduced. In [15], extended S_b-metric spaces were introduced

Definition 2 ([15]). *Let X be a non empty set and a function $\theta : X^3 \to [1, \infty)$. If the function $S_\theta : X^3 \to [0, \infty)$ satisfies the following conditions for all $x, y, z, t \in X$:*

1. $S_\theta(x, y, z) = 0$ *implies* $x = y = z$;
2. $S_\theta(x, y, z) \leq \theta(x, y, z)[S_\theta(x, x, t) + S_\theta(y, y, t) + S_\theta(z, z, t)]$,

then the pair (X, S_θ) is called an extended S_b-metric spaces.

First, note that, if $\theta(x, y, z) = s \geq 1$, then we have an S_b-metric spaces, which leads us to conclude that every S_b-metric spaces is an extended S_b-metric spaces, but the converse is not always true.

Definition 3 ([15]). *Let (X, S_θ) be an extended S_b-metric space. Then,*

(i) *A sequence $\{x_n\}$ is called convergent if and only if there exists $z \in X$ such that $S_\theta(x_n, x_n, z)$ goes to 0 as n goes toward ∞. In this case, we write $\lim_{n \to \infty} x_n = z$.*
(ii) *A sequence $\{x_n\}$ is called a Cauchy sequence if and only if $S_\theta(x_n, x_n, x_m)$ goes to 0 as n, m goes toward ∞.*

(iii) (X, S_θ) is said to be a complete extended S_b-metric space if every Cauchy sequence $\{x_n\}$ converges to a point $x \in X$

(iv) Define the diameter of a subset Y of X by

$$diam(Y) := Sup\{S_\theta(x,y,z) \mid x,y,z \in Y\}.$$

In the extended S_b-metric spaces, we define a ball as follows:

$$B(x,\epsilon) = \{y \in X \mid S_\theta(x,x,y) \leq \epsilon\}.$$

Next, we present some examples of extended S_b-metric spaces.

Example 1. *Let $X = C([a,b], (-\infty, \infty))$ be the set of all continuous real valued functions on $[a,b]$. Define*

$$S_\theta : X^3 \to [0, \infty); \; S_\theta(x(t), y(t), z(t)) = \sup_{t \in [a,b]} |\max\{x(t), y(t)\} - z(t)|^2,$$

and

$$\theta : X^3 \to [1, \infty); \; \theta(x(t), y(t), z(t)) = \max\{|x(t)|, |y(t)|\} + |z(t)| + 1.$$

It is not difficult to see that (X, S_θ) is a complete extended S_b-metric spaces.

2. Extended Partial S_b-metric spaces

In this section, as a generalization of the extended S_b-metric spaces, we introduce extended partial S_b-metric spaces, along with its topology.

Definition 4. *Let X be a non empty set and a function $\theta : X^3 \to [1, \infty)$. If the function $S_\theta : X^3 \to [0, \infty)$ satisfies the following conditions for all $x, y, z, t \in X$:*

1. $x = y = z$ if and only if $S_\theta(x,y,z) = S_\theta(x,x,x) = S_\theta(y,y,y) = S_\theta(z,z,z)$;
2. $S_\theta(x,x,x) \leq S_\theta(x,y,z)$,
3. $S_\theta(x,y,z) \leq \theta(x,y,z)[S_\theta(x,x,t) + S_\theta(y,y,t) + S_\theta(z,z,t)]$,

then the pair (X, S_θ) is called an extended partial S_b-metric spaces.

First, note that, if $\theta(x,y,z) = s \geq 1$, then we have a partial S_b-metric spaces, which leads us to conclude that every S_b-metric spaces is an extended S_b-metric spaces, but the converse is not always true.

Definition 5. *Let (X, S_θ) be a extended partial S_b-metric space. Then,*

- *A sequence $\{x_n\}$ is called convergent if and only if there exists $z \in X$ such that $S_\theta(x_n, x_n, z)$ goes to $S_\theta(z,z,z)$ as n goes toward ∞. In this case, we write $\lim_{n \to \infty} x_n = z$.*
- *A sequence $\{x_n\}_{n=0}^\infty$ of elements in X is called S_θ-Cauchy if $\lim_{n,m} S_\theta(x_n, x_n, x_m)$ exists and is finite.*
- *The extended partial S_θ-metric spaces (X, S_θ) is called complete if, for each S_θ-Cauchy sequence $\{x_n\}_{n=0}^\infty$, there exists $z \in X$ such that*

$$S_\theta(z,z,z) = \lim_n S_\theta(z,z,x_n) = \lim_{n,m} S_\theta(x_n, x_n, x_m).$$

- *A sequence $\{x_n\}_n$ in an extended partial S_b-metric spaces (X, S_θ) is called 0-Cauchy if*

$$\lim_{n,m} S_\theta(x_n, x_n, x_m) = 0.$$

- *We say that (X, S_θ) is 0-complete if every 0-Cauchy in X converges to a point $x \in X$ such that $S_\theta(x,x,x) = 0$.*

Note that every extended S_b-metric spaces is an extended partial S_b-metric spaces, but the converse is not always true. The following example is an example of an extended partial S_b-metric spaces which is not extended S_b-metric spaces.

Example 2. Let $X = C([a,b], (-\infty, \infty))$ be the set of all continuous real valued functions on $[a,b]$. Define

$$S_\theta : X^3 \to [0, \infty); \quad S_\theta(x(t), y(t), z(t)) = \sup_{t \in [a,b]} |\max\{x(t), y(t), z(t)\}|^2,$$

and

$$\theta : X^3 \to [1, \infty); \quad \theta(x(t), y(t), z(t)) = \max\{|x(t)|, |y(t)|\} + |z(t)| + 1.$$

First, note that, for all $x(t), y(t), z(t)$ and $f(t) \in X$, we have

$$S_\theta(x(t), y(t), z(t)) = \sup_{t \in [a,b]} |\max\{x(t), y(t), z(t)\}|^2$$
$$\leq \sup_{t \in [a,b]} |\max\{x(t), x(t), f(t)\}|^2 + \sup_{t \in [a,b]} |\max\{y(t), y(t), f(t)\}|^2$$
$$+ \sup_{t \in [a,b]} |\max\{z(t), z(t), f(t)\}|^2$$
$$\leq S_\theta(x(t), x(t), f(t)) + S_\theta(y(t), y(t), f(t)) + S_\theta(z(t), z(t), f(t))$$
$$\leq \theta(x(t), y(t), z(t))[S_\theta(x(t), x(t), f(t)) + S_\theta(y(t), y(t), f(t)) + S_\theta(z(t), z(t), f(t))].$$

In the last inequality, we used the fact that, for all $x(t), y(t), z(t) \in X$, we have $\theta(x(t), y(t), z(t)) \geq 1$. Hence, (X, S_θ) is an extended partial S_b-metric spaces, but it is not an extended S_b-metric spaces, since the self distance is not zero.

In the extended partial S_b-metric spaces, we define a ball as follows:

$$B(x, \epsilon) = \{y \in X \mid S_\theta(x, x, y) \leq \epsilon\}.$$

Definition 6. An extended partial S_b-metric spaces (X, S_θ) is said to be symmetric if it satisfies the following condition:

$$S_\theta(x, x, y) = S_\theta(y, y, x) \text{ for all } x, y \in X.$$

Theorem 1. Let (X, S_θ) be a complete symmetric extended partial S_b-metric spaces such that S_θ is continuous, and let T be a continuous self mapping on X satisfying the following condition:

$$S_\theta(Tx, Ty, Tz) \leq k S_\theta(x, y, z) \text{ for all } x, y, z \in X,$$

where $0 < k < 1$ and for every $x_0 \in X$ we have $\lim_{n,m \to \infty} \theta(T^n x, T^n x, T^m) < \frac{1}{k}$. Then, T has a unique fixed point say u. In addition, for every $y \in X$, we have $\lim_{n \to \infty} T^n y = u$.

Proof. Since X is a nonempty set, pick $x_0 \in X$ and define the sequence $\{x_n\}$ as follows:

$$x_1 = Tx_0, x_2 = Tx_1 = T^2 x_0, \cdots, x_n = T^n x_0, \cdots$$

Note that, by (1), we have

$$S_\theta(x_n, x_n, x_{n+1}) \leq k S_\theta(x_{n-1}, x_{n-1}, x_n) \leq \cdots \leq k^n S_\theta(x_0, x_0, x_1).$$

Now, pick two natural numbers $n < m$. Hence, by the triangle inequality of the extended partial S_b-metric space, we deduce

$$S_\theta(x_n, x_n, x_m) \leq \theta(x_n, x_n, x_m) 2k^n S_\theta(x_0, x_0, x_1) + \theta(x_n, x_n, x_m)\theta(x_{n+1}, x_{n+1}, x_m) 2k^{n+1} S_\theta(x_0, x_0, x_1)$$
$$+ \cdots + \theta(x_n, x_n, x_m) \cdots \theta(x_{m-1}, x_{m-1}, x_m) 2k^{m-1} S_\theta(x_0, x_0, x_1)$$
$$\leq S_\theta(x_0, x_0, x_1)[\theta(x_1, x_1, x_m)\theta(x_2, x_2, x_m) \cdots \theta(x_{n-1}, x_{n-1}, x_m)\theta(x_n, x_n, x_m) 2k^n$$
$$+ \theta(x_1, x_1, x_m)\theta(x_2, x_2, x_m) \cdots \theta(x_n, x_n, x_m)\theta(x_{n+1}, x_{n+1}, x_m) 2k^{n+1}$$
$$+ \cdots + \theta(x_1, x_1, x_m)\theta(x_2, x_2, x_m) \cdots \theta(x_{m-2}, x_{m-2}, x_m)\theta(x_{m-1}, x_{m-1}, x_m) 2k^{m-1}].$$

By the hypothesis of the theorem, we have

$$\lim_{n,m \to \infty} \theta(x_n, x_n, x_m)(k) < 1.$$

Therefore, by the Ratio test, the series $\sum_{n=1}^\infty 2k^n \prod_{i=1}^n \theta(x_i, x_i, x_m)$ converges. Now, let

$$A = \sum_{n=1}^\infty 2k^n \prod_{i=1}^n \theta(x_i, x_i, x_m) \text{ and } A_n = \sum_{j=1}^n 2k^j \prod_{i=1}^j \theta(x_i, x_i, x_m).$$

Hence, for $m > n$, we deduce that

$$S_\theta(x_n, x_n, x_m) \leq S_\theta(x_0, x_0, x_1)[A_{m-1} - A_{n-1}].$$

Taking the limit as $n, m \to \infty$, we conclude that $\{x_n\}$ is a Cauchy sequence. Since X is complete, $\{x_n\}$ converges to some $u \in X$, such that

$$S_\theta(u, u, u) = \lim_{n \to \infty} S_\theta(u, u, x_n) = \lim_{n,m \to \infty} S_\theta(x_n, x_n, x_m).$$

Now, using the fact that T and S_θ are continuous, we deduce that

$$S_\theta(x_{n+1}, x_{n+1}, x_{n+1}) = S_\theta(x_{n+1}, x_{n+1}, Tx_n) = S_\theta(Tx_n, Tx_n, Tx_n).$$

Taking the limit in the above equalities, we can easily conclude that $Tu = u$. Hence, u is a fixed point of T. To show uniqueness, assume that there exists $v \neq u \in X$ such that $Tu = u$ and $Tv = v$. Thus,

$$S_\theta(u, u, v) = S_\theta(Tu, Tu, Tv)$$
$$\leq k S_\theta(u, u, v)$$
$$< S_\theta(u, u, v),$$

which leads us to a contradiction. Therefore, T has a unique fixed point in X as desired. □

Now, we present the following example as an application of Theorem 1.

Example 3. Let $X = C([a, b], (-\infty, \infty))$ be the set of all continuous real valued functions on $[a, b]$. Define

$$S_\theta : X^3 \to [0, \infty); \ S_\theta(x(t), y(t), z(t)) = \sup_{t \in [a,b]} |\max\{x(t), y(t), z(t)\}|^2,$$

and

$$\theta : X^3 \to [1, \infty); \ \theta(x(t), y(t), z(t)) = \max\{|x(t)|, |y(t)|\} + |z(t)| + 1.$$

Now, let T be a self mapping on X defined by

$$Tx = \frac{x}{2}.$$

Note that

$$S_\theta(Tx, Ty, Tz) \leq \frac{1}{3} S_\theta(x, y, z).$$

In this case, we have $k = \frac{1}{3}$. On the other hand, it is not difficult to see that, for every $x \in X$, we have

$$T^n x = \frac{x}{2^n}.$$

Thus, it is not difficult to see that

$$\lim_{n,m \to \infty} \theta(T^n x, T^n x, T^m x) < \frac{3}{2}.$$

Therefore, all the conditions of Theorem 1 are satisfied and hence T has a unique fixed point which is in this case 0.

Theorem 2. *Let (X, S_θ) be a symmetric complete extended partial S_b-metric spaces such that S_θ is continuous and T be a continuous self mapping on X satisfying*

$$S_\theta(Tx, Ty, Tz) \leq \psi[S_\theta(x, y, z)] \text{ for all } x, y, z \in X,$$

where ψ is a comparison function (i.e., $\psi : [0, +\infty) \longrightarrow [0, +\infty)$ is an increasing function such that $\lim_{n \to \infty} \psi^n(t) = 0$ for each fixed $t > 0$.) In addition, assume that there exists $s > 1$ such that, for every $x_0 \in X$ and $x \in X$, we have

$$\lim_{n \to \infty} \theta(x_n, x_n, x) < \frac{s}{2}.$$

Then, T has a unique fixed point in X.

Proof. Let $x \in X$ and $\epsilon > 0$. Let n be a natural number such that $\psi^n(\epsilon) < \frac{\epsilon}{2s}$.
Let $F = T^n$ and $x_k = F^k(x)$ for $k \in \mathbb{N}$. Then, for $x, y \in X$ and $\alpha = \psi^n$, we have

$$S_\theta(Fx, Fx, Fy) \leq \psi^n(S_\theta(x, x, y))$$
$$= \alpha(S_\theta(x, x, y)).$$

Hence, for $k \in \mathbb{N}$ $S_\theta(x_{k+1}, x_{k+1}, x_k)$ goes to 0 as k goes toward ∞. Therefore, let k be such that

$$S_\theta(x_{k+1}, x_{k+1}, x_k) < \frac{\epsilon}{2s}.$$

Note that $x_k \in B(x_k, \epsilon)$. Therefore, $B(x_k, \epsilon) \neq \emptyset$. Hence, for all $z \in B(x_k, \epsilon)$, we have

$$S_\theta(Fz, Fz, Fx_k) \leq \alpha(S_\theta(x_k, x_k, z))$$
$$\leq \alpha(\epsilon) = \psi^n(\epsilon) < \frac{\epsilon}{2s} < \frac{\epsilon}{s}.$$

Since $S_\theta(Fx_k, Fx_k, Fx_k) = S_\theta(x_{k+1}, x_{k+1}, x_k) < \frac{\epsilon}{2s}$, thus

$$S_\theta(x_k, x_k, Fz) \leq \theta(x_k, x_k, Fz)[S_\theta(x_k, x_k, x_{k+1}) + S_\theta(x_k, x_k, x_{k+1}) + S_\theta(Fz, Fz, x_{k+1})]$$
$$= \theta(x_k, x_k, Fz)[2S_\theta(x_k, x_k, x_{k+1}) + S_\theta(Fz, Fz, x_{k+1})]$$
$$\leq \theta(x_k, x_k, Fz)[2\frac{\epsilon}{2s} + \frac{\epsilon}{s}].$$

Now, taking the limit of the above inequality as $k \to \infty$, we get

$$S_\theta(x_k, x_k, F_z) \leq \epsilon.$$

Hence, F maps $B(x_k, \epsilon)$ to itself. Since $x_k \in B(x_k, \epsilon)$, we have $Fx_k \in B(x_k, \epsilon)$. By repeating this process, we get

$$F_{x_k}^m \in B(x_k, \epsilon) \text{ for all } m \in \mathbb{N}.$$

That is, $x_l \in B(x_k, \epsilon)$ for all $l \geq k$. Hence,

$$S_\theta(x_m, x_m, x_l) < \epsilon \text{ for all } m, l > k.$$

Therefore, $\{x_k\}$ is a Cauchy sequence and, by the completeness of X, there exists $u \in X$ such that x_k converges to u as k goes toward ∞. Moreover, $u = \lim_{k \to \infty} x_{k+1} = \lim_{k \to \infty} x_k = F(u)$.

Thus, F has u as a fixed point. We prove now the uniqueness of the fixed point for F. Since $\alpha(t) = \psi^n(t) < t$ for any $t > 0$, let u and u_1 be two fixed points of F:

$$\begin{aligned} S_\theta(u, u, u_1) &= S_\theta(F_u, F_u, F_{u_1}) \\ &\leq \psi^n(S_\theta(u, u, u_1)) \\ &= \alpha(S_\theta(u, u, u_1)) \\ &< S_\theta(u, u, u_1). \end{aligned}$$

Thus, $S_\theta(u, u, u_1) = 0$, that is $u = u_1$ and hence, F has a unique fixed point in X. On the other hand, $T^{nk+r}(x) = F^k(T^r(x))$ goes to u as k goes toward ∞. Hence, $T^m x$ goes to u as m goes toward ∞ for every x. That is $u = \lim_{m \to \infty} Tx_m = T(u)$. Therefore, T has a fixed point. □

Theorem 3. *Let (X, S_θ) be a complete symmetric extended partial S_b-metric spaces such that S_θ is continuous, and let T be a continuous self mapping on X satisfying the following condition:*

$$S_\theta(Tx, Ty, Tz) \leq \lambda[S_\theta(x, x, Tx) + S_\theta(y, y, Ty) + S_\theta(z, z, Tz)] \text{ for all } x, y, z \in X, \quad (1)$$

where $\lambda \in [0, \frac{1}{3})$, $\lambda \neq \frac{1}{3\theta(T^n x, T^n x, T^m)}$ for every $x_0 \in X$. Then, T has a unique fixed point $u \in X$ and $S_\theta(u, u, u) = 0$.

Proof. We first prove that if T has a fixed point, then it is unique. We must show that, if $u \in X$ is a fixed point of T, that is, $Tu = u$ then $S_\theta(u, u, u) = 0$.
From (1), we obtain

$$\begin{aligned} S_\theta(u, u, u) = S_\theta(Tu, Tu, Tu) &\leq \lambda[S_\theta(u, u, Tu) + S_\theta(u, u, Tu) + S_\theta(u, u, Tu)] \\ &= 3\lambda S_\theta(u, u, Tu) \text{ since } \lambda \in [0, \frac{1}{3}), \text{ we have} \\ &< S_\theta(u, u, u), \end{aligned}$$

which implies that we must have $S_\theta(u, u, u) = 0$ Suppose $u, v \in X$ be two fixed points, that is, $Tu = u$ and $Tv = v$. Then, we have $S_\theta(u, u, u) = S_\theta(v, v, v) = 0$. Relation (1) gives

$$\begin{aligned} S_\theta(u, u, v) &= S_\theta(Tu, Tu, Tv) \\ &\leq \lambda[S_\theta(u, u, Tu) + S_\theta(u, u, Tu) + S_\theta(v, v, Tv)] \\ &= 2\lambda S_\theta(u, u, u) + \lambda S_\theta(v, v, v) \\ &= 0. \end{aligned}$$

Therefore, $u = v$. Thereby, we get the uniqueness of the fixed point if it exists. For the existence of the fixed point, let $x_0 \in X$ arbitrary, set $x_n = T^n x_0$ and $S_{b_n} = S_\theta(x_n, x_n, x_{n+1})$. We can assume $S_{b_n} > 0$ for all $n \in \mathbb{N}$; otherwise, x_n is a fixed point of T for at least one $n \geq 0$. For all n, we obtain from (1)

$$\begin{aligned} S_{b_n} &= S_\theta(x_n, x_n, x_{n+1}) = S_\theta(Tx_{n-1}, Tx_{n-1}, Tx_n) \\ &\leq \lambda[2S_\theta(x_{n-1}, x_{n-1}, Tx_{n-1}) + S_\theta(x_n, x_n, Tx_n)] \\ &= \lambda[2S_\theta(x_{n-1}, x_{n-1}, x_n) + S_\theta(x_n, x_n, x_{n+1})] \\ &= \lambda[2S_{b_{n-1}} + S_{b_n}]. \end{aligned}$$

Therefore, $(1 - \lambda)S_{b_n} \leq 2\lambda S_{b_{n-1}}$. Thus,

$$S_{b_n} \leq \frac{2\lambda}{1 - \lambda} S_{b_{n-1}}, \quad \lambda \in [0, \frac{1}{3}). \tag{2}$$

Let $\beta = \frac{2\lambda}{1 - \lambda} < 1$. By repeating this process, we obtain

$$S_{b_n} \leq \beta^n b_0.$$

Therefore, $\lim_{n \to \infty} S_{b_n} = 0$. Let us prove that $\{x_n\}$ is a Cauchy sequence. It follows from (1) that, for $n, m \in \mathbb{N}$:

$$\begin{aligned} S_\theta(x_n, x_n, x_m) &= S_\theta(Tx_{n-1}, Tx_{n-1}, Tx_{m-1}) \\ &\leq \lambda[2S_\theta(x_{n-1}, x_{n-1}, Tx_{n-1}) + S_\theta(x_{m-1}, x_{m-1}, Tx_{m-1})] \\ &= \lambda[2S_\theta(x_{n-1}, x_{n-1}, x_n) + S_\theta(x_{m-1}, x_{m-1}, x_m)] \\ &= \lambda[2S_{b_{n-1}} + S_{b_{m-1}}]. \end{aligned}$$

Thus, for every $\epsilon > 0$, as $\lim_{n \to \infty} S_{b_n} = 0$, we can find $n_0 \in \mathbb{N}$ such that $S_{b_{n-1}} < \frac{\epsilon}{4}$ and $S_{b_{m-1}} < \frac{\epsilon}{2}$ for all $n, m > n_0$. Then, we obtain $2S_{b_{n-1}} + S_{b_{m-1}} \leq 2\frac{\epsilon}{4} + \frac{\epsilon}{2} = \epsilon$. As $\lambda < 1$, it follows that $S_\theta(x_n, x_n, x_m) < \epsilon$ for all $n, m > n_0$. Thus, $\{x_n\}$ is a Cauchy sequence in X and $\lim_{n \to \infty} S_\theta(x_n, x_n, x_m) = 0$. By completeness of X, there exists $u \in X$ such that

$$\lim_{n,m \to \infty} S_\theta(x_n, x_n, u) = S_\theta(u, u, u) = 0.$$

Now, we shall prove that $Tu = u$. For any $n \in \mathbb{N}$,

$$\begin{aligned} S_\theta(u, u, Tu) &\leq \theta(u, u, Tu)[2S_\theta(u, u, x_{n+1}) + S_\theta(Tu, Tu, Tx_n)] \\ &\leq \theta(u, u, Tu)[2S_\theta(u, u, x_{n+1}) + \lambda(2S_\theta(u, u, Tu) + S_\theta(x_n, x_n, Tx_n))]. \end{aligned}$$

Therefore, $(1 - 2\theta(u, u, Tu)\lambda)S_\theta(u, u, Tu) \leq 2\theta(u, u, Tu)S_\theta(u, u, x_{n+1}) + \theta(u, u, Tu)\lambda S_\theta(x_n, x_n, Tx_n)$, giving

$$S_\theta(u, u, Tu) \leq \frac{2\theta(u, u, Tu)}{1 - 2\theta(u, u, Tu)\lambda} S_\theta(u, u, x_{n+1}) + \frac{\theta(u, u, Tu)\lambda}{1 - 2\theta(u, u, Tu)\lambda} S_\theta(x_n, x_n, Tx_n).$$

Since S_θ and T are continuous, we have $S_\theta(x_n, x_n, Tx_n)$ goes to $S_\theta(u, u, Tu)$, and n goes toward ∞. Therefore, we obtain

$$S_\theta(u, u, Tu) \leq \frac{2\theta(u, u, Tu)}{1 - 2\theta(u, u, Tu)\lambda} S_\theta(u, u, u) + \frac{\theta(u, u, Tu)\lambda}{1 - 2\theta(u, u, Tu)\lambda} S_\theta(u, u, Tu)$$

$$\left(1 - \frac{\theta(u, u, Tu)\lambda}{1 - 2\theta(u, u, Tu)\lambda}\right) S_\theta(u, u, Tu) \leq \frac{2\theta(u, u, Tu)}{1 - 2\theta(u, u, Tu)\lambda} S_\theta(u, u, u)$$

$$S_\theta(u, u, Tu) \leq \frac{2\theta(u, u, Tu)}{1 - 3\theta(u, u, Tu)\lambda} S_\theta(u, u, u).$$

As $\lambda \neq \dfrac{1}{3\theta(u, u, Tu)}$ and from (2), we obtain $S_\theta(u, u, Tu) = 0$ and then $Tu = u$. □

In closing, we would like to present to the reader the following open questions:

Question 1. *Is it possible to omit the continuity of T in Theorem 1, and obtain a unique fixed point?*

Question 2. *If we omit the symmetry condition of the extended partial S_b-metric spaces in Theorem 2, is it possible to prove the existence of a fixed point?*

Funding: The author would like to thank Prince Sultan University for funding this work through research group Nonlinear Analysis Methods in Applied Mathematics (NAMAM) group number RG-DES-2017-01-17.

Conflicts of Interest: The authors declare no conflict of interest.

References

1. Czerwik, S. Contraction mappings in b-metric spaces. *Acta Math. Inform. Univ. Ostrav.* **1993**, *1*, 5–11.
2. Souayah, N.; Mlaiki, N. A fixed point theorem in S_b metric spaces. *J. Math. Comput. Sci.* **2016**, *16*, 131–139. [CrossRef]
3. Dung, N.V.; Hang, V.T.L. Remarks on partial b-metric spaces and fixed point theorems. *Matematicki Vesnik* **2017**, *69*, 231–240.
4. Ge, X.; Lin, S. A note on partial b-metric spaces. *Mediterr. J. Math.* **2016**, *13*, 1273–1276. [CrossRef]
5. Haghi, R.H.; Rezapour, S.; Shahzad, N. Be careful on partial metric fixed point results. *Topol. Appl.* **2013**, *60*, 450–454. [CrossRef]
6. Kiek, W.; Shahzad, N. *Fixed Point Theory in Distance Spaces*; Springer International Publishing: Basel, Switzerland, 2014.
7. Shukla, S. Partial b-mretric spaces and fixed point theorems. *Mediterr. J. Math.* **2014**, *11*, 703–711. [CrossRef]
8. Yumnam, R.; Tatjana, D.; Stojan, R. A note on the paper, A fixed point theorems in S_b-metric spaces. *Filomat* **2017**, *31*, 3335–3346.
9. Banach, S. Sur les operations dans les ensembles et leur application aux equation sitegrales. *Fundam. Math.* **1922**, *3*, 133–181. [CrossRef]
10. Khan, M.S.; Singh, Y.; Maniu, G.; Postolache, M. On generalized convex contractions of type-2 in b-metric and 2-metric spaces. *J. Nonlinear Sci. Appl.* **2017**, *10*, 2902–2913. [CrossRef]
11. Mlaiki, N.; Mukheimer, A.; Rohen, Y.; Souayah, N.; Abdeljawad, T. Fixed point theorems for-contractive mapping in Sb-metric spaces. *J. Math. Anal.* **2017**, *8*, 40–46.
12. Mukheimer, A. α-ψ-φ-contractive mappings in ordered partial b-metric spaces. *J. Nonlinear Sci. Appl.* **2014**, *7*, 168–179. [CrossRef]
13. Bakhtin, A. The contraction mapping principle in almost metric spaces. *Funct. Anal. Gos. Ped. Inst. Unianowsk* **1989**, *30*, 26–37.

14. Sedghi, S.; Shobe, N.; Aliouche, A. A generalization of fixed point theorems in S-metric spaces. *Mat. Vesnik.* **2012**, *64*, 258–266.
15. Mlaiki, N. Extended S_b-metric spaces. *J. Math. Anal.* **2018**, *9*, 124–135.

© 2018 by the authors. Licensee MDPI, Basel, Switzerland. This article is an open access article distributed under the terms and conditions of the Creative Commons Attribution (CC BY) license (http://creativecommons.org/licenses/by/4.0/).

Article

On the Fixed-Circle Problem and Khan Type Contractions

Nabil Mlaiki [1,*], Nihal Taş [2] and Nihal Yılmaz Özgür [2]

1. Department of Mathematics and General Sciences, Prince Sultan University, 11586 Riyadh, Saudi Arabia
2. Department of Mathematics, Balıkesir University, 10145 Balıkesir, Turkey; nihaltas@balikesir.edu.tr (N.T.); nihal@balikesir.edu.tr (N.Y.Ö.)
* Correspondence: nmlaiki@psu.edu.sa

Received: 18 October 2018; Accepted: 5 November 2018; Published: 8 November 2018

Abstract: In this paper, we consider the fixed-circle problem on metric spaces and give new results on this problem. To do this, we present three types of F_C-Khan type contractions. Furthermore, we obtain some solutions to an open problem related to the common fixed-circle problem.

Keywords: fixed circle; common fixed circle; fixed-circle theorem

MSC: Primary: 47H10; Secondary: 54H25, 55M20, 37E10

---

1. Introduction

Recently, the fixed-circle problem has been considered for metric and some generalized metric spaces (see [1–6] for more details). For example, in [1], some fixed-circle results were obtained using the Caristi type contraction on a metric space. Using Wardowski's technique and some classical contractive conditions, new fixed-circle theorems were proved in [5,6]. In [2,3], the fixed-circle problem was studied on an S-metric space. In [7], a new fixed-circle theorem was proved using the modified Khan type contractive condition on an S-metric space. Some generalized fixed-circle results with geometric viewpoint were obtained on S_b-metric spaces and parametric N_b-metric spaces (see [8,9] for more details, respectively). Also, it was proposed to investigate some fixed-circle theorems on extended M_b-metric spaces [10]. On the other hand, an application of the obtained fixed-circle results was given to discontinuous activation functions on metric spaces (see [1,4,11]). Hence it is important to study new fixed-circle results using different techniques.

Let (X, d) be a metric space and $C_{x_0,r} = \{x \in X : d(x, x_0) = r\}$ be any circle on X. In [5], it was given the following open problem.

Open Problem CC: What is (are) the condition(s) to make any circle $C_{x_0,r}$ as the common fixed circle for two (or more than two) self-mappings?

In this paper, we give new results to the fixed-circle problem using Khan type contractions and to the above open problem using both of Khan and Ćirić type contractions on a metric space. In Section 2, we introduce three types of F_C-Khan type contractions and obtain new fixed-circle results. In Section 3, we investigate some solutions to the above Open Problem CC. In addition, we construct some examples to support our theoretical results.

2. New Fixed-Circle Theorems

In this section, using Khan type contractions, we give new fixed-circle theorems (see [12–15] for some Khan type contractions used to obtain fixed-point theorems). At first, we recall the following definitions.

Definition 1 ([16]). *Let $\mathbb{F}$ be the family of all functions $F : (0, \infty) \to \mathbb{R}$ such that*
  (F1) *F is strictly increasing,*
  (F2) *For each sequence $\{\alpha_n\}_{n=1}^{\infty}$ of positive numbers, $\lim_{n \to \infty} \alpha_n = 0$ if and only if $\lim_{n \to \infty} F(\alpha_n) = -\infty$,*
  (F3) *There exists $k \in (0,1)$ such that $\lim_{\alpha \to 0^+} \alpha^k F(\alpha) = 0$.*

Definition 2 ([16]). *Let (X,d) be a metric space. A mapping $T : X \to X$ is said to be an F-contraction on (X,d), if there exist $F \in \mathbb{F}$ and $\tau \in (0, \infty)$ such that*

$$d(Tx, Ty) > 0 \implies \tau + F(d(Tx, Ty)) \leq F(d(x,y)),$$

for all $x, y \in X$.

Definition 3 ([15]). *Let $\mathbb{F}_k$ be the family of all increasing functions $F : (0, \infty) \to \mathbb{R}$, that is, for all $x, y \in (0, \infty)$, if $x < y$ then $F(x) \leq F(y)$.*

Definition 4 ([15]). *Let (X,d) be a metric space and $T : X \to X$ be a self-mapping. T is said to be an F-Khan-contraction if there exist $F \in \mathbb{F}_k$ and $t > 0$ such that for all $x, y \in X$ if $\max\{d(Ty, x), d(Tx, y)\} \neq 0$ then $Tx \neq Ty$ and*

$$t + F(d(Tx, Ty)) \leq F\left(\frac{d(Tx,x)d(Ty,x) + d(Ty,y)d(Tx,y)}{\max\{d(Ty,x), d(Tx,y)\}}\right),$$

and if $\max\{d(Ty, x), d(Tx, y)\} = 0$ then $Tx = Ty$.

Now we modify the definition of an F-Khan-contractive condition, which is used to obtain a fixed point theorem in [15], to get new fixed-circle results. Hence, we define the notion of an F_C-Khan type I contractive condition as follows.

Definition 5. *Let (X,d) be a metric space and $T : X \to X$ be a self-mapping. T is said to be an F_C-Khan type I contraction if there exist $F \in \mathbb{F}_k$, $t > 0$ and $x_0 \in X$ such that for all $x \in X$ if the following condition holds*

$$\max\{d(Tx_0, x_0), d(Tx, x)\} \neq 0, \tag{1}$$

then

$$t + F(d(Tx, x)) \leq F\left(h\frac{d(Tx,x)d(Tx_0,x) + d(Tx_0,x_0)d(Tx,x_0)}{\max\{d(Tx_0,x_0), d(Tx,x)\}}\right),$$

where $h \in \left[0, \frac{1}{2}\right)$ and if $\max\{d(Tx_0, x_0), d(Tx, x)\} = 0$ then $Tx = x$.

One of the consequences of this definition is the following proposition.

Proposition 1. *Let (X,d) be a metric space. If a self-mapping T on X is an F_C-Khan type I contraction with $x_0 \in X$ then we get $Tx_0 = x_0$.*

Proof. Let $Tx_0 \neq x_0$. Then using the hypothesis, we find

$$\max\{d(Tx_0, x_0), d(Tx, x)\} \neq 0$$

and

$$\begin{aligned}
t + F(d(Tx_0, x_0)) &\leq F\left(h\frac{d(Tx_0,x_0)d(Tx_0,x_0) + d(Tx_0,x_0)d(Tx_0,x_0)}{d(Tx_0,x_0)}\right) \\
&= F(2hd(Tx_0, x_0)) < F(d(Tx_0, x_0)).
\end{aligned}$$

This is a contradiction since $t > 0$ and so it should be $Tx_0 = x_0$. □

Consequently, the condition (1) can be replaced with $d(Tx, x) \neq 0$ and so $Tx \neq x$. Considering this, now we give a new fixed-circle theorem.

Theorem 1. *Let (X, d) be a metric space, $T : X \to X$ be a self-mapping and*

$$r = \inf \{d(Tx, x) : Tx \neq x\}. \tag{2}$$

If T is an F_C-Khan type I contraction with $x_0 \in X$ then $C_{x_0, r}$ is a fixed circle of T.

Proof. Let $x \in C_{x_0, r}$. Assume that $Tx \neq x$. Then we have $d(Tx, x) \neq 0$ and by the F_C-Khan type I contractive condition, we obtain

$$\begin{aligned}
t + F(d(Tx, x)) &\leq F\left(h \frac{d(Tx, x)d(Tx_0, x) + d(Tx_0, x_0)d(Tx, x_0)}{\max\{d(Tx_0, x_0), d(Tx, x)\}}\right) \\
&= F(hr) \leq F(hd(Tx, x)) < F(d(Tx, x)),
\end{aligned}$$

a contradiction since $t > 0$. Therefore, we have $Tx = x$ and so T fixes the circle $C_{x_0, r}$. □

Corollary 1. *Let (X, d) be a metric space, $T : X \to X$ be a self-mapping and r be defined as in (2). If T is an F_C-Khan type I contraction with $x_0 \in X$ then T fixes the disc $D_{x_0, r} = \{x \in X : d(x, x_0) \leq r\}$.*

We recall the following theorem.

Theorem 2 ([12]). *Let (X, d) be a metric space and $T : X \to X$ satisfy*

$$d(Tx, Ty) \leq \begin{cases} k \frac{d(x, Tx)d(x, Ty) + d(y, Ty)d(y, Tx)}{d(x, Ty) + d(y, Tx)} & \text{if } d(x, Ty) + d(y, Tx) \neq 0 \\ 0 & \text{if } d(x, Ty) + d(y, Tx) = 0 \end{cases}, \tag{3}$$

where $k \in [0, 1)$ and $x, y \in X$. Then T has a unique fixed point $x^* \in X$. Moreover, for all $x \in X$, the sequence $\{T^n x\}_{n \in \mathbb{N}}$ converges to x^*.

We modify the inequality (3) using Wardowski's technique to obtain a new fixed-point theorem. We give the following definition.

Definition 6. *Let (X, d) be a metric space and $T : X \to X$ be a self-mapping. T is said to be an F_C-Khan type II contraction if there exist $F \in \mathbb{F}_k$, $t > 0$ and $x_0 \in X$ such that for all $x \in X$ if $d(Tx_0, x_0) + d(Tx, x) \neq 0$ then $Tx \neq x$ and*

$$t + F(d(Tx, x)) \leq F\left(h \frac{d(Tx, x)d(Tx_0, x) + d(Tx_0, x_0)d(Tx, x_0)}{d(Tx_0, x_0) + d(Tx, x)}\right),$$

where $h \in \left[0, \frac{1}{2}\right)$ and if $d(Tx_0, x_0) + d(Tx, x) = 0$ then $Tx = x$.

An immediate consequence of this definition is the following result.

Proposition 2. *Let (X, d) be a metric space. If a self-mapping T on X is an F_C-Khan type II contraction then we get $Tx_0 = x_0$.*

Proof. Let $Tx_0 \neq x_0$. Then using the hypothesis, we find

$$d(Tx_0, x_0) + d(Tx, x) \neq 0$$

and

$$t + F(d(Tx_0, x_0)) \leq F\left(h\frac{d(Tx_0, x_0)d(Tx_0, x_0) + d(Tx_0, x_0)d(Tx_0, x_0)}{2d(Tx_0, x_0)}\right)$$
$$= F(hd(Tx_0, x_0)) < F(d(Tx_0, x_0)),$$

which is a contradiction since $t > 0$. Hence it should be $Tx_0 = x_0$. □

Theorem 3. *Let (X, d) be a metric space, $T : X \to X$ be a self-mapping and r be defined as in (2). If T is an F_C-Khan type II contraction with $x_0 \in X$ then $C_{x_0, r}$ is a fixed circle of T.*

Proof. Let $x \in C_{x_0, r}$. Assume that $Tx \neq x$. Then using Proposition 2, we get

$$d(Tx_0, x_0) + d(Tx, x) = d(Tx, x) \neq 0.$$

By the F_C-Khan type II contractive condition, we obtain

$$t + F(d(Tx, x)) \leq F\left(h\frac{d(Tx, x)d(Tx_0, x) + d(Tx_0, x_0)d(Tx, x_0)}{d(Tx_0, x_0) + d(Tx, x)}\right)$$
$$= F(hr) \leq F(hd(Tx, x)) < F(d(Tx, x)),$$

a contradiction since $t > 0$. Therefore, we have $Tx = x$ and T fixes the circle $C_{x_0, r}$. □

Corollary 2. *Let (X, d) be a metric space, $T : X \to X$ be a self-mapping and r be defined as in (2). If T is an F_C-Khan type II contraction with $x_0 \in X$ then T fixes the disc $D_{x_0, r}$.*

In the following theorem, we see that the F_C-Khan type I and F_C-Khan type II contractive conditions are equivalent.

Theorem 4. *Let (X, d) be a metric space and $T : X \to X$ be a self-mapping. T satisfies the F_C-Khan type I contractive condition if and only if T satisfies the F_C-Khan type II contractive condition.*

Proof. Let the F_C-Khan type I contractive condition be satisfied by T. Using Proposition 1 and Proposition 2, we get

$$t + F(d(Tx, x)) \leq F\left(h\frac{d(Tx, x)d(Tx_0, x) + d(Tx_0, x_0)d(Tx, x_0)}{\max\{d(Tx_0, x_0), d(Tx, x)\}}\right)$$
$$= F\left(h\frac{d(Tx, x)d(Tx_0, x)}{d(Tx, x)}\right)$$
$$= F(hd(Tx_0, x))$$
$$= F\left(h\frac{d(Tx, x)d(Tx_0, x) + d(Tx_0, x_0)d(Tx, x_0)}{d(Tx_0, x_0) + d(Tx, x)}\right).$$

Using the similar arguments, the converse statement is clear. Consequently, the F_C-Khan type I contractive and the F_C-Khan type II contractive conditions are equivalent. □

Remark 1. *By Theorem 4, we see that Theorem 1 and Theorem 3 are equivalent.*

Now we give an example.

Example 1. Let $X = \mathbb{R}$ be the metric space with the usual metric $d(x,y) = |x - y|$. Let us define the self-mapping $T : \mathbb{R} \to \mathbb{R}$ as

$$Tx = \begin{cases} x & \text{if } |x| < 6 \\ x + 1 & \text{if } |x| \geq 6 \end{cases},$$

for all $x \in \mathbb{R}$. The self-mapping T is both of an F_C-Khan type I and an F_C-Khan type II contraction with $F = \ln x$, $t = \ln 2$, $x_0 = 0$ and $h = \frac{1}{3}$. Indeed, we get

$$d(Tx, x) = 1 \neq 0,$$

for all $x \in \mathbb{R}$ such that $|x| \geq 6$. Then we have

$$\ln 2 \leq \ln\left(\frac{1}{3}|x|\right)$$
$$\implies \ln 2 + \ln 1 \leq \ln(hd(x, 0)) = \ln(hd(x, x_0))$$
$$\implies t + F(d(Tx, x)) \leq F\left(h\frac{d(Tx, x)d(Tx_0, x) + d(Tx_0, x_0)d(Tx, x_0)}{\max\{d(Tx_0, x_0), d(Tx, x)\}}\right)$$

and

$$\ln 2 \leq \ln\left(\frac{1}{3}|x|\right)$$
$$\implies \ln 2 + \ln 1 \leq \ln(hd(x, 0)) = \ln(hd(x, x_0))$$
$$\implies t + F(d(Tx, x)) \leq F\left(h\frac{d(Tx, x)d(Tx_0, x) + d(Tx_0, x_0)d(Tx, x_0)}{d(Tx_0, x_0) + d(Tx, x)}\right).$$

Also we obtain

$$r = \min\{d(Tx, x) : Tx \neq x\} = 1.$$

Consequently, T fixes the circle $C_{0,1} = \{-1, 1\}$ and the disc $D_{0,1} = \{x \in X : |x| \leq 1\}$. Notice that the self-mapping T has other fixed circles. The above results give us only one of these circles. Also, T has infinitely many fixed circles.

Now we consider the case if $T : X \to X$ is a self-mapping, then for all $x, y \in X$,

$$x \neq y \implies d(Ty, x) + d(Tx, y) \neq 0.$$

Definition 7. Let (X, d) be a metric space and $T : X \to X$ be a self-mapping. Then T is called a C-Khan type contraction if there exists $x_0 \in X$ such that

$$d(Tx, x) \leq h\frac{d(Tx, x)d(Tx_0, x) + d(Tx_0, x_0)d(Tx, x_0)}{d(Tx_0, x) + d(Tx, x_0)}, \qquad (4)$$

where $h \in [0, 1)$ for all $x \in X - \{x_0\}$.

We can give the following fixed-circle result.

Theorem 5. Let (X, d) be a metric space, $T : X \to X$ be a self-mapping and $C_{x_0,r}$ be a circle on X. If T satisfies the C-Khan type contractive condition (4) for all $x \in C_{x_0,r}$ with $Tx_0 = x_0$, then T fixes the circle $C_{x_0,r}$.

Proof. Let $x \in C_{x_0,r}$. Suppose that $Tx \neq x$. Using the C-Khan type contractive condition with $Tx_0 = x_0$, we find

$$d(Tx,x) \leq h\frac{d(Tx,x)d(Tx_0,x) + d(Tx_0,x_0)d(Tx,x_0)}{d(Tx_0,x) + d(Tx,x_0)}$$
$$= \frac{hrd(Tx,x)}{r + d(Tx,x_0)}$$
$$\leq \frac{hrd(Tx,x)}{r} = hd(Tx,x),$$

which is a contradiction since $h < 1$. Consequently, T fixes the circle $C_{x_0,r}$. $\square$

Theorem 6. *Let (X,d) be a metric space, $x_0 \in X$ and $T : X \to X$ be a self-mapping. If T is a C-Khan type contraction for all $x \in X - \{x_0\}$ with $Tx_0 = x_0$, then T is the identity map I_X on X.*

Proof. Let $x \in X - \{x_0\}$ be any point. If $Tx \neq x$ then using the C-Khan type contractive condition (4) with $Tx_0 = x_0$, we find

$$d(Tx,x) \leq h\frac{d(Tx,x)d(Tx_0,x) + d(Tx_0,x_0)d(Tx,x_0)}{d(Tx_0,x) + d(Tx,x_0)}$$
$$= h\frac{d(Tx,x)d(x_0,x)}{d(x_0,x) + d(Tx,x_0)}$$
$$\leq h\frac{d(Tx,x)d(x_0,x) + d(Tx,x)d(Tx,x_0)}{d(x_0,x) + d(Tx,x_0)}$$
$$= h\frac{d(Tx,x)[d(x_0,x) + d(Tx,x_0)]}{d(x_0,x) + d(Tx,x_0)}$$
$$= hd(Tx,x),$$

which is a contradiction since $h < 1$. Consequently, we have $Tx = x$ and hence T is the identity map I_X on X. $\square$

Example 2. *Let $X = \mathbb{R}$ be the usual metric space and consider the circle $C_{0,3} = \{-3,3\}$. Let us define the self-mapping $T : \mathbb{R} \to \mathbb{R}$ as*

$$Tx = \begin{cases} \frac{-9x+8}{2x-9} & \text{if} \quad x \in \{-3,3\} \\ 0 & \text{if} \quad x \in \mathbb{R} - \{-3,3\} \end{cases},$$

for all $x \in \mathbb{R}$. Then the self-mapping T satisfies the C-Khan type contractive condition for all $x \in C_{0,3}$ and $T0 = 0$. Consequently, $C_{0,3}$ is a fixed circle of T.

3. Common Fixed-Circle Results

Recently, it was obtained some coincidence and common fixed-point theorems using Wardowski's technique and the Ćirić type contractions (see [17] for more details). In this section, we extend the notion of a Khan type F_C-contraction to a pair of maps to obtain a solution to the Open Problem CC. At first, we give the following definition.

Definition 8. *Let (X,d) be a metric space and $T,S : X \to X$ be two self-mappings. A pair of self-mappings (T,S) is called a Khan type $F_{T,S}$-contraction if there exist $F \in \mathbb{F}_k$, $t > 0$ and $x_0 \in X$ such that for all $x \in X$ if the following condition holds*

$$\max\{d(Tx_0,x_0), d(Sx_0,x_0)\} \neq 0,$$

then

$$t + F(d(Tx,Sx)) \leq F\left(h\frac{d(Tx,Sx)d(Tx,x_0) + d(Tx_0,Sx_0)d(Sx,x_0)}{\max\{d(Tx_0,x_0), d(Sx_0,x_0)\}}\right),$$

where $h \in \left[0, \frac{1}{2}\right)$ and if $\max\{d(Tx_0,x_0), d(Sx_0,x_0)\} = 0$ then $Tx = Sx$.

An immediate consequence of this definition is the following proposition.

Proposition 3. *Let (X, d) be a metric space and $T, S : X \to X$ be two self-mappings. If the pair of self-mappings (T, S) is a Khan type $F_{T,S}$-contraction with $x_0 \in X$ then x_0 is a coincidence point of T and S, that is, $Tx_0 = Sx_0$.*

Proof. We prove this proposition under the following cases:
Case 1: Let $Tx_0 = x_0$ and $Sx_0 \neq x_0$. Then using the hypothesis, we get

$$\max\{d(Tx_0, x_0), d(Sx_0, x_0)\} = d(Sx_0, x_0) \neq 0$$

and so

$$t + F(d(Tx_0, Sx_0)) \leq F\left(h\frac{d(Tx_0, Sx_0)d(Tx_0, x_0) + d(Tx_0, Sx_0)d(Sx_0, x_0)}{d(Sx_0, x_0)}\right)$$
$$= F(hd(Tx_0, Sx_0)),$$

which is a contradiction since $h \in \left[0, \frac{1}{2}\right)$ and $t > 0$.
Case 2: Let $Tx_0 \neq x_0$ and $Sx_0 = x_0$. By the similar arguments used in the proof of Case 1, we get a contradiction.
Case 3: Let $Tx_0 = x_0$ and $Sx_0 = x_0$. Then we get $Tx_0 = Sx_0$.
Case 4: Let $Tx_0 \neq x_0$, $Sx_0 \neq x_0$ and $Tx_0 \neq Sx_0$. Using the hypothesis, we obtain

$$\max\{d(Tx_0, x_0), d(Sx_0, x_0)\} \neq 0$$

and so

$$t + F(d(Tx_0, Sx_0)) \leq F\left(h\frac{d(Tx_0, Sx_0)d(Tx_0, x_0) + d(Tx_0, Sx_0)d(Sx_0, x_0)}{\max\{d(Tx_0, x_0), d(Sx_0, x_0)\}}\right). \quad (5)$$

Assume that $d(Tx_0, x_0) > d(Sx_0, x_0)$. Using the inequality (5), we get

$$t + F(d(Tx_0, Sx_0)) \leq F\left(h\frac{d(Tx_0, Sx_0)d(Tx_0, x_0) + d(Tx_0, Sx_0)d(Sx_0, x_0)}{d(Tx_0, x_0)}\right)$$
$$= F\left(hd(Tx_0, Sx_0) + h\frac{d(Tx_0, Sx_0)d(Sx_0, x_0)}{d(Tx_0, x_0)}\right)$$
$$< F(2hd(Tx_0, Sx_0)) < F(d(Tx_0, Sx_0)),$$

which is a contradiction. Suppose that $d(Tx_0, x_0) < d(Sx_0, x_0)$. Using the inequality (5), we find

$$t + F(d(Tx_0, Sx_0)) < F(d(Tx_0, Sx_0)),$$

which is a contradiction. Consequently, x_0 is a coincidence point of T and S, that is, $Tx_0 = Sx_0$. □

Now we use the following number given in [17] (see Definition 3.1 on page 183):

$$M(x, y) = \max\left\{d(Sx, Sy), d(Sx, Tx), d(Sy, Ty), \frac{d(Sx, Ty) + d(Sy, Tx)}{2}\right\}. \quad (6)$$

We give the following definition.

Definition 9. *Let (X, d) be a metric space and $T, S : X \to X$ be two self-mappings. A pair of self-mappings (T, S) is called a Ćirić type $F_{T,S}$-contraction if there exist $F \in \mathbb{F}_k$, $t > 0$ and $x_0 \in X$ such that for all $x \in X$*

$$d(Tx, x) > 0 \implies t + F(d(Tx, x)) \leq F(M(x, x_0)).$$

We get the following proposition.

Proposition 4. Let (X,d) be a metric space and $T, S : X \to X$ be two self-mappings. If the pair of self-mappings (T, S) is both a Khan type $F_{T,S}$-contraction and a Ćirić type $F_{T,S}$-contraction with $x_0 \in X$ then x_0 is a common fixed point of T and S, that is, $Tx_0 = Sx_0 = x_0$.

Proof. By the Khan type $F_{T,S}$-contractive property and Proposition 3, we know that x_0 is a coincidence point of T and S, that is, $Tx_0 = Sx_0$. Now we prove that x_0 is a common fixed point of T and S. Let $Tx_0 \neq x_0$. Then using the Ćirić type $F_{T,S}$-contractive condition, we get

$$\begin{aligned}
t + F(d(Tx_0, x_0)) &\leq F(M(x_0, x_0)) \\
&= F\left(\max \left\{ \begin{array}{l} d(Sx_0, Sx_0), d(Sx_0, Tx_0), d(Sx_0, Tx_0), \\ \frac{d(Sx_0, Tx_0) + d(Sx_0, Tx_0)}{2} \end{array} \right\} \right) \\
&= F(d(Sx_0, Tx_0)) = F(0),
\end{aligned}$$

which is a contradiction because of the definition of F. Therefore it should be $Tx_0 = x_0$. Consequently, x_0 is a common fixed point of T and S, that is, $Tx_0 = Sx_0 = x_0$. $\square$

Notice that we get a coincidence point result for a pair of self-mappings using the Khan type $F_{T,S}$-contractive condition by Proposition 3. We obtain a common fixed-point result for a pair of self-mappings using the both of Khan type $F_{T,S}$-contractive condition and the Ćirić type $F_{T,S}$-contractive condition by Proposition 4.

We prove the following common fixed-circle theorem as a solution to the Open Problem CC.

Theorem 7. Let (X,d) be a metric space, $T, S : X \to X$ be two self-mappings and r be defined as in (2). If $d(Tx, x_0) = d(Sx, x_0) = r$ for all $x \in C_{x_0, r}$ and the pair of self-mappings (T, S) is both a Khan type $F_{T,S}$-contraction and a Ćirić type $F_{T,S}$-contraction with $x_0 \in X$ then $C_{x_0, r}$ is a common fixed circle of T and S, that is, $Tx = Sx = x$ for all $x \in C_{x_0, r}$.

Proof. Let $x \in C_{x_0, r}$. We show that x is a coincidence point of T and S. Using Proposition 4, we get

$$\max\{d(Tx_0, x_0), d(Sx_0, x_0)\} = 0$$

and so by the definition of the Khan type $F_{T,S}$-contraction we obtain

$$Tx = Sx.$$

Now we prove that $C_{x_0, r}$ is a common fixed circle of T and S. Assume that $Tx \neq x$. Using Proposition 4 and the hypothesis Ćirić type $F_{T,S}$-contractive condition, we find

$$\begin{aligned}
t + F(d(Tx, x)) &\leq F(M(x, x_0)) \\
&= F\left(\max \left\{ \begin{array}{l} d(Sx, Sx_0), d(Sx, Tx), d(Sx_0, Tx_0), \\ \frac{d(Sx, Tx_0) + d(Sx_0, Tx)}{2} \end{array} \right\} \right) \\
&= F\left(\max \left\{ d(Sx, x_0), d(Sx, Tx), \frac{d(Sx, x_0) + d(x_0, Tx)}{2} \right\} \right) \\
&= F(\max\{r, d(Sx, Tx), r\}) = F(r),
\end{aligned}$$

which contradicts with the definition of r. Consequently, we have $Tx = x$ and so $C_{x_0, r}$ is a common fixed circle of T and S. $\square$

Corollary 3. Let (X,d) be a metric space, $T, S : X \to X$ be two self-mappings and r be defined as in (2). If $d(Tx, x_0) = d(Sx, x_0) = r$ for all $x \in C_{x_0, r}$ and the pair of self-mappings (T, S) is both a Khan type

$F_{T,S}$-contraction and a Ćirić type $F_{T,S}$-contraction with $x_0 \in X$ then T and S fix the disc $D_{x_0,r}$, that is, $Tx = Sx = x$ for all $x \in D_{x_0,r}$.

We give an illustrative example.

Example 3. Let $X = [1, \infty) \cup \{-1, 0\}$ be the metric space with the usual metric. Let us define the self-mappings $T : X \to X$ and $S : X \to X$ as

$$Tx = \begin{cases} x^2 & \text{if } x \in \{0, 1, 3\} \\ -1 & \text{if } x = -1 \\ x + 1 & \text{otherwise} \end{cases}$$

and

$$Sx = \begin{cases} \frac{1}{x} & \text{if } x \in \{-1, 1\} \\ 3x & \text{if } x \in \{0, 3\} \\ x + 1 & \text{otherwise} \end{cases},$$

for all $x \in X$. The pair of the self-mappings (T, S) is both a Khan type $F_{T,S}$-contraction and a Ćirić type $F_{T,S}$-contraction with $F = \ln x$, $t = \ln \frac{3}{2}$ and $x_0 = 0$. Indeed, we get

$$\max\{d(T0, 0), d(S0, 0)\} = 0$$

and so $Tx = Sx$. Therefore, the pair (T, S) is a Khan type $F_{T,S}$-contraction. Also we get

$$d(T3, 3) = 6 \neq 0,$$

for $x = 3$ and

$$d(Tx, x) = 1 \neq 0,$$

for all $x \in X \setminus \{-1, 0, 1, 3\}$. Then we have

$$\ln \frac{3}{2} \leq \ln 9$$
$$\implies \ln \frac{3}{2} + \ln 6 \leq \ln 9$$
$$\implies \ln \frac{3}{2} + \ln(d(T3, 3)) \leq \ln(M(3, 0))$$

and

$$\ln \frac{3}{2} \leq \ln |x + 1|$$
$$\implies \ln \frac{3}{2} + \ln 1 \leq \ln |x + 1|$$
$$\implies \ln \frac{3}{2} + \ln(d(Tx, x)) \leq \ln(M(x, 0)).$$

Hence the pair (T, S) is a Ćirić type $F_{T,S}$-contraction. Also we obtain

$$r = \min\{d(Tx, x) : Tx \neq x\} = \min\{1, 6\} = 1.$$

Consequently, T fixes the circle $C_{0,1} = \{-1, 1\}$ and the disc $D_{0,1}$.

In closing, we want to bring to the reader attention the following question, under what conditions we can prove the results in [18–20] in fixed circle?

Author Contributions: All authors contributed equally in writing this article. All authors read and approved the final manuscript.

Funding: The first author would like to thank Prince Sultan University for funding this work through research group Nonlinear Analysis Methods in Applied Mathematics (NAMAM) group number RG-DES-2017-01-17.

Conflicts of Interest: The authors declare no conflict of interest.

References

1. Özgür, N.Y.; Taş, N. Some fixed-circle theorems on metric spaces. *Bull. Malays. Math. Sci. Soc.* **2017**. [CrossRef]
2. Özgür, N.Y.; Taş, N.; Çelik, U. New fixed-circle results on S-metric spaces. *Bull. Math. Anal. Appl.* **2017**, *9*, 10–23.
3. Özgür, N.Y.; Taş, N. Fixed-circle problem on S-metric spaces with a geometric viewpoint. *arXiv* **2017**, arXiv:1704.08838.
4. Özgür, N.Y.; Taş, N. Some fixed-circle theorems and discontinuity at fixed circle. *AIP Conf. Proc.* **2018**. [CrossRef]
5. Taş, N.; Özgür, N.Y.; Mlaiki, N. New fixed-circle results related to F_c-contractive and F_c-expanding mappings on metric spaces. **2018**, submitted for publication.
6. Taş, N.; Özgür, N.Y.; Mlaiki, N. New types of F_c-contractions and the fixed-circle problem. *Mathematics* **2018**, *6*, 188. [CrossRef]
7. Taş, N. Various types of fixed-point theorems on S-metric spaces. *J. BAUN Inst. Sci. Technol.* **2018**. [CrossRef]
8. Özgür, N.Y.; Taş, N. Generalizations of metric spaces: From the fixed-point theory to the fixed-circle theory. In *Applications of Nonlinear Analysis*; Rassias T., Ed.; Springer: Cham, Switzerland, 2018; Volume 134.
9. Taş, N.; Özgür, N.Y. Some fixed-point results on parametric N_b-metric spaces. *Commun. Korean Math. Soc.* **2018**, *33*, 943–960.
10. Mlaiki, N.; Özgür, N.Y.; Mukheimer, A.; Taş, N. A new extension of the M_b-metric spaces. *J. Math. Anal.* **2018**, *9*, 118–133.
11. Taş, N.; Özgür, N.Y. A new contribution to discontinuity at fixed point. *arXiv* **2017**, arXiv:1705.03699.
12. Fisher, B. On a theorem of Khan. *Riv. Math. Univ. Parma* **1978**, *4*, 135–137.
13. Khan, M.S. A fixed point theorem for metric spaces. *Rend. Inst. Math. Univ. Trieste* **1976**, *8*, 69–72. [CrossRef]
14. Piri, H.; Rahrovi, S.; Kumam, P. Generalization of Khan fixed point theorem. *J. Math. Comput. Sci.* **2017**, *17*, 76–83. [CrossRef]
15. Piri, H.; Rahrovi, S.; Marasi, H.; Kumam, P. A fixed point theorem for F-Khan-contractions on complete metric spaces and application to integral equations. *J. Nonlinear Sci. Appl.* **2017**, *10*, 4564–4573. [CrossRef]
16. Wardowski, D. Fixed points of a new type of contractive mappings in complete metric spaces. *Fixed Point Theory Appl.* **2012**, *2012*, 94. [CrossRef]
17. Tomar, A.; Sharma, R. Some coincidence and common fixed point theorems concerning F-contraction and applications. *J. Int. Math. Virtual Inst.* **2018**, *8*, 181–198.
18. Kadelburg, Z.; Radenovic, S. Notes on some recent papers concerning Fcontractions in b-metric spaces. *Constr. Math. Anal.* **2018**, *1*, 108–112.
19. Satish, S.; Stojan, R.; Zoran, K. Some fixed point theorems for F-generalized contractions in 0-orbitally complete partial metric spaces. *Theory Appl. Math. Comput. Sci.* **2014**, *4*, 87–98.
20. Lukacs, A.; Kajanto, S. Fixed point therorems for various types of F-contractions in complete b-metric spaces. *Fixed Point Theory* **2018**, *19*, 321–334 [CrossRef]

© 2018 by the authors. Licensee MDPI, Basel, Switzerland. This article is an open access article distributed under the terms and conditions of the Creative Commons Attribution (CC BY) license (http://creativecommons.org/licenses/by/4.0/).

Article

Common Fixed Point Theorems for Generalized Geraghty (α, ψ, ϕ)-Quasi Contraction Type Mapping in Partially Ordered Metric-Like Spaces

Haitham Qawaqneh [1,*,†], Mohd Noorani [1,†], Wasfi Shatanawi [2,3,†] and Habes Alsamir [1,†]

[1] Department of Mathematics, University Kebangsaan Malaysia, Bangi 43600, Malaysia; msn@ukm.my (M.N.); h.alsamer@gmail.com (H.A.)
[2] Department of Mathematics, Hashemite University, Zarqa 1315, Jordan; swasfi@hu.edu.jo
[3] Department of Mathematics and General Courses, Prince Sultan University, Riyadh 11586, Saudi Arabia; wshatanawi@psu.edu.sa
* Correspondence: Haitham.math77@gmail.com; Tel.: +60-128813450
† These authors contributed equally to this work.

Received: 18 September 2018; Accepted: 15 October 2018; Published: 25 October 2018

Abstract: The aim of this paper is to establish the existence of some common fixed point results for generalized Geraghty (α, ψ, ϕ)-quasi contraction self-mapping in partially ordered metric-like spaces. We display an example and an application to show the superiority of our results. The obtained results progress some well-known fixed (common fixed) point results in the literature. Our main results cannot be specifically attained from the corresponding metric space versions. This paper is scientifically novel because we take Geraghty contraction self-mapping in partially ordered metric-like spaces via $\alpha-$admissible mapping. This opens the door to other possible fixed (common fixed) point results for non-self-mapping and in other generalizing metric spaces.

Keywords: common fixed point; metric-like space; α-Geraghty contraction; triangular α-admissible mapping

1. Introduction

Fixed point theory occupies a central role in the study of solving nonlinear equations of kinds $Sx = x$, where the function S is characterized on abstract space X. It is outstanding that the Banach contraction principle is a standout amongst essential and principal results in the fixed point theorem. It ensures the existence of fixed points for certain self-maps in a complete metric space and provides a helpful technique to find those fixed points. Many authors studied and extended it in many generalizations of metric spaces with new contractive mappings, for example, see References [1–3] and the references therein.

Otherwise, Hitzler and Seda [4] introduce the notation of metric-like (dislocated) metric space as a generalization of a metric space, they introduced variants of the Banach fixed point theorem in such space. Metric like spaces were revealed by Amini-Harandi [5] who proved the existence of fixed point results. This interesting subject has been mediated by certain authors, for example, see References [6–8]. In partial metric spaces and partially ordered metric-like spaces, the usual contractive condition is weakened and many researchers apply their results to problems of existence and uniqueness of solutions for some boundary value problems of differential and Integral equations, for example, see References [9–22] and the references therein.

Additionally, Geraghty [23] characterized a kind of the set of functions $\mathfrak{S}$ to be classified as the functions $\beta:[0, \infty) \to [0, 1)$ such that if $\{t_n\}$ is a sequence in $[0, +\infty)$ with $\beta(t_n) \to 1$, then $t_n \to 0$.

By using the function $\beta \in \mathfrak{S}$, Geraghty [23] presented the following exceptional theorem

Theorem 1. *Suppose (Y,d) is a complete metric space. Assume that $T:Y \to Y$ and $\beta:[0,\infty) \to [0,1)$ are functions such that for all $u,v \in Y$,*

$$d(Tu, Tv) \leq \beta(d(u,v))d(u,v), \tag{1}$$

where $\beta \in \mathcal{S}$, then T has a fixed point and has to be unique.

The main results of Geraghty have engaged many of authors, see References [24–26] and the references therein.

Recently, Amini-Harandi and Emami [27] reconsidered Theorem 1 as the framework of partially ordered metric spaces and they presented taking into account existence theorem.

Theorem 2. *Let (Y,d) be a partially ordered complete metric space. Assume $S:Y \to Y$ is a mapping such that there exists $u_0 \in Y$ with $u_0 \preceq Su_0$ and $\alpha \in \mathcal{F}$ such that*

$$d(Su, Sv) \leq \alpha(d(u,v))d(u,v), \text{ for any } u,v \in Y \text{ with } u \succeq v. \tag{2}$$

Hence, S has a fixed point supported that either S is continuous or Y is such that if an increasing sequence $\{u_n\} \to u$, then $u_n \leq u$ for all n.

In 2015, Karapinar [28] demonstrated the following specific results:

Theorem 3. *[28] Let (Y, σ) be a complete metric-like space. Assume that $S:Y \to Y$ is a mapping. If there exists $\beta \in \mathfrak{S}$ such that*

$$\sigma(Su, Sv) \leq \beta(\sigma(u,v))\sigma(u,v) \tag{3}$$

for all $u, v \in Y$, then S has a unique fixed point $u^* \in Y$ with $\sigma(u^*, u^*) = 0$.

The notion of quasi-contraction presented by Reference [29], is known as one of the foremost common contractive self-mappings.

A mapping $S:Y \to Y$ is expressed to be a quasi contraction if there exists $0 \leq \lambda < 1$ such that

$$d(Su, Sv) \leq \lambda \max\{d(u,v).d(u,fv), d(u,fv), d(fu,v), d(u,fv)\}, \tag{4}$$

for any $u, v \in Y$.

In this paper, we show the generalized Geraghty (α, ψ, ϕ)-quasi contraction type mapping in partially ordered metric like space, then we present some fixed and common fixed point theorems for such mappings in an ordered complete metric-like space. We investigate this new contractive mapping as a generalized weakly contractive mapping in our main results, then we display an example and an application to support our obtained results.

2. Preliminaries

In this section, we review a few valuable definitions and assistant results that will be required within the following sections.

Definition 1. *[5] Let Y be a nonempty set. A function $\sigma:Y \times Y \to [0,\infty)$ is expressed to be a metric-like space on X if for any $u,v,z \in Y$, the accompanying stipulations satisfied:*

(σ_1) $\sigma(u,v) = 0 \Rightarrow u = v$,
(σ_2) $\sigma(u,v) = \sigma(v,u)$,
(σ_3) $\sigma(u,z) \leq \sigma(u,v) + \sigma(v,z)$.

The pair (Y, σ) is called a metric-like space.

Obviously, we can consider that every metric space and partial metric space could be a metric-like space. However, this assertion isn't valid.

Example 1. *[5] Let $Y = \{0,1\}$ and*

$$\sigma(u,v) = \begin{cases} 2, & \text{if } u = v = 0; \\ 1, & \text{otherwise.} \end{cases} \quad (5)$$

We note that $\sigma(0,0) \not\leq \sigma(0,1)$. So, (Y,σ) is a metric-like space and at the same time it is not a partial metric space.

Additonally, each metric-like σ on Y create a topology τ_σ on Y whose use as a basis of the group of open σ-balls

$$B_\sigma(Y,\epsilon) = \{u \in Y : |\sigma(u,v) - \sigma(u,u)| < \epsilon\}, \text{ for all } u,v \in Y \text{ and } \epsilon > 0.$$

Let (Y,σ) be a metric-like space and $f:Y \to Y$ be a continuous mapping. Then

$$\lim_{n \to \infty} u_n = u \Rightarrow \lim_{n \to \infty} fu_n = fu.$$

A sequence $\{u_n\}$ of elements of Y is considered σ-Cauchy if the limit $\lim_{n,m \to \infty} \sigma(u_n, u_m)$ exists as a finite number. The metric-like space (Y,σ) is considered complete if for each σ-Cauchy sequence $\{u_n\}$, there is some $u \in Y$ such that

$$\lim_{n \to \infty} \sigma(u_n, u) = \sigma(u,u) = \lim_{n,m \to \infty} \sigma(u_n, u_m).$$

Remark 1. *[30] Let $Y = \{0,1\}$, and $\sigma(u,v) = 1$ for each $u,v \in Y$ and $u_n = 1$ for each $n \in \mathbb{N}$. Then, it is easy to see that $u_n \to 0$ and $u_n \to 1$ and so in metric-like spaces the limit of a convergent sequence is not necessarily unique.*

Lemma 1. *[30] Let (Y,σ) be a metric-like space. Let $\{u_n\}$ be a sequence in Y such that $u_n \to u$ where $u \in Y$ and $\sigma(u,u) = 0$. Then, for all $u,v \in Y$, we have $\lim_{n \to \infty} \sigma(u_n, v) = \sigma(u,v)$.*

Example 2. *[5] Let $Y = \mathbb{R}$ and $\sigma:Y \times Y \to [0, +\infty)$ be defined by*

$$\sigma(u,v) = \begin{cases} 2n, & \text{if } u = v = 0; \\ n, & \text{otherwise.} \end{cases}$$

Then, we can consider (Y,σ) to be a metric-like space, but it does not satisfy the conditions of the partial metric space, as $\sigma(0,0) \not\leq \sigma(0,1)$.

Samet et al. [31] displayed the definition of α-admissible mapping as followings:

Definition 2. *[31] Let $S:X \to X$ and $\alpha:X \times X \to [0,\infty)$ are two functions. Then, S is called α-admissible if $\forall u,v \in X$ with $\alpha(u,v) \geq 1$ implies $\alpha(fu, fv) \geq 1$.*

Definition 3. *[32] Let $S,T:X \to X$ be two mappings and $\alpha:X \times X \to \mathbb{R}$ be a function. We consider that the pair (S,T) is α-admissible if*

$$u,v \in X, \ \alpha(u,v) \geq 1 \Rightarrow \alpha(Su, Tv) \geq 1 \text{ and } \alpha(Tu, Sv) \geq 1$$

Definition 4. *[33] Let $S:X \to X$ and $\alpha:X \times X \to [0,\infty)$. Then, S is called a triangular α-admissible mapping if*

(1) S is α-admissible,
(2) $\alpha(u,z) \geq 1$ and $\alpha(z,v) \geq 1$ imply $\alpha(u,v) \geq 1$.

Definition 5. *[32] Let $S, T: X \to X$ and $\alpha: X \times X \to [0, \infty)$. Then, (S, T) is called a triangular α-admissible mapping if*

(1) *The pair (S, T) is α-admissible,*
(2) *$\alpha(u,z) \geq 1$ and $\alpha(z,v) \geq 1$ imply $\alpha(u,v) \geq 1$.*

Let Ψ indicate the set of functions $\psi:[0,\infty) \to [0,\infty)$ that approve the following stipulations:

(1) ψ is strictly continuous increasing,
(2) $\psi(t) = 0 \Leftrightarrow t = 0$.

and Φ indicates the set of all continuous functions $\phi:[0,\infty) \to [0,\infty)$ with $\phi(t) > \psi(t)$ for all $t > 0$ and $\phi(0) = 0$.

Definition 6. *[12] Let $(X, d, \preceq)$ be a partially ordered metric space. Assume $f, g: X \to X$ are two mappings. Then:*

(1) *For all $x, y \in X$ are said to be comparable if $x \preceq y$ or $y \preceq x$ holds,*
(2) *f is said to be nondecreasing if $x \preceq y$ implies $fx \preceq fy$,*
(3) *f, g are called weakly increasing if $fx \preceq gfx$ and $gx \preceq fgx$ for all $x \in X$,*
(4) *f is called weakly increasing if f and I are weakly increasing, where I is denoted to the identity mapping on X.*

3. Main Results

In this section, we present the notation of generalized Geraghty (α, ψ, ϕ)-quasi contraction self-mappings in partially ordered metric-like space. Then, we present some fixed and common fixed point theorems for such self-mappings. We investigate this new contractive self-mapping as a generalized weakly contractive self-mapping which is a generalization of the results of Reference [34]. Results of this kind are amongst the most useful in fixed point theory and it's applications.

Definition 7. *Let (X, σ) be a partially ordered metric-like space and $S, T: X \to X$ be two mappings. Then, we consider that the pair (S, T) is generalized Geraghty (α, ψ, ϕ)-quasi contraction self-mapping if there exist $\alpha: X \times X \to [0, \infty)$, $\beta \in \mathfrak{S}$, $\psi \in \Psi$ and $\phi:[0,\infty) \to [0,\infty)$ are continuous functions with $\phi(t) \leq \psi(t)$ for all $t > 0$ such that*

$$\alpha(x,y)\psi(\sigma(Sx, Ty)) \leq \lambda \beta(\psi(M_{x,y}))\phi(M_{x,y}), \tag{6}$$

holds for all elements $x, y \in X$ and $0 \leq \lambda < 1$, where

$$M_{x,y} = \max\{\sigma(x,y), \sigma(x, Sx), \sigma(y, Ty), \sigma(Sx, y), \sigma(x, Ty)\}.$$

The following two lemmas will be utilized proficiently within the verification of our fundamental result.

Lemma 2. *If $\psi \in \Psi$ and $\phi:[0,\infty) \to [0,\infty)$ are continuous function that satisfy the condition $\psi(t) > \phi(t)$ for all $t > 0$, then $\phi(0) = 0$.*

Proof. From the assumption $\phi(t) < \psi(t)$, since ψ and ϕ are continuous, we have

$$0 \leq \phi(0) = \lim_{t \to 0} \phi(t) \leq \lim_{t \to 0} \psi(t) = \psi(0) = 0.$$

□

Lemma 3. *Let $S, T: X \to X$ be two mappings and $\alpha: X \times X \to [0, \infty)$ be a function such that S, T are triangular α-admissible. Suppose that there exists $x_0 \in X$ such that $\alpha(x_0, Sx_0) \geq 1$. Define a sequence $\{x_n\}$ in X by $Sx_{2n} = x_{2n+1}$ and $Tx_{2n+1} = x_{2n+2}$. Then $\alpha(x_n, x_m) \geq 1$ for all $m, n \in \mathbb{N}$ with $n < m$.*

Proof. Since $\alpha(x_0, Sx_0) \geq 1$ and S, T are α-admissible, we get

$$\alpha(x_0, x_1) = \alpha(x_0, Sx_0) \geq 1.$$

By triangular α-admissibility, we get

$$\alpha(Sx_0, Tx_1) = \alpha(x_1, x_2) \geq 1$$

and

$$\alpha(TSx_0, STx_1) = \alpha(x_2, x_3) \geq 1.$$

Again, since $\alpha(x_2, x_3) \geq 1$, then

$$\alpha(Sx_2, Tx_3) = \alpha(x_3, x_4) \geq 1$$

and

$$\alpha(TSx_2, STx_3) = \alpha(x_4, Sx_5) \geq 1.$$

By proceeding the above process, we conclude that $\alpha(x_n, x_{n+1}) \geq 1$ for all $n \in \mathbb{N} \cup \{0\}$. Now, we prove that $\alpha(x_n, x_m) \geq 1$, for all $m, n \in \mathbb{N}$ with $n < m$. Since

$$\begin{cases} \alpha(x_n, x_{n+1}) \geq 1, \\ \alpha(x_{n+1}, x_{n+2}) \geq 1, \end{cases}$$

then, we have

$$\alpha(x_n, x_{n+2}) \geq 1.$$

Again, since

$$\begin{cases} \alpha(x_n, x_{n+2}) \geq 1 \\ \alpha(x_{n+2}, x_{n+3}) \geq 1, \end{cases}$$

we deduce that

$$\alpha(x_n, x_{n+3}) \geq 1.$$

By continuing this process, we have

$$\alpha(x_n, x_m) \geq 1$$

for all $n \in \mathbb{N}$ with $m > n$. □

Lemma 4. *Let $(X, \preceq, \sigma)$ be a partially ordered metric-like space. Assume S, T are two self-mappings of X which the pair (S, T) is generalized (α, ψ, ϕ)-quasi contraction self-mappings. Fix $x_1 \in X$ and define a sequence $\{x_n\}$ by $x_{2n+1} = Sx_{2n}$ and $x_{2n+2} = Tx_{2n+1}$ for all $n \in \mathbb{N}$. If $\lim_{n \to \infty} \sigma(x_n, x_{n+1}) = 0$ and the sequence $\{x_n\}$ is nondecreasing, then $\{x_n\}$ is a Cauchy sequence.*

Proof. Since S, T are a generalized (α, ψ, ϕ)-quasi contraction non-self mapping, then there exist $\psi \in \Psi, \phi \in \Phi$ such that

$$\alpha(x,y)\psi(\sigma(Sx, Ty)) \leq \lambda\beta(\psi(M_{x,y}))\phi(M_{x,y}), \quad (7)$$

holds for all elements $x, y \in X$ and $0 \leq \lambda < 1$, where

$$M_{x,y} = \max\{\sigma(x,y), \sigma(x, Sx), \sigma(y, Ty), \sigma(Sx, y), \sigma(x, Ty)\}.$$

Now, we show that the sequence $\{x_n\}$ is Cauchy sequence. Assume, for contradiction's sake, that $\{x_n\}$ isn't Cauchy sequence. Therefore, there exist $\epsilon > 0$ and two subsequences $\{n_k\}$ and $\{m_k\}$ of the sequence $\{x_n\}$ such that $\sigma(x_{2n_k}, x_{2m_k}), \sigma(x_{2n_k-1}, x_{2m_k})$ and $\sigma(x_{2n_k}, x_{2m_k+1})$ converge to ϵ^+ when $k \to \infty$.

$$n_k > m_k > k, \ \sigma(x_{2n_k}, x_{2m_k-2}) < \epsilon, \ \sigma(x_{2n_k}, x_{2m_k}) \geq \epsilon. \quad (8)$$

By the above inequalities and triangle inequality property, we imply that

$$\begin{aligned}
\epsilon &\leq \sigma(x_{2n_k}, x_{2m_k}) \leq \sigma(x_{2n_k}, x_{2m_k-2}) + \sigma(x_{2m_k-2}, x_{2m_k-1}) + \sigma(x_{2m_k-1}, x_{2m_k}) \\
&< \epsilon + \sigma(x_{2m_k-2}, x_{2m_k-1}) + \sigma(x_{2m_k-1}, x_{2m_k}).
\end{aligned}$$

In view of $\lim_{n \to \infty} \sigma(x_n, x_{n+1}) = 0$ and letting $k \to \infty$ in the above inequalities, we obtain

$$\lim_{k \to \infty} \sigma(x_{2n_k}, x_{2m_k}) = \epsilon. \quad (9)$$

By the triangle inequality, we have

$$\begin{aligned}
\sigma(x_{2n_k}, x_{2m_k}) &\leq \sigma(x_{2n_k}, x_{2n_k+1}) + \sigma(x_{2n_k+1}, x_{2m_k}) \\
&\leq \sigma(x_{2n_k}, x_{2n_k+1}) + \sigma(x_{2n_k+1}, x_{2m_k+1}) + \sigma(x_{2m_k+1}, x_{2m_k}) \\
&\leq \sigma(x_{2n_k}, x_{2n_k+1}) + \sigma(x_{2n_k+1}, x_{2n_k+2}) + \sigma(x_{2n_k+2}, x_{2m_k}) + 2\sigma(x_{2m_k}, x_{2m_k+1}) \\
&\leq 2\sigma(x_{2n_k}, x_{2n_k+1}) + 2\sigma(x_{2m_k+2}, x_{2m_k+1}) + \sigma(x_{2n_k}, x_{2m_k}) + 2\sigma(x_{2m_k}, x_{2m_k+1}).
\end{aligned}$$

Taking the limit as $k \to \infty$ in the above inequalities and using Equation (9), we get

$$\lim_{k \to \infty} \sigma(x_{2n_k}, x_{2m_k}) = \lim_{k \to \infty} \sigma(x_{2n_k+1}, x_{2m_k}) = \lim_{k \to \infty} \sigma(x_{2n_k+1}, x_{2m_k+1}) = \epsilon. \quad (10)$$

Since $x_{n_k+1} \preceq x_{m_k}$ and $\alpha(x_{n_k+1}, x_{m_k}) \geq 1$ for all $k \in \mathbb{N}$, so by substituting x with x_{n_k+1} and y with x_{m_k} in Equation (7), it follows that

$$\psi(\sigma(x_{n_k+1}, x_{m_k})) \leq \alpha(x_{n_k+1}, x_{m_k})\psi(\sigma(Sx_{n_k}, Tx_{m_k-1})) \leq \lambda\beta(\psi(M_{x,y}))\phi(M_{x,y}), \quad (11)$$

holds for all elements $x, y \in X$ and $0 \leq \lambda < 1$, where

$$\begin{aligned}
M_{x_{n_k}, x_{m_k-1}} &= \max\{\sigma(x_{n_k}, x_{m_k-1}), \sigma(x_{n_k}, Sx_{n_k}), \sigma(x_{m_k-1}, Tx_{m_k-1}), \\
&\quad \sigma(Sx_{n_k}, x_{m_k-1}), \sigma(x_{n_k}, Tx_{m_k-1})\} \\
&= \max\{\sigma(x_{n_k}, x_{m_k-1}), \sigma(x_{n_k}, x_{n_k+1}), \sigma(x_{m_k-1}, x_{m_k}), \\
&\quad \sigma(x_{n_k+1}, x_{m_k-1}), \sigma(x_{n_k}, x_{m_k})\}.
\end{aligned}$$

Taking the limit as $k \to \infty$ of the above inequality and applying Equations (9), (10), we get

$$\lim_{k \to \infty} M_{x_{2n_k}, x_{2m_k}} = \epsilon. \quad (12)$$

Letting $k \to \infty$ in Equation (11) and using $\phi \in \Phi$, $\beta \in \mathcal{S}$ and Equation (12), we deduce that

$$\psi(\epsilon) \leq \lambda \beta(\psi(\epsilon))\phi(\epsilon)$$
$$< \lambda \phi(\epsilon)$$
$$< \lambda \psi(\epsilon).$$

This is possible only if $\epsilon = 0$. Which contradicts the positivity of ϵ. Therefore, we get the desired result. □

Theorem 4. *Let (X, σ) be a partially ordered metric like space. Assume that $S, T: X \to X$ are two self-mappings fulfilling the following conditions:*

(1) *(S, T) is triangular α-admissible and there exists an $x_0 \in X$ such that $\alpha(x_0, Sx_0) \geq 1$,*
(2) *the pair (S, T) is weakly increasing,*
(3) *the pair (S, T) is a generalized Geraghty (α, ψ, ϕ)-quasi contraction non-self mapping,*
(4) *S and T are σ-continuous mappings.*

Then, the pair (S, T) has a common fixed point $z \in X$ with $\sigma(z, z) = 0$. Moreover, assume that if $x_1, x_2 \in X$ such $\sigma(x_1, x_1) = \sigma(x_2, x_2) = 0$ implies that x_1 and x_2 are comparable elements. Then the common fixed point of the pair (S, T) is unique.

Proof. Let $x_0 \in X$ such that $\alpha(x_0, Sx_0) \geq 1$. Define the sequence $\{x_n\}$ in X as follows:

$$x_{2n+1} = Sx_{2n} \; x_{2n+2} = Tx_{2n+1} \; \text{for all } n \geq 0. \tag{13}$$

Suppose that $x_{2n} \neq x_{2n+1}$ for all $n \in \mathbb{N}_0$. Then, $\sigma(x_{2n}, x_{2n+1}) > 0$ for all $n \in \mathbb{N}_0$. Indeed, if $x_{2n} \neq x_{2n+1}$, which is a contradiction. By using the assumption of Equations (1), (2), and Lemma 3, we have

$$\alpha(x_n, x_{n+1}) \geq 1 \tag{14}$$

for all $n \in \mathbb{N} \cup \{0\}$.

Since the pair (S, T) is weakly increasing, we have

$$x_1 = Sx_0 \preceq TSx_0 = x_2 = Sx_1 \preceq \ldots x_{2n} \preceq TSx_{2n} = x_{2n+2} \preceq \ldots.$$

Thus, $x_n \preceq x_{n+1}$, for all $n \in \mathbb{N}$. Since $\alpha(x_{2n}, x_{2n+1}) \geq 1$, by applying Equation (6), we obtain

$$\psi(\sigma(x_{2n+1}, x_{2n+2})) = \psi(\sigma(Sx_{2n}, Tx_{2n+1}))$$
$$\leq \alpha(x_{2n}, x_{2n+1})\psi(\sigma(Sx_{2n}, Tx_{2n+1}))$$
$$\leq \lambda\beta(\psi(M_{x_{2n}, x_{2n+1}}))\phi(M_{x_{2n}, x_{2n+1}}). \tag{15}$$

Set $\sigma_n = \sigma(x_{2n+1}, x_{2n+2})$. We have

$$\psi(\sigma_n) = \psi(\sigma(x_{2n+1}, x_{2n+2})) \tag{16}$$
$$\leq \lambda\beta(\psi(M_{x_{2n}, x_{2n+1}}))\phi(M_{x_{2n}, x_{2n+1}}). \tag{17}$$

For the rest, for each n assume that $(\sigma_n \neq 0)$.

$$\begin{aligned}
M_{x_{2n}, x_{2n+1}} &= \max\{\sigma(x_{2n}, x_{2n+1}), \sigma(x_{2n}, Sx_{2n}), \sigma(x_{2n+1}, Tx_{2n+1}), \sigma(Sx_{2n}, x_{2n+1}), \sigma(x_{2n}, Tx_{2n+1})\} \\
&= \max\{\sigma(x_{2n}, x_{2n+1}), \sigma(x_{2n}, x_{2n+1}), \sigma(x_{2n+1}, x_{2n+2}), \sigma(x_{2n+1}, x_{2n+1}), \sigma(x_{2n}, x_{2n+2})\} \\
&= \max\{\sigma(x_{2n}, x_{2n+1}), \sigma(x_{2n+1}, x_{2n+2}), \sigma(x_{2n}, x_{2n+2})\} \\
&= \max\{\sigma_{n-1}, \sigma_n, \sigma_{n-1} + \sigma_n\}
\end{aligned}$$

If for some $n \in \mathbb{N}$, $\max\{\sigma_{n-1}, \sigma_n, \sigma_{n-1} + \sigma_n\} = \sigma_n$ then from Equation (16), we find that $\psi(\sigma_n) < \lambda \psi(\sigma_n)$ which is a contradiction with respect to $0 \leq \lambda < 1$. We deduce $\max\{\sigma_{n-1}, \sigma_n, \sigma_{n-1} + \sigma_n\} = \max\{\sigma_{n-1}, \sigma_{n-1} + \sigma_n\}$. Therefore Equation (16) becomes

$$\psi(\sigma_n) < \lambda \psi(\max\{\sigma_{n-1}, \sigma_{n-1} + \sigma_n\}).$$

Put

$$\gamma = \max\{\lambda, \frac{\lambda}{1-\lambda}\}.$$

Thus,

$$\psi(\sigma_n) \leq \gamma \beta(\psi(\sigma_{n-1})) \phi(\sigma_{n-1}), \text{ for all } n \in \mathbb{N}_0. \tag{18}$$

It is clear that $\gamma < 1$. Therefore, the sequence $\{\sigma(x_n, x_{n+1})\}$ is a decreasing sequence. Thus, there exists $r \geq 0$ such that

$$\lim_{n \to \infty} \sigma(x_n, x_{n+1}) = r.$$

Now, we show that $r = 0$. Presume to the contrary, that is $r > 0$. Since $\beta \in \mathcal{S}$ and by using the condition of Theorem 4 and taking the limit as $k \to \infty$ in Equation (18), we conclude

$$\psi(r) \leq \lambda \beta(\psi(r)) \phi(r) < \lambda \phi(r) < \lambda \psi(r),$$

which could be a contradiction. So $r = 0$. Then,

$$\lim_{n \to \infty} \sigma(x_n, x_{n+1}) = 0.$$

Lemma 4 implies that $\{x_n\}$ is a Cauchy sequence and from the completeness of (X, σ), then there exists a $x^* \in X$ in order that

$$\lim_{n \to \infty} \sigma(x_n, x^*) = \sigma(x^*, x^*) = \lim_{n, m \to \infty} \sigma(x_n, x_m). \tag{19}$$

Whereas, S and T are continuous, we conclude

$$\lim_{n \to \infty} \sigma(x_{n+1}, Tx^*) = \lim_{n \to \infty} \sigma(Sx_n, Tx^*) = \sigma(Sx^*, Tx^*), \tag{20}$$

$$\lim_{n \to \infty} \sigma(Sx^*, x_{n+1}) = \lim_{n \to \infty} \sigma(Sx^*, Tx_n) = \sigma(Sx^*, Tx^*). \tag{21}$$

By Lemma 1 and Equation (19), we obtain that

$$\lim_{n \to \infty} \sigma(x_{n+1}, Tx^*) = \sigma(x^*, Tx^*) \tag{22}$$

and

$$\lim_{n \to \infty} \sigma(Sx^*, x_{n+1}) = \sigma(Sx^*, x^*). \tag{23}$$

By merging Equations (20) and (22), we deduce that $\sigma(x^*, Tx^*) = \sigma(Sx^*, x^*)$. In addition, by Equations (21) and (23), we deduce that $\sigma(Sx^*, x^*) = \sigma(Sx^*, Tx^*)$. So

$$\sigma(x^*, Tx^*) = \sigma(Sx^*, x^*) = \sigma(Sx^*, Tx^*). \tag{24}$$

Presently, we display that $\sigma(x^*, Tx^*) = 0$. Assume the opposite, that is, $\sigma(x^*, Tx^*) > 0$, we get

$$\begin{aligned}\psi(\sigma(x^*, Tx^*)) &= \psi(\sigma(Sx^*, Tx^*)) \\ &\leq \lambda \beta(\psi(M_{x^*, x^*})) \phi(M_{x^*, x^*}),\end{aligned} \tag{25}$$

where

$$M_{x^*,x^*} = \max\{\sigma(x^*,x^*), \sigma(x^*,Sx^*), \sigma(x^*,Tx^*), \sigma(Sx^*,x^*), \sigma(x^*,Tx^*),\}$$
$$= \max\{\sigma(x^*,Tx^*), \sigma(x^*,Sx^*)\}$$
$$= \max\{\sigma(x^*,Tx^*), \sigma(x^*,Tx^*)\}.$$

Therefore, from Equation (25), we get

$$\begin{aligned}\psi(\sigma(x^*,Tx^*)) &\leq \beta(\psi(\sigma(x^*,Tx^*)))\phi(\sigma(x^*,Tx^*))\\ &< \lambda\phi(\sigma(x^*,Tx^*))\\ &< \lambda\psi(\sigma(x^*,Tx^*))\end{aligned} \qquad (26)$$

Since $\psi \in \Psi$, we have $\sigma(x^*,Tx^*) < \lambda\sigma(x^*,Tx^*)$ which is a discrepancy. Thus, we have $\sigma(x^*,Tx^*) = 0$. Hence, $Tx^* = x^*$. From Equation (24), we deduce that $\sigma(x^*,Sx^*) = 0$. Therefore, $Sx^* = x^*$. Hence, x^* is a common fixed point of S and T. To demonstrate the uniqueness of the common fixed point, we suppose that $\tilde{x}$ is another fixed point of S and T. Directly, we prove that $\sigma(\tilde{x},\tilde{x}) = 0$. Assume the antithesis, that is, $\sigma(\tilde{x},\tilde{x}) > 0$. Since $\tilde{x} \preceq \tilde{x}$, we get

$$\begin{aligned}\psi(\sigma(\tilde{x},\tilde{x})) &= \psi(\sigma(S\tilde{x},T\tilde{x}))\\ &\leq \lambda\beta(\psi(\sigma(\tilde{x},\tilde{x})))\phi(\sigma(\tilde{x},\tilde{x}))\\ &< \lambda\phi(\sigma(\tilde{x},\tilde{x}))\\ &< \lambda\psi(\sigma(\tilde{x},\tilde{x}))\end{aligned}$$

which is a discrepancy. Thus, $\sigma(\tilde{x},\tilde{x}) = 0$. Therefore, by the further conditions on X, we deduce that x^* and $\tilde{x}$ are comparable. Presently, suppose that $\sigma(x^*,\tilde{x}) \neq 0$. Then

$$\begin{aligned}\psi(\sigma(x^*,\tilde{x})) &= \psi(\sigma(Sx^*,T\tilde{x}))\\ &\leq \lambda(\psi(\sigma(x^*,\tilde{x})))\phi(\sigma(x^*,\tilde{x}))\\ &< \lambda\phi(\sigma(x^*,\tilde{x}))\end{aligned}$$

which is a discrepancy with the condition of Theorem 4. Therefore, $\sigma(x^*,\tilde{x}) = 0$. Hence, $x^* = \tilde{x}$. Thus, S and T have a unique common fixed point. □

It is additionally conceivable to expel the continuity of S and T by exchanging a weaker condition.

$(\mathcal{C})$ If $\{x_n\}$ is a nondecreasing sequence in X such that $\alpha(x_n, x_{n+1}) \geq 1$ for all $n \in \mathbb{N} \cup \{0\}$ and $x_n \to u \in X$ as $n \to \infty$, then there exists a subsequence $\{x_{n_l}\}$ of $\{x_n\}$ such that $x_{n_l} \preceq u$ for all l.

Theorem 5. *Let (X, σ) be a partially ordered metric-like space. Assume that $S, T{:}X \to X$ are two self-mappings fulfilling the following conditions:*

(1) *the pair (S,T) is triangular α-admissible,*
(2) *there exists an $x_0 \in X$ such that $\alpha(x_0, Sx_0) \geq 1$,*
(3) *the pair (S,T) is a generalized Geraghty (α, ψ, ϕ)-quasi contraction non-self mapping,*
(4) *the pair (S,T) is weakly increasing,*
(5) *$(\mathcal{C})$ holds.*

Then, the pair (S,T) has a common fixed point $v \in X$ with $\sigma(v,v) = 0$. Moreover, suppose that if $x_1, x_2 \in X$ such $\sigma(x_1,x_1) = \sigma(x_2,x_2) = 0$ implies that x_1 and x_2 are comparable. Then, the common fixed point of the pair (S,T) is unique.

Proof. Here, we define $\{x_n\}$ as in the proof of Theorem 4. Clearly $\{x_n\}$ is a Cauchy sequence in X, then there exists $v \in X$ in order that

$$\lim_{n \to \infty} x_n = v \tag{27}$$

As a result of the condition of Equation (5), there exists a subsequence $\{x_{n_l}\}$ of $\{x_n\}$ in order that $x_{n_l} \preceq v$ for all l. Therefore, x_{n_l} and v are comparable. In addition, from Equation (13) on taking limit as $n \to \infty$ and using Equation (27), we get

$$\lim_{n \to +\infty} x_n = v.$$

$$\lim_{n \to \infty} Sx_{2n_l} = \lim_{n \to \infty} x_{2n_l+1} = v, \lim_{n \to \infty} Tx_{2n_l+1} = \lim_{n \to \infty} x_{2n_l+2} = v. \tag{28}$$

From the definition of α yields that $\alpha(x_{n_l}, v) \geq 1$ for all l. Now by applying Equation (6), we have

$$\begin{aligned}
\psi(\sigma(x_{2n_l+1}, Tv)) &= \psi(\sigma(Sx_{2n}, Tv)) \\
&\leq \lambda \beta(\psi(M_{x_{2n_l}, v}))\phi(M_{x_{2n_l}, v}) \\
&< \lambda \phi(M_{x_{2n_l}, v}) \\
&< \lambda \psi(M_{x_{2n_l}, v})
\end{aligned} \tag{29}$$

where

$$M_{x_{2n_l}, v} = \max\{\sigma(x_{2n}, v), \sigma(x_{2n}, Sx_{2n}), \sigma(v, Tv), \sigma(Sx_{2n}, v), \sigma(x_{2n}, Tv)\}$$

Letting $l \to +\infty$ and using Equations (27) and (28), we have

$$\lim_{l \to \infty} M_{x_{2n_l}, v} = \max\{\sigma(v, Sv), \sigma(v, Tv)\} \tag{30}$$

Case I: Assume that $\lim_{l \to \infty} M_{x_{2n_l}, v} = \sigma(v, Tv)$.
From Equation (30) and letting $l \to \infty$ in Equation (29). Then, we have

$$\psi(\sigma(v, Tv)) < \lambda \psi(\sigma(v, Tv)).$$

Regarding the concept of ψ, we deduce that $\sigma(v, Tv) < \lambda \sigma(v, Tv)$ which is a discrepancy. Hence, we get that $\sigma(v, Tv) = 0$. As a result of (σ_1), we have $v = Tv$.
Case II: Assume that $\lim_{l \to \infty} M_{x_{2n_l}, v} = \sigma(v, Sv)$. Then, arguing like above, we get $v = Sv$. Thus, $v = Sv = Tv$. Uniqueness of the fixed point is follows from the Theorem 4. This completes the proof. □

If we set $S = T$ and $M(x, y) = \max\{\sigma(x, y), \sigma(x, Tx), \sigma(y, Ty), \sigma(Tx, y), \sigma(x, Ty)\}$ in Theorems 4 and 5, then we obtain the following corollaries.

Corollary 1. *Let (X, σ) be a partially ordered metric-like space and $\alpha: X \times X \to [0, \infty)$ a function. Assume that $S: X \to X$ holds the following:*

(1) *there exists $\psi \in \Psi, \beta \in \mathfrak{S}$ and a continuous function $\phi : [0, \infty) \to [0, \infty)$ are continuous functions with $\phi(t) < \psi(t)$ for all $t > 0$ such that*

$$\alpha(x, y)\psi(\sigma(Sx, Sy)) \leq \lambda \beta(\psi(M_{x,y}))\phi(M_{x,y}), \tag{31}$$

holds for all comparable elements $x, y \in X$ and $0 \leq \lambda < 1$,
(2) *S is triangular α-admissible and there exists an $x_0 \in X$ such that $\alpha(x_0, Sx_0) \geq 1$,*
(3) *$Sx \preceq S(Sx)$ for all $x, y \in X$,*

(4) T is σ-continuous mappings.

Then, S has an unique fixed point $v \in X$ with $\sigma(v,v) = 0$.

Corollary 2. *Let (X, σ) be a partially ordered metric-like space and $\alpha: X \times X \to [0, \infty)$ a function. Assume that $S: X \to X$ holds the following:*

(1) *there exists $\psi \in \Psi, \beta \in \mathfrak{S}$ and a continuous function $\phi : [0, \infty) \to [0, \infty)$ are continuous functions with $\phi(t) < \psi(t)$ for all $t > 0$ such that*

$$\alpha(x,y)\psi(\sigma(Sx, Sy)) \leq \lambda \beta(\psi(M_{x,y}))\phi(M_{x,y}), \qquad (32)$$

holds for all comparable elements $x, y \in X$ and $0 \leq \lambda < 1$,
(2) *S is triangular α-admissible and there exists an $x_0 \in X$ such that $\alpha(x_0, Sx_0) \geq 1$,*
(3) *$Sx \preceq S(Sx)$ for all $x, y \in X$,*
(4) *$(\mathcal{C})$ holds.*

Then, S has an unique fixed point $v \in X$ with $\sigma(v,v) = 0$.

If we take $\alpha(x,y) = 1$ in Theorems 4 and 5, we have the following corollaries.

Corollary 3. *Let (X, σ) be a partially ordered metric-like space. Assume $S, T: X \to X$ are two mappings holding the following:*

(1) *there exists $\psi \in \Psi, \beta \in \mathfrak{S}$ and a continuous function $\phi : [0, \infty) \to [0, \infty)$ are continuous functions with $\phi(t) < \psi(t)$ for all $t > 0$ such that*

$$\psi(\sigma(Sx, Ty)) \leq \lambda \beta(\psi(M_{x,y}))\phi(M_{x,y}), \qquad (33)$$

holds for all comparable elements $x, y \in X$ and $0 \leq \lambda < 1$, where

$$M_{x,y} = \max\{\sigma(x,y), \sigma(x, Sx), \sigma(y, Ty), \sigma(Sx, y), \sigma(x, Ty)\}.$$

(2) *the pair (S, T) is weakly increasing,*
(3) *S and T are σ-continuous mappings.*

Then, the pair S, T has an unique common fixed point $v \in X$ with $\sigma(v,v) = 0$.

Corollary 4. *Let (X, σ) be a partially ordered metric-like space, Assume $S, T: X \to X$ are two mappings holding the following:*

(1) *there exists $\psi \in \Psi, \beta \in \mathfrak{S}$ and a continuous function $\phi : [0, \infty) \to [0, \infty)$ are continuous functions with $\phi(t) < \psi(t)$ for all $t > 0$ such that*

$$\psi(\sigma(Tx, Ty)) \leq \lambda \beta(\psi(M_{x,y}))\phi(M_{x,y}), \qquad (34)$$

holds for all comparable elements $x, y \in X$ and $0 \leq \lambda < 1$, where

$$M_{x,y} = \max\{\sigma(x,y), \sigma(x, Sx), \sigma(y, Ty), \sigma(Sx, y), \sigma(x, Ty)\},$$

(2) *the pair (S, T) is weakly increasing,*
(3) *the pair (S, T) is a generalized (α, ψ, ϕ)-quasi contraction non-self,*
(4) *$(\mathcal{C})$ holds.*

Then, the pair S, T has an unique common fixed point $v \in X$ with $\sigma(v,v) = 0$.

4. Consequences

If we put $M_{x,y} = \sigma(x,y)$, then, by Theorems 4 and 5, we get the following corollaries as an expansion of results from the literature.

Corollary 5. *Let (X,σ) be a partially ordered metric like space and $\alpha: X \times X \to [0,\infty)$ be a function. Suppose that $S,T: X \to X$ are two self-mappings holding the following:*

(1) *(S,T) is triangular α-admissible and there exists an $x_0 \in X$ such that $\alpha(x_0, Sx_0) \geq 1$,*
(2) *there exists $\psi \in \Psi, \beta \in \mathfrak{S}$ and a continuous function $\phi : [0,\infty) \to [0,\infty)$ are continuous functions with $\phi(t) < \psi(t)$ for all $t > 0$ in order that*

$$\psi(\sigma(Sx, Ty)) \leq \lambda \beta(\psi(\sigma(x,y)))\phi(\sigma(x,y)), \tag{35}$$

satisfies for $x,y \in X$ and $0 \leq \lambda < 1$,
(3) *the pair (S,T) is weakly increasing,*
(4) *the pair (S,T) is σ-continuous mappings.*

Then, the pair (S,T) has an unique common fixed point $v \in X$ with $\sigma(v,v) = 0$.

Corollary 6. *Let (X,σ) be a partially ordered metric-like space. Assume $S,T: X \to X$ are two mappings holding the following:*

(1) *(S,T) is triangular α-admissible and there exists an $x_0 \in X$ such that $\alpha(x_0, Sx_0) \geq 1$,*
(2) *there exists $\psi \in \Psi, \beta \in \mathfrak{S}$ and a continuous function $\phi : [0,\infty) \to [0,\infty)$ are continuous functions with $\phi(t) \leq \psi(t)$ for all $t > 0$ in order that*

$$\psi(\sigma(Sx, Ty)) \leq \lambda \beta(\psi(\sigma(x,y)))\phi(\sigma(x,y)), \tag{36}$$

satisfies for $x,y \in X$ and $0 \leq \lambda \leq 1$,
(3) *the pair (S,T) is weakly increasing,*
(4) *$(\mathcal{C})$ holds.*

Then, the pair (S,T) has an unique common fixed point $v \in X$ with $\sigma(v,v) = 0$.

Corollary 7. *Let (X,σ) be a partially ordered metric-like space. Assume $\alpha: X \times X \to [0,\infty)$ is a function and $S: X \to X$ is a mapping holding the following:*

(1) *S is triangular α-admissible and there exists an $x_0 \in X$ such that $\alpha(x_0, Sx_0) \geq 1$.*
(2) *there exists $\psi \in \Psi, \beta \in \mathfrak{S}$ and a continuous function $\phi : [0,\infty) \to [0,\infty)$ are continuous functions with $\phi(t) < \psi(t)$ for all $t > 0$ in order that*

$$\alpha(x,y)\psi(\sigma(Sx, Sy)) \leq \lambda \beta(\psi(\sigma(x,y)))\phi(\sigma(x,y)), \tag{37}$$

holds for all comparable elements $x,y \in X$ and $0 \leq \lambda < 1$,
(3) *$S \preceq S(Sx)$,*
(4) *the pair (S,T) is σ-continuous mappings.*

Then, S has an unique fixed point $v \in X$ with $\sigma(v,v) = 0$.

Corollary 8. *Let (X,σ) be a partially ordered metric-like space. Assume $\alpha: X \times X \to [0,\infty)$ is a function and $S: X \to X$ is a mapping holding the following:*

(1) *S is triangular α-admissible and there exists an $x_0 \in X$ such that $\alpha(x_0, Sx_0) \geq 1$,*

(2) there exists $\psi \in \Psi, \beta \in \mathfrak{S}$ and a continuous function $\phi : [0,\infty) \to [0,\infty)$ are continuous functions with $\phi(t) < \psi(t)$ for all $t > 0$ in order that

$$\alpha(x,y)\psi(\sigma(Sx,Sy)) \leq \lambda\beta(\psi(\sigma(x,y))\phi(\sigma(x,y)), \qquad (38)$$

satisfies for $x, y \in X$ and $0 \leq \lambda < 1$,
(3) $S \preceq S(Sx)$,
(4) (C) holds.

Then S has an unique fixed point $v \in X$ with $\sigma(v,v) = 0$.

Example 3. Let $X = \{0, 1, 2\}$ and specify the partial order $\preceq$ on X in order that

$$\preceq := \{(0,0), (1,1), (2,2), (0,2), (2,1), (0,1)\}.$$

Take into consideration that the function $S : X \to X$ specified as

$$S = \begin{pmatrix} 0 & 1 & 2 \\ 1 & 1 & 0 \end{pmatrix}, \qquad (39)$$

which increasing with respect to $\preceq$. Let $x_0 = 0$. Hence, $S(x_0) = 1$ and $S(S(X_0)) = S(1) = 1$. Characterize to begin with the metric like space σ on X by $\sigma(0,1) = 1, \sigma(0,2) = \frac{5}{2}, \sigma(1,2) = \frac{3}{2}$ and $\sigma(x,x) = 0$. Then, (X, σ) is a complete metric-like space. Let $\beta \in \mathcal{S}$ is given by $\beta(t) = \frac{e^t}{2}, \psi(t) = t, \lambda = \frac{1}{2}$ and $\phi(t) = \frac{2}{3}t$. Define a function $\alpha : X \times X \to [0,\infty)$ in order that

$$\alpha(x,y) = \begin{cases} 1 & \text{if } x \in \{0,1,2\} \\ 0 & \text{if otherwise}. \end{cases}$$

Note that $S \in X$ and is continuous. S is α-admissible mapping. Indeed, $\alpha(Sx, Sy) = 1$. If $(x,y) = (0,1)$, then $\alpha(0,1) = 1$ and

$$\begin{aligned} M_{0,1} &= \max\{\sigma(0,1), \sigma(0,S0), \sigma(1,S1), \sigma(S0,1), \sigma(0,S1)\} \\ &= \max\{\sigma(0,1), \sigma(0,1), \sigma(1,1), \sigma(1,1), \sigma(0,1)\} \\ &= \max\{1, 1, 0, 1, 0\} = 1. \end{aligned}$$

$\sigma(S0, S1) = \sigma(1,1) = 0$. Now

$$0 = \alpha(0,1)\psi(\sigma(\sigma(S0,S1))) \leq \beta(\psi(M_{0,1}))\phi(M_{0,1}) = \frac{1}{2}\beta(1) \simeq \phi(1) = \frac{1}{2} \times \frac{e}{2} \times \frac{2}{3} = \frac{e}{6}$$

holds.
If $(x,y) = (0,2)$, then $\alpha(0,2) = 1$ and

$$\begin{aligned} M_{0,2} &= \max\{\sigma(0,2), \sigma(0,S0), \sigma(2,S2), \sigma(S0,2), \sigma(0,S2)\} \\ &= \max\{\sigma(0,2), \sigma(0,1), \sigma(2,0), \sigma(1,2), \sigma(0,0)\} \\ &= \max\{\frac{5}{2}, 1, \frac{5}{2}, \frac{3}{2}, 0\} = \frac{5}{2}. \end{aligned}$$

$\sigma(S0, S2) = \sigma(1,0) = \frac{5}{2}$. Now

$$\frac{5}{2} = \alpha(0,2)\psi(\sigma(\sigma(S0,S2))) \leq \beta(\psi(M_{0,2}))\phi(M_{0,2}) = \frac{1}{2}\beta(\frac{e^{\frac{5}{2}}}{2}) \times \frac{2}{3} \times \frac{5}{2} = \frac{5e^{\frac{5}{2}}}{12}$$

holds. Similarly, for the case $(x = 1, y = 2)$, it is simple to examine that the contractive condition in Corollary 1 is satisfied.

All conditions (1)–(4) of Corollary 1 are satisfied. Hence S has a unique fixed point $x = 1$.

5. Application

The aim of this section is to give the existence of fixed points of an integral equation, where we can apply the obtained result of Corollary 1 to get a common solution.

We consider X with the partial order $\preceq$ presented by:

$$x \preceq y \Leftrightarrow x(t) \preceq y(t) \; for \; all \; t \in [0,1].$$

Let $X = C(I, \mathbb{R})$ be the set of continuous functions specified on $I = [0,1]$. The metric-like space $\sigma : X \times X \to [0, \infty)$ presented by

$$\sigma(x,y) = \sup_{t \in [0,1]} | x(t) - y(t) |,$$

for all $x, y \in X$. Since (X, σ) is a complete metric-like space. We consider the integral equation

$$x(t) = g(t) + \int_0^1 P(t,r) f(r, x(r)) dr; \; t \in [0,1] \tag{40}$$

for all $x \in X$.

We suppose that $f:[0,1] \times \mathbb{R} \to \mathbb{R}$ and $g:[0,1] \to \mathbb{R}$ are two continuous functions. Suppose that $P:[0,1] \times [0,1] \to [0, \infty)$ in order that

$$Sx(t) = g(t) + \int_0^1 P(t,r) f(r, x(r)) dr; \; t \in [0,1] \tag{41}$$

for all $x \in X$. Then, a solution of Equation (40) is a fixed point of S.

Now, We will prove the following Theorem with our obtained results.

Theorem 6. *Assume that the following conditions are satisfied:*

(i) There exists $\zeta : X \times X \to [0,1)$ such that for all $r \in [0,1]$ and for all $x, y \in X$

$$0 \leq | f(r, x(r)) - f(r, y(r)) | \leq \zeta(x,y) | x(r) - y(r) |,$$

(ii) there exists $\beta : [0, \infty) \to [0,1)$ such that

$$\lim_{n \to \infty} \beta(t_n) = 1 \Rightarrow \lim_{n \to \infty} t_n = 0,$$

and

$$\| \int_0^1 P(t,r) \zeta(x,y) dr \|_\infty \leq (\frac{1}{4} \beta(\| x - y \|_\infty)).$$

Then the integral Equation (41) has a unique solution in X.

Proof. By conditions (i) and (ii), we get

$$
\begin{aligned}
|S(x)(t) - S(y)(t)| &= \left| \int_0^1 P(t,r)[f(r,x(r)) - f(r,x(r))]dr \right| \\
&\leq \int_0^1 P(t,r) |f(r,x(r)) - f(r,y(r))| \, dr \\
&\leq \int_0^1 P(t,r)\zeta(x,y) |f(r,x(r)) - f(r,y(r))| \, dr \\
&\leq \int_0^1 P(t,r)\zeta(x,y) \|x-y\|_\infty \, dr \\
&\leq \sigma(x,y) \int_0^1 P(t,r)\zeta(x,y) dr \\
&\leq \frac{1}{4}\beta(\sigma(x,y))\sigma(x,y) \\
&= \frac{1}{2}\beta(\sigma(x,y))\frac{1}{2}\sigma(x,y) \\
&= \frac{1}{2}\beta(\sigma(x,y))\phi(\sigma(x,y)).
\end{aligned}
$$

At that point, we have

$$\|S(x)(t) - S(y)(t)\|_\infty \leq \frac{1}{2}\beta(\sigma(x,y))\phi(\sigma(x,y)).$$

for all $x, y \in X$.
Thus, we obtain

$$\sigma(Sx, Sy) \leq \frac{1}{2}\beta(\sigma(x,y))\phi(\sigma(x,y)), \; for all \, x, y \in X.$$

Lastly, we specify $\beta: X \times X \to [0, \infty)$ such that

$$\alpha(x,y) = \begin{cases} 1 & \text{if } x, y \in X, \\ 0 & \text{if } otherwise. \end{cases}$$

Then, we have

$$\alpha(x,y)\sigma(Sx, Sy) \leq \frac{1}{2}\beta(\sigma(x,y))\sigma(x,y).$$

Obviously, $\alpha(x,y) = 1$ and $\alpha(Sx, Sy) = 1$ for all $x, y, z \in X$. Therefore, S is triangular α–admissible mapping.

Hence, the hypotheses of Corollary 1 hold with $\psi(t) = t, \lambda = \frac{1}{2}$ and $\phi(t) = \frac{t}{2}$. Thus, S has a unique fixed point, that is, the integral Equation (40) has a unique solution in X. □

6. Conclusions

We have introduced some common fixed point results for generalized (α, ψ, ϕ)-quasi contraction self-mapping in partially ordered metric-like spaces. We have generalized weakly contractive mapping as we used quasi contraction self-mapping, α-admissible mapping, triangular α-admissible mapping and ψ, ϕ as strictly increasing and continuous functions. We have provided an example and application to show the superiority of our results over corresponding (common) fixed point results. Alternatively, we suggest finding new results by replacing the single-valued mapping with multi-valued mapping. Furthermore, we suggest generalizing more results in other spaces like b-metric space, metric-like space, and others. Otherwise, we suggest using our main results for non-self-mapping to establish the existence of an optimal approximate solution.

Author Contributions: All authors contributed equally to the main text.

Funding: This research was funded by UKM Grant DIP-2017-011 and Ministry of Education, Malaysia grant FRGS/1/2017/STG06/UKM/01/1 for financial support.

Acknowledgments: The author would like to thank the anonymous reviewers and editor for their valuable comments.

Conflicts of Interest: The authors declare no conflict of interest.

References

1. Qawaqneh, H.; Noorani, M.S.M.; Shatanawi, W.; Alsamir, H. Common fixed points for pairs of triangular α-admissible mappings. *J. Nonlinear Sci. Appl.* **2017**, *10*, 6192–6204. [CrossRef]
2. Qawaqneh, H.; Noorani, M.S.M.; Shatanawi, W.; Abodayeh, K.; Alsamir, H. Fixed point for mappings under contractive condition based on simulation functions and cyclic (α, β)-admissibility. *J. Math. Anal.* **2018**, *9*, 38–59. [CrossRef]
3. Qawaqneh, H.; Noorani, M.S.M.; Shatanawi, W. Fixed point for mappings under contractive condition based on simulation functions and cyclic (α, β)-admissibility. *J. Math. Anal.* **2018**, *11*, 702–716. [CrossRef]
4. Hitzler, P.; Seda, A.K. Dislocated topologies. *J. Electr. Eng.* **2000**, *12*, 4223–4229.
5. Amini-Harandi, A. Metric-like spaces, partial metric spaces and fixed points. *Fixed Point Theory Appl.* **2012**, *2012*, 204. [CrossRef]
6. Chen, C.; Dong, J.; Zhu, C. Some fixed point theorems in *b*-metric-like spaces. *Fixed Point Theory Appl.* **2015**, *2015*. [CrossRef]
7. Roshan, J.R.; Parvaneh, V.; Sedghi, S.; Shbkolaei, N.; Shatanawi, W. Common fixed points of almost generalized contractive mappings in ordered *b*-metric spaces. *Fixed Point Theory Appl.* **2013**, *2013*, 159. [CrossRef]
8. Zoto, K. Some generalizations for (α, ψ, ϕ)-contractions in *b*-metric-like spaces and an application. *Fixed Point Theory Appl.* **2017**, *2017*, 26. [CrossRef]
9. Aydi, H.; Felhi, A. Best proximity points for cyclic Kannan-Chatterjea- Ciric non-self contractions on metric-like spaces. *J. Nonlinear Sci. Appl.* **2016**, *9*, 2458–2466. [CrossRef]
10. Aydi, H.; Felhi, A.; Sahmim, S. On common fixed points for (α, ψ)-contractions and generalized cyclic contractions in b-metric-like spaces and consequences. *J. Nonlinear Sci. Appl.* **2016**, *9*, 2492–2510. [CrossRef]
11. Aydi, H.; Felhi, A. On best proximity points for various α-proximal contractions on metric-like spaces. *J. Nonlinear Sci. Appl.* **2016**, *9*, 5202–5218. [CrossRef]
12. Aydi, H.; Felhi, A.; Afshari, H. New Geraghty non-self contractions on metric-like spaces. *J. Nonlinear Sci. Appl.* **2017**, *10*, 780–788. [CrossRef]
13. Aydi, H.; Felhi, A.; Sahmim, S. Common fixed points via implicit contractions on b-metric-like spaces. *J. Nonlinear Sci. Appl.* **2017**, *10*, 1524–1537. [CrossRef]
14. Alsamir, H; Noorani, M.S.M.; Shatanawi, W. On fixed points of (η, θ)-quasicontraction mappings in generalized metric spaces. *J. Nonlinear Sci. Appl.* **2016**, *9*, 4651–4658. [CrossRef]
15. Shatanawi, W.; Nashine, H.K. A generalization of Banachs contraction principle for nonlinear contraction in a partial metric space. *J. Nonlinear Sci. Appl.* **2012**, *5*, 37–43. [CrossRef]
16. Aydi, H.; Shatanawi, W.; Vetro, C. On generalized weak G-contraction mapping in G-metric spaces. *Comput. Math. Appl.* **2011**, *62*, 4223–4229. [CrossRef]
17. Cho, S. Fixed point theorems for generalized weakly contractive mappings in metric spaces with application. *Fixed Point Theory Appl.* **2018**, *2018*, 3. [CrossRef]
18. Zoto, K.; Radenovic, S.; Ansari, A.H. On some fixed point results for (s, p, α)-contractive mappings in *b*-metric-like spaces and applications to integral equations. *Open Math.* **2018**, *16*, 235–249. [CrossRef]
19. Mohammadi, B.; Golkarmanesh, F.; Parvaneh, V. Common fixed point results via implicit contractions for multi-valued mappings on *b*-metric like spaces. *Cogent Math. Stat.* **2018**, *5*, 1493761. [CrossRef]
20. Karapinar, E.; Czerwik, S.; Aydi, H. Meir-Keeler Contraction Mappings in Generalized-Metric Spaces. *J. Funct. Spaces* **2018**, *2018*, 3264620. [CrossRef]
21. Liang, M.; Zhu, C.; Wu, Z.; Chen, C. Some New Coupled Coincidence Point and Coupled Fixed Point Results in Partially Ordered Metric-Like Spaces and an Application. *J. Funct. Spaces* **2018**, *2018*, 1378379. [CrossRef]

22. Hussain, N.; Roshan, J.R.; Parvaneh, V.; Kadelburg, Z. Fixed Points of Contractive Mappings in-Metric-Like Spaces. *Sci. World J.* **2014**, *2014*, 471827. [CrossRef] [PubMed]
23. Geraghty, M. On contractive mappings. *Proc. Am. Math. Soc.* **1973**, *40*, 604–608. [CrossRef]
24. Karapinar, E. A Discussion on α-ψ-Geraghty Contraction Non-self Mappings. *Filomat* **2014**, *28*, 761–766. [CrossRef]
25. Karapinar, E. α-ψ-Geraghty Contraction Non-self Mappings and Some Related Fixed Point Results. *Filomat* **2014**, *28*, 37–48. [CrossRef]
26. Karapinar, E.; Pitea, A. On alpha-psi-Geraghty contraction type mappings on quasi-Branciari metric spaces. *J. Nonlinear Convex Anal.* **2016**, *17*, 1291–1301.
27. Amini-Harandi, A.; Emami, H. A fixed point theorem for contraction non-self maps in partially ordered metric spaces and application to ordinary differential equations. *Nonlinear Anal.* **2010**, *72*, 2238–2242. [CrossRef]
28. karapinar, E.; Alsulami, H.; Noorwali, M. Some extensions for Geragthy non-self contractive mappings. *Fixed Point Theory Appl.* **2015**, *2015*, 303. [CrossRef]
29. Ćirič, L. A generalization of Banach contraction principle. *Proc. Am. Math. Soc.* **1974**, *45*, 267–273. [CrossRef]
30. Isik, H.; Turkoglu, D. Fixed point theorems for weakly contractive mappings in partially ordered metric-like spaces. *Fixed Point Theory Appl.* **2013**, *2013*, 51. [CrossRef]
31. Samet, B.; Vetro, C.; Vetro, P. Fixed point theorems for a α-ψ-contractive non-self mappings. *Nonlinear Anal.* **2012**, *75*, 2154–2165. [CrossRef]
32. Abdeljawad, T. Meir-Keeler a-contractive fixed and common fixed point theorems. *Fixed Point Theory Appl.* **2013**, *2013*, 19. [CrossRef]
33. Karapinar, E.; Kumam, P.; Salimi, P. On α-ψ-Meir-Keeler contractive mappings. *Fixed Point Theory Appl.* **2013**, *2013*, 94. [CrossRef]
34. Alsamir, H.; Noorani, M.S.M.; Shatanawi, W.; Abodayeh, K. Common Fixed Point Results for Generalized (ψ, β)-Geraghty Contraction Non-self Mapping in Partially Ordered Metric-like Spaces with Application. *Filomat* **2017**, *31*, 5497–5509. [CrossRef]

© 2018 by the authors. Licensee MDPI, Basel, Switzerland. This article is an open access article distributed under the terms and conditions of the Creative Commons Attribution (CC BY) license (http://creativecommons.org/licenses/by/4.0/).

Review

Differential Equations for Classical and Non-Classical Polynomial Sets: A Survey †

Paolo Emilio Ricci

Section of Mathematics, International Telematic University UniNettuno, Corso Vittorio Emanuele II, 39, 00186 Roma, Italy; paoloemilioricci@gmail.com
† Dedicated to Prof. Dr. Hari M. Srivastava.

Received: 8 March 2019; Accepted: 20 April 2019; Published: 25 April 2019

Abstract: By using the monomiality principle and general results on Sheffer polynomial sets, the differential equation satisfied by several old and new polynomial sets is shown.

Keywords: differential equations; Sheffer polynomial sets; generating functions; monomiality principle

MSC: 33C99; 12E10; 11B83

1. Introduction

In this survey article, a uniform method is presented for constructing the differential equations satisfied by several sets of classical and non classical polynomials. This has been done by starting from the basic elements of the relevant generating functions, using the monomiality principle by G. Dattoli [1] and a general result by Y. Ben Cheikh [2]. Of course, the polynomials considered in this paper are only examples for showing that the method works, but obviously this technique can be theoretically extended to every polynomial set.

This method has been recently applied in several articles (see [3–9]), which include works in collaboration with several authors. The most outstanding of them is Prof. Dr. Hari M. Srivastava, to whom this article is dedicated.

The derived differential equations are generally of infinite order, but they reduce to finite order when applied to polynomials.

It is worth noting that the differential equations for Sheffer polynomial sets have been studied even with different methods (see [10–13]), but here we use only elements directly connected with the theory of polynomials.

We start recalling, in Section 2, the definitions relevant to Sheffer polynomials, the G. Dattoli monomiality principle, and a general result by Y. Ben Cheikh.

The classical polynomial sets, considered in Section 3, are the Bernoulli, Euler, Genocchi and Mittag–Leffler polynomials. In Section 4, we show some new polynomial sets derived from non-classical generating functions.

2. Sheffer Polynomials

The Sheffer polynomials $\{s_n(x)\}$ are introduced [14] by means of the exponential generating function [15] of the type:

$$A(t)\exp(xH(t)) = \sum_{n=0}^{\infty} s_n(x)\frac{t^n}{n!}, \qquad (1)$$

where

$$A(t) = \sum_{n=0}^{\infty} a_n \frac{t^n}{n!}, \qquad (a_0 \neq 0),$$

$$H(t) = \sum_{n=0}^{\infty} h_n \frac{t^n}{n!}, \qquad (h_0 = 0). \qquad (2)$$

According to a different characterization (see [16], p. 18), the same polynomial sequence can be defined by means of the pair $(g(t), f(t))$, where $g(t)$ is an invertible series and $f(t)$ is a delta series:

$$g(t) = \sum_{n=0}^{\infty} g_n \frac{t^n}{n!}, \qquad (g_0 \neq 0),$$

$$f(t) = \sum_{n=0}^{\infty} f_n \frac{t^n}{n!}, \qquad (f_0 = 0, f_1 \neq 0). \qquad (3)$$

Denoting by $f^{-1}(t)$ the compositional inverse of $f(t)$ (i.e., such that $f\left(f^{-1}(t)\right) = f^{-1}\left(f(t)\right) = t$), the exponential generating function of the sequence $\{s_n(x)\}$ is given by

$$\frac{1}{g[f^{-1}(t)]} \exp\left(xf^{-1}(t)\right) = \sum_{n=0}^{\infty} s_n(x)\frac{t^n}{n!}, \qquad (4)$$

so that

$$A(t) = \frac{1}{g[f^{-1}(t)]}, \qquad H(t) = f^{-1}(t). \qquad (5)$$

When $g(t) \equiv 1$, the Sheffer sequence corresponding to the pair $(1, f(t))$ is called the associated Sheffer sequence $\{\sigma_n(x)\}$ for $f(t)$, and its exponential generating function is given by

$$\exp\left(xf^{-1}(t)\right) = \sum_{n=0}^{\infty} \sigma_n(x)\frac{t^n}{n!}. \qquad (6)$$

A list of known Sheffer polynomial sequences and their associated ones can be found in [17].

Shift Operators and Differential Equation

We recall that a polynomial set $\{p_n(x)\}$ is called quasi-monomial if and only if there exist two operators $\hat{P}$ and $\hat{M}$ such that

$$\hat{P}\left(p_n(x)\right) = np_{n-1}(x), \qquad \hat{M}\left(p_n(x)\right) = p_{n+1}(x), \qquad (n = 1, 2, \ldots). \qquad (7)$$

$\hat{P}$ is called the *derivative* operator and $\hat{M}$ the *multiplication* operator, as they act in the same way as classical operators on monomials.

This definition traces back to a paper by J.F. Steffensen [18] recently improved by G. Dattoli and widely used in several applications [19,20].

Y. Ben Cheikh proved that every polynomial set is quasi-monomial under the action of suitable derivative and multiplication operators. In particular, in the same article, the following result is proved, as a particular case of Corollary 3.2 in [2]:

Theorem 1. *Let $(p_n(x))$ denote a Sheffer polynomial set, defined by the generating function*

$$A(t)\exp(xH(t)) = \sum_{n=0}^{\infty} p_n(x) \frac{t^n}{n!}, \qquad (8)$$

where

$$A(t) = \sum_{n=0}^{\infty} \tilde{a}_n t^n, \qquad (\tilde{a}_0 \neq 0), \qquad (9)$$

and

$$H(t) = \sum_{n=0}^{\infty} \tilde{h}_n t^{n+1}, \qquad (\tilde{h}_0 \neq 0). \qquad (10)$$

Denoting, as before, by $f(t)$ the compositional inverse of $H(t)$, the Sheffer polynomial set $\{p_n(x)\}$ is quasi-monomial under the action of the operators

$$\hat{P} = f(D_x), \qquad \hat{M} = \frac{A'[f(D_x)]}{A[f(D_x)]} + xH'[f(D_x)], \qquad (11)$$

where prime denotes the ordinary derivatives with respect to t.

Furthermore, according to the monomiality principle, the quasi-monomial polynomials $\{p_n(x)\}$ satisfy the differential equation

$$\hat{M}\hat{P}\, p_n(x) = n\, p_n(x). \qquad (12)$$

3. Differential Equations of Classical Polynomials

3.1. Bernoulli Polynomials

The Bernoulli polynomials are defined by the generating function

$$G(t,x) = \frac{t}{e^t - 1} e^{xt}, \qquad (13)$$

so that

$$A(t) = \frac{t}{e^t - 1} = \sum_{k=0}^{\infty} b_k \frac{t^k}{k!},$$

$$G(t,x) = \sum_{k=0}^{\infty} B_k(x) \frac{t^k}{k!} = \sum_{k=0}^{\infty} \left[\sum_{h=0}^{k} \binom{k}{h} b_{k-h} x^h \right] \frac{t^k}{k!}, \qquad (14)$$

$$B_k(x) = \sum_{h=0}^{k} \binom{k}{h} b_{k-h} x^h,$$

where b_k are the Bernoulli numbers.

Differential Equation of the $B_k(x)$

Note that, recalling that $B_n(1) = (-1)^n b_n$, the following expansion holds:

$$t \frac{A'(t)}{A(t)} = \frac{e^t - te^t - 1}{e^t - 1} = 1 - \frac{te^t}{e^t - 1} = 1 - \sum_{n=0}^{\infty} B_n(1) \frac{t^n}{n!} = $$
$$= \sum_{n=1}^{\infty} (-1)^{n+1} b_n \frac{t^n}{n!}. \qquad (15)$$

The shift operators for the Bernoulli polynomials are given by

$$\hat{P} = D_x,$$

$$\hat{M} = \frac{e^{D_x} - D_x e^{D_x} - 1}{D_x (e^{D_x} - 1)}. \qquad (16)$$

Therefore, by using the factorization method, we find

Theorem 2. *The Bernoulli polynomials $\{B_n(x)\}$ satisfy the differential equation*

$$\left(\frac{e^{D_x} - D_x e^{D_x} - 1}{e^{D_x} - 1} + x D_x \right) B_n(x) = n B_n(x), \qquad (17)$$

that is

$$\left(\sum_{k=1}^{\infty} \frac{(-1)^{k+1} b_k}{k!} D_x^k + x D_x \right) B_n(x) = n B_n(x), \qquad (18)$$

or, in equivalent form:

$$\left(\sum_{k=1}^{n} \frac{(-1)^{k+1} b_k}{k!} D_x^k + x D_x \right) B_n(x) = n B_n(x). \qquad (19)$$

Proof. It is sufficient to expand in series the operator (17). Equation (19) follows because, for any fixed n, the series expansion in Equation (18) reduces to a finite sum when applied to a polynomial of degree n. □

3.2. Euler Polynomials

The Euler polynomials are defined by the generating function

$$G(t,x) = \frac{2}{e^t+1} e^{xt}, \qquad (20)$$

so that

$$A(t) = \frac{2}{e^t+1} = \sum_{k=0}^{\infty} e_k \frac{t^k}{k!},$$

$$G(t,x) = \sum_{k=0}^{\infty} E_k(x)\frac{t^k}{k!} = \sum_{k=0}^{\infty}\left[\sum_{h=0}^{k}\binom{k}{h} e_{k-h} x^h\right]\frac{t^k}{k!}, \qquad (21)$$

$$E_k(x) = \sum_{h=0}^{k}\binom{k}{h} e_{k-h} x^h,$$

where e_k are the Euler numbers.

Differential Equation of the $E_k(x)$

Note that the following expansion holds:

$$\frac{A'(t)}{A(t)} = -\frac{e^t}{e^t+1} = -1+\frac{1}{e^t+1} = -1+\frac{1}{2}\sum_{n=0}^{\infty} e_n \frac{t^n}{n!} =$$
$$= -\frac{1}{2} + \frac{1}{2}\sum_{n=1}^{\infty} e_n \frac{t^n}{n!} = \sum_{n=0}^{\infty} c_n \frac{t^n}{n!}, \qquad (22)$$

where $c_0 = -1/2$, and $c_n = e_n/2$.

The shift operators for the Euler polynomials are given by

$$\hat{P} = D_x,$$
$$\hat{M} = -\frac{e^{D_x}}{e^{D_x}+1}. \qquad (23)$$

Therefore, by using the factorization method, we find

Theorem 3. *The Euler polynomials $\{E_n(x)\}$ satisfy the differential equation*

$$\left(-\frac{e^{D_x} D_x}{e^{D_x}+1} + x D_x\right) E_n(x) = n E_n(x). \qquad (24)$$

that is

$$\left(\sum_{k=0}^{\infty} \frac{c_k}{k!} D_x^{k+1} + x D_x\right) E_n(x) = n E_n(x), \qquad (25)$$

or, in equivalent form:

$$\left(\sum_{k=0}^{n-1} \frac{c_k}{k!} D_x^{k+1} + x D_x\right) E_n(x) = n E_n(x). \qquad (26)$$

Proof. It is sufficient to expand in series the operator (24). Equation (26) follows because, for any fixed n, the series expansion in Equation (25) reduces to a finite sum when applied to a polynomial of degree n. □

3.3. Genocchi Polynomials

The Genocchi polynomials are defined by the generating function

$$G(t,x) = \frac{2t}{e^t+1} e^{xt}, \tag{27}$$

so that

$$A(t) = \frac{2t}{e^t+1} = \sum_{k=0}^{\infty} g_k \frac{t^k}{k!},$$

$$G(t,x) = \sum_{k=0}^{\infty} G_k(x) \frac{t^k}{k!} = \sum_{k=0}^{\infty} \left[\sum_{h=0}^{k} \binom{k}{h} g_{k-h} x^h \right] \frac{t^k}{k!}, \tag{28}$$

$$G_k(x) = \sum_{h=0}^{k} \binom{k}{h} g_{k-h} x^h,$$

where g_k are the Genocchi numbers.

Differential Equation of the $G_k(x)$

Note that the following expansion holds:

$$t\frac{A'(t)}{A(t)} = \frac{e^t - te^t + 1}{e^t + 1} = 1 - \frac{te^t}{e^t+1} = 1 - \frac{1}{2}t + \frac{1}{2}\sum_{n=2}^{\infty} e_n \frac{t^{n+1}}{n!} =$$

$$= \sum_{n=0}^{\infty} d_n \frac{t^n}{n!}, \tag{29}$$

where $d_0 = 1$, $d_1 = -1/2$, and $d_k = e_k/2$, ($k \geq 2$).

The shift operators for the Genocchi polynomials are given by

$$\hat{P} = D_x,$$

$$\hat{M} = \frac{e^{D_x} - D_x e^{D_x} + 1}{D_x(e^{D_x}+1)}, \tag{30}$$

so that the Genocchi polynomials satisfy the differential equation

$$\left(\frac{e^{D_x} - D_x e^{D_x} + 1}{e^{D_x}+1} + x D_x \right) G_n(x) = n\, G_n(x). \tag{31}$$

Therefore, by using the factorization method, we find

Theorem 4. *The Genocchi polynomials $\{G_n(x)\}$ satisfy the differential equation*

$$\left(\frac{e^{D_x} - D_x e^{D_x} + 1}{e^{D_x}+1} + x D_x \right) G_n(x) = n\, G_n(x), \tag{32}$$

that is

$$\left(\sum_{k=0}^{\infty} \frac{d_k}{k!} D_x^k + x D_x\right) G_n(x) = n G_n(x), \tag{33}$$

or, in equivalent form:

$$\left(\sum_{k=0}^{n} \frac{d_k}{k!} D_x^k + x D_x\right) G_n(x) = n G_n(x). \tag{34}$$

Proof. It is sufficient to expand in series the operator (32). Equation (34) follows because, for any fixed n, the series expansion in Equation (33) reduces to a finite sum when applied to a polynomial of degree n. □

3.4. The Mittag–Leffler Polynomials

We recall that the Mittag–Leffler polynomials [21] are a special case of associated Sheffer polynomials, defined by the generating function

$$A(t) = 1, \qquad H(t) = \log \frac{1+t}{1-t},$$

$$G(t,x) = \left(\frac{1+t}{1-t}\right)^x = \exp\left(x \log \frac{1+t}{1-t}\right) = \sum_{n=0}^{\infty} \mathcal{M}_n(x) \frac{t^n}{n!}. \tag{35}$$

Therefore, we have

$$\frac{A'(t)}{A(t)} = 0, \qquad H'(t) = \frac{2}{1-t^2}, \qquad H^{-1}(t) = f(t) = \frac{e^t - 1}{e^t + 1}, \tag{36}$$

so that, for the Mittag-Leffer polynomials, we find the shift operators:

$$\hat{P} = \frac{e^{D_x} - 1}{e^{D_x} + 1} = \tanh\left(\frac{D_x}{2}\right),$$

$$\hat{M} = x \frac{(e^{D_x} + 1)^2}{2 e^{D_x}} = x \left[1 + \cosh(D_x)\right]. \tag{37}$$

3.5. Differential Equation of the $\mathcal{M}_n(x)$

In the present case, according to the identity:

$$[1 + \cosh x] \tanh(x/2) = \sinh x,$$

we can write

$$\hat{M}\hat{P} = x \frac{e^{2D_x} - 1}{2 e^{D_x}} = x \sinh(D_x), \tag{38}$$

so that we have the theorem

Theorem 5. The Mittag–Leffler polynomials $\{\mathcal{M}_n(x)\}$ satisfy the differential equation

$$x \sinh(D_x) \mathcal{M}_n(x) = n \mathcal{M}_n(x), \tag{39}$$

that is

$$x \sum_{k=0}^{\infty} \frac{D_x^{2k+1}}{(2k+1)!} \mathcal{M}_n(x) = n \mathcal{M}_n(x), \tag{40}$$

or

$$x \sum_{k=0}^{\left[\frac{n-1}{2}\right]} \frac{D_x^{2k+1}}{(2k+1)!} \mathcal{M}_n(x) = n \mathcal{M}_n(x), \tag{41}$$

where $\left[\frac{n-1}{2}\right]$ denotes the integral part of $(n-1)/2$.

Proof. It is sufficient to expand in series the operator (39). Equation (41) follows because, for any fixed n, the series expansion in Equation (40) reduces to a finite sum when applied to a polynomial of degree n. □

4. Differential Equations of Non-Classical Polynomials

4.1. Euler-Type Polynomials

Here, we introduce a Sheffer polynomial set connected with the classical Euler polynomials. Assuming:

$$A(t) = \frac{1}{\cosh t}, \quad H(t) = \sinh t, \tag{42}$$

we consider the Euler-type polynomials $\tilde{E}_n(x)$, defined by the generating function

$$G(t,x) = \frac{1}{\cosh t} \exp\left[x \sinh t\right] = \sum_{k=0}^{\infty} \tilde{E}_k(x) \frac{t^k}{k!}. \tag{43}$$

Note that the Euler numbers are recovered, since we have:

$$G(t,0) = \frac{2}{e^t + e^{-t}} = \sum_{k=0}^{\infty} \tilde{E}_k(0) \frac{t^k}{k!}, \tag{44}$$

so that $\tilde{E}_n(0) = E_n$.

In what follows, we use the expansions

$$\sinh t = \sum_{k=0}^{\infty} \frac{t^{2k+1}}{(2k+1)!} = \sum_{k=0}^{\infty} \left(\frac{1+(-1)^{k+1}}{2}\right) \frac{t^k}{k!}, \tag{45}$$

$$\cosh t = \sum_{k=0}^{\infty} \frac{t^{2k}}{(2k)!} = \sum_{k=0}^{\infty} \left(\frac{1+(-1)^k}{2}\right) \frac{t^k}{k!}. \tag{46}$$

Note that, in our case, we are dealing with a Sheffer polynomial set, so that, since we have $\psi(t) = e^t$, the operator σ defined by Equation (6) simply reduces to the derivative operator D_x. Furthermore, we have:

$$A(t) = \frac{1}{\cosh t}, \qquad \frac{A'(t)}{A(t)} = -\tanh t,$$

$$H(t) = \sinh t = \sum_{k=0}^{\infty} \frac{t^{2k+1}}{(2k+1)!}, \qquad \left(\tilde{h}_k = \left(\frac{1+(-1)^{k+1}}{2}\right)\frac{1}{k!}\right),$$

$$H'(t) = \cosh t, \qquad f(t) = H^{-1}(t) = \log(t + \sqrt{t^2+1}),$$

so that we have the theorem

Theorem 6. *The Euler-type polynomial set $\{\tilde{E}_n(x)\}$ is quasi-monomial under the action of the operators*

$$\hat{P} = \log(D_x + \sqrt{D_x^2+1}), \qquad \hat{M} = -\tanh(\operatorname{arcsinh} D_x) + x\operatorname{arcsinh} D_x \qquad (47)$$

(by $\operatorname{arcsinh} t = \log(t + \sqrt{t^2+1})$, we denote the inverse of the function $\sinh t$), i.e.,

$$\hat{P} = \sum_{k=0}^{\infty} (-1)^k \frac{(2k)!}{4^k (k!)^2 (2k+1)} D_x^{2k+1},$$

$$\hat{M} = -\frac{D_x}{\sqrt{1+D_x^2}} + x\sqrt{1+D_x^2} = (xD_x^2 - D_x + x)(1+D_x^2)^{-1/2}, \qquad (48)$$

$$\hat{M} = (xD_x^2 - D_x + x)\sum_{k=0}^{\infty} \binom{-1/2}{k} D_x^{2k}.$$

There is no problem about the convergence of the above series, since they reduce to finite sums when applied to polynomials.

4.2. Differential Equation of the $\tilde{E}_n(x)$

In the present case, we have

Theorem 7. *The Euler-type polynomials $\{\tilde{E}_n(x)\}$ satisfy the differential equation*

$$\left\{\left[(xD_x^2 - D_x + x)\sum_{k=0}^{\infty} \binom{-1/2}{k} D_x^{2k}\right]\right.$$

$$\left.\sum_{k=0}^{\infty} (-1)^k \frac{(2k)!}{4^k (k!)^2 (2k+1)} D_x^{2k+1}\right\} \tilde{E}_n(x) = n\tilde{E}_n(x), \qquad (49)$$

i.e.,

$$(xD_x^2 - D_x + x) \sum_{k=0}^{\left[\frac{n-1}{2}\right]} \sum_{h=0}^{k} (-1)^h \binom{-1/2}{k-h} \frac{(2h)!}{4^h (h!)^2 (2h+1)} D_x^{2k+1} \tilde{E}_n(x) \qquad (50)$$

$$= n \tilde{E}_n(x).$$

Note that, for any fixed n, the Cauchy product of series expansions in Equation (49) reduces to a finite sum, with upper limit $\left[\frac{n-1}{2}\right]$, when it is applied to a polynomial of degree n, because the successive addends vanish.

Remark 1. *The first few Euler-type polynomials are as follows:*

$$\tilde{E}_0(x) = 1,$$
$$\tilde{E}_1(x) = x,$$
$$\tilde{E}_2(x) = x^2 - 1,$$
$$\tilde{E}_3(x) = x^3 - 2x,$$
$$\tilde{E}_4(x) = x^4 - 2x^2 + 5,$$
$$\tilde{E}_5(x) = x^5 + 16x,$$
$$\tilde{E}_6(x) = x^6 + 5x^4 + 31x^2 - 61,$$
$$\tilde{E}_7(x) = x^7 + 14x^5 + 56x^3 - 272x,$$
$$\tilde{E}_8(x) = x^8 + 28x^6 + 126x^4 - 692x^2 + 1385,$$
$$\tilde{E}_9(x) = x^9 + 48x^7 + 336x^5 - 1280x^3 + 7936x,$$
$$\tilde{E}_{10}(x) = x^{10} + 75x^8 + 882x^6 - 1490x^4 + 25,261x^2 - 50,521.$$

5. Adjointness for Sheffer Polynomial Sequences

According to the above considerations, Sheffer polynomials are characterized both by the ordered couples $(A(t), H(t))$, or by $(g(t), f(t))$.

Definition 1. *Adjoint Sheffer polynomials are defined by interchanging the ordered couple $(A(t), H(t))$ with $(g(t), f(t))$, when writing the generating function.*

Here and in the following the *tilde* "~" above the symbol of a polynomial set stands for the adjective *"adjoint"* (see e.g., [4]).

5.1. Adjoint Hahn Polynomials

Assuming:

$$A(t) = \sec t, \qquad H(t) = \tan t, \qquad (51)$$

we consider the adjoint Hahn $\tilde{R}_n(x)$, defined by the generating function

$$G(t, x) = \sec t \, \exp(x \tan t) = \sum_{n=0}^{\infty} \tilde{R}_n(x) \frac{t^n}{n!}. \qquad (52)$$

It is a Sheffer set.

We have:

$$\frac{\partial G}{\partial x} = \frac{1}{\cos^3 t} \exp(x \tan t) = \frac{1}{\cos t} G(t,x).$$

Note that, in this case, we have:

$$A(t) = \sec t, \qquad H(t) = \tan t,$$

$$H'(t) = \sec^2 t, \qquad f(t) = H^{-1}(t) = \arctan t,$$

$$\frac{A'(t)}{A(t)} = \tan t,$$

so that we have the theorem

Theorem 8. *The adjoint Hahn polynomial set* $\{\tilde{R}_n(x)\}$ *is quasi-monomial under the action of the operators*

$$\hat{P} = \arctan D_x, \tag{53}$$

$$\hat{M} = \tan(\arctan D_x) + x \sec^2(\arctan D_x),$$

i.e.,

$$\hat{P} = \arctan D_x = \sum_{k=0}^{\infty} \frac{(-1)^k}{2k+1} D_x^{2k+1}, \tag{54}$$

$$\hat{M} = D_x + x(1+D_x^2) = x D_x^2 + D_x + x.$$

5.2. Differential Equation of the $\tilde{R}_n(x)$

In the present case, we have

Theorem 9. *The Sheffer-type adjoint Hahn polynomials* $\{\tilde{R}_n(x)\}$ *satisfy the differential equation*

$$\left(x D_x^2 + D_x + x\right) \sum_{k=0}^{[\frac{n-1}{2}]} \frac{(-1)^k}{2k+1} D_x^{2k+1} \tilde{R}_n(x) = n \tilde{R}_n(x). \tag{55}$$

Note that, for any fixed n, in Equation (55), a finite sum appears, with upper limit $\left[\frac{n-1}{2}\right]$, instead of a complete series expansion since, when this series is applied to a polynomial of degree n, the subsequent addends vanish.

Remark 2. *The first few values of the adjoint Hahn polynomials are as follows:*

$$\tilde{R}_0(x) = 1,$$
$$\tilde{R}_1(x) = x,$$
$$\tilde{R}_2(x) = x^2 + 1,$$
$$\tilde{R}_3(x) = x^3 + 5x,$$
$$\tilde{R}_4(x) = x^4 + 14x^2 + 5,$$
$$\tilde{R}_5(x) = x^5 + 30x^3 + 61x,$$
$$\tilde{R}_6(x) = x^6 + 55x^4 + 331x^2 + 61,$$
$$\tilde{R}_7(x) = x^7 + 91x^5 + 1211x^3 + 1385x,$$
$$\tilde{R}_8(x) = x^8 + 140x^6 + 3486x^4 + 12,284x^2 + 1385,$$
$$\tilde{R}_9(x) = x^9 + 204x^7 + 8526x^5 + 68,060x^3 + 50,521x,$$
$$\tilde{R}_{10}(x) = x^{10} + 285x^8 + 18,522x^6 + 281,210x^4 + 663,061x^2 + 50,521.$$

Remark 3. *Table of adjoint Hahn numbers*

$$\tilde{R}_0(0) = 1 \qquad \tilde{R}_1(0) = 0 \qquad \tilde{R}_2(0) = 1,$$
$$\tilde{R}_3(0) = 0 \qquad \tilde{R}_4(0) = 5 \qquad \tilde{R}_{2k+1}(0) = 0, \ \forall k \geq 2,$$
$$\tilde{R}_6(0) = 61 \qquad \tilde{R}_8(0) = 1385 \qquad \tilde{R}_{10}(0) = 50,521.$$

Note that the sequence $\{1, 1, 5, 61, 1385, 50,521, \ldots\}$ appears in the Encyclopedia of Integer Sequences [22] under #A000364—Euler (or secant numbers): $a(n)$ = number of downup permutations of $[2n]$.

Example 1. *$a(2) = 5$ counts 4231, 4132, 3241, 3142, 2143. - David Callan, Nov 21, 2011.*

5.3. Adjoint Bernoulli Polynomials of the Second Kind

Assuming

$$A(t) = \frac{t}{e^t - 1}, \qquad H(t) = e^t - 1, \tag{56}$$

we consider the adjoint Bernoulli polynomials of the second kind $\{\tilde{b}_k(x)\}$, defined by the generating function

$$G(t,x) = \frac{t}{e^t - 1} \exp\left[x(e^t - 1)\right] = \sum_{k=0}^{\infty} \tilde{b}_k(x) \frac{t^k}{k!}. \tag{57}$$

Note that, in our case, we are dealing with a Sheffer polynomial set, so that, since we have $\psi(t) = e^t$, the operator σ defined by Equation (6) simply reduces to the derivative operator D_x. Furthermore, we have

$$A(t) = \frac{t}{e^t - 1}, \qquad H(t) = e^t - 1 = \sum_{k=1}^{\infty} \frac{t^k}{k!}, \qquad (\tilde{h}_k = 1/(k+1)!) \,,$$

$$H'(t) = e^t, \qquad f(t) = H^{-1}(t) = \log(t+1),$$

$$\frac{A'(t)}{A(t)} = \frac{e^t - te^t - 1}{t(e^t - 1)} = \frac{1}{t} - \frac{1}{e^t - 1} - 1.$$

so that we have the theorem

Theorem 10. *The adjoint Bernoulli polynomials of the second kind* $\{\tilde{b}_n(x)\}$ *are quasi-monomials under the action of the operators*

$$\hat{P} = \log(D_x + 1),$$

$$\hat{M} = \frac{1}{\log(D_x + 1)} - \frac{1}{D_x} - 1 + x(D_x + 1),$$
(58)

that is

$$\hat{P} = \log(D_x + 1) = \sum_{k=1}^{\infty} \frac{(-1)^{k+1}}{k} D_x^k,$$

$$\hat{M} = \frac{1}{\log(D_x + 1)} + \left(x - \frac{1}{D_x}\right)(D_x + 1).$$
(59)

5.4. Differential Equation of the $\tilde{b}_n(x)$

In the present case, we have

$$\hat{M}\hat{P} = 1 + \left(x - \frac{1}{D_x}\right)(D_x + 1)\log(D_x + 1),$$
(60)

so that we have the theorem

Theorem 11. *The adjoint Bernoulli polynomials of the second kind* $\{\tilde{b}_n(x)\}$ *satisfy the differential equation*

$$\left[1 + \left(x - \frac{1}{D_x}\right)(D_x + 1)\log(D_x + 1)\right]\tilde{b}_n(x) = n\tilde{b}_n(x),$$
(61)

that is

$$\left[1 + (xD_x - 1)(D_x + 1)\sum_{k=0}^{n}\frac{(-1)^k}{k+1}D_x^k\right]\tilde{b}_n(x) = n\tilde{b}_n(x),$$
(62)

because, for any fixed n, the series expansion in Equation (61) reduces to a finite sum when it is applied to a polynomial of degree n.

Note that, in this case, due to the presence of the operator D_x^{-1}, it is necessary to consider derivatives up to the order $n + 1$.

Remark 4. *The first few values of the adjoint Bernoulli polynomials of the second kind are as follows:*

$\tilde{b}_0(x) = 1,$
$\tilde{b}_1(x) = x - \frac{1}{2},$
$\tilde{b}_2(x) = x^2 + \frac{1}{6},$
$\tilde{b}_3(x) = x^3 + \frac{3}{2}x^2,$
$\tilde{b}_4(x) = x^4 + 4x^3 + 2x^2 - \frac{1}{30},$
$\tilde{b}_5(x) = x^5 + \frac{15}{2}x^4 + \frac{35}{3}x^3 + \frac{5}{2}x^2,$
$\tilde{b}_6(x) = x^6 + 12x^5 + \frac{75}{2}x^4 + 30x^3 + 3x^2 + \frac{1}{42},$
$\tilde{b}_8(x) = x^8 + 24x^7 + \frac{560}{3}x^6 + 560x^5 + 602x^4 + 168x^3 + 4x^2 - \frac{1}{30},$
$\tilde{b}_{10}(x) = x^{10} + 40x^9 + \frac{1155}{2}x^8 + 3780x^7 + 11,585x^6 + 15,540x^5 + \frac{15,125}{2}x^4 + 850x^3 +$
$\qquad + 5x^2 + \frac{5}{66},$
$\tilde{b}_{12}(x) = x^{12} + 60x^{11} + 1386x^{10} + 15,840x^9 + \frac{191,961}{2}x^8 + 307,692x^7 + 493,460x^6 +$
$\qquad + 349,800x^5 + 85,503x^4 + 4092x^3 + 6x^2 - \frac{691}{2730}.$

Note that for $x = 0$ the generating function becomes

$$G(t,0) = \frac{t}{e^t - 1} = \sum_{n=0}^{\infty} \tilde{b}_n(0) \frac{t^n}{n!}, \tag{63}$$

so that $\tilde{b}_n(0) = B_n$, namely the nth classical Bernoulli number.

6. Conclusions

In this survey article, it has been shown that the common belonging of some polynomial sets to the Sheffer class allows to construct, in a uniform way, the differential equations they verify. This follows from the fact that it is possible to construct their shift operators, on the basis of general results due to G. Dattoli and Y. Ben Cheikh.

The equations derived in such a way are, in general, of infinite order, but they reduce to finite-order equations when they are applied to polynomials of the considered set. This means that the order of the equation increases with the degree of the polynomial, in a similar way to what happens for the order of the recurrence they verify.

Both classic and other polynomials—the so-called associated Sheffer polynomials—have been examined. In fact, it has been shown that, for the polynomials of the Sheffer class, the differential equation follows from the basic elements of their generating function, in a constructive way, using a simple and general method linked to the monomiality principle.

Funding: This research received no external funding.

Conflicts of Interest: The author declares no conflicts of interest.

References

1. Dattoli, G. Hermite-Bessel and Laguerre-Bessel functions: A by-product of the monomiality principle. In *Advanced Special Functions and Applications, Proceedings of the Melfi School on Advanced Topics in Mathematics and Physics, Melfi, Italy, 9–12 May 1999*; Cocolicchio, D., Dattoli, G., Srivastava, H.M., Eds.; Aracne Editrice: Rome, Italy, 2000; pp. 147–164.
2. Cheikh, Y.B. Some results on quasi-monomiality. *Appl. Math. Comput.* **2003**, *141*, 63–76.

3. Bretti, G.; Natalini, P.; Ricci, P.E. New sets of Euler-type polynomials. *J. Anal. Number Theor.* **2018**, *6*, 51–54. [CrossRef]
4. Natalini, P.; Bretti, G.; Ricci, P.E. Adjoint Hermite and Bernoulli polynomials. *Bull. Allahabad Math. Soc.* **2018**, *33*, 251–264.
5. Natalini, P.; Ricci, P.E. New Bell-Sheffer polynomial sets. *Axioms* **2018**, *7*, 71. [CrossRef]
6. Natalini, P.; Ricci, P.E. Bell-Sheffer exponential polynomials of the second kind. *Georgian Math. J.* **2018** (to appear).
7. Ricci, P.E. Logarithmic-Sheffer polynomials of the second kind. *Tbilisi Math. J.* **2018**, *11*, 95–106. [CrossRef]
8. Ricci, P.E. Mittag–Leffler polynomial differential equation. *Jñānābha* **2018**, *48*, 1–6.
9. Srivastava, H.M.; Ricci, P.E.; Natalini, P. A Family of Complex Appell Polynomial Sets. *Real Acad. Sci. Exact. Fis Nat. Ser. A Math.* **2018**. [CrossRef]
10. Costabile, F.A.; Longo, E. An algebraic approach to Sheffer polynomial sequences. *Integral Transform. Spec. Funct.* **2014**, *25*, 295–311. [CrossRef]
11. Costabile, F.A.; Longo, E. An algebraic exposition of Umbral calculus with applications to general interpolation problem. A Survey. *Publ. Inst. Math. Nouv. Ser.* **2014**, *96*, 67–83. [CrossRef]
12. Youn, H.; Yang, Y. Differential equations and recursive formulas of Sheffer polynomial sequences. *ISRN Discret. Math.* **2011**, *2011*, 476462. [CrossRef]
13. Yang, S.L. Recurrence relation for the Sheffer sequences. *Linear Algebra Appl.* **2012**, *437*, 2986–2996. [CrossRef]
14. Sheffer, I.M. Some properties of polynomials sets of zero type. *Duke Math. J.* **1939**, *5*, 590–622. [CrossRef]
15. Srivastava, H.M.; Manocha, H.L. *A Treatise on Generating Functions*; Halsted Press (Ellis Horwood Limited, Chichester), John Wiley and Sons: New York, NY, USA; Chichester, UK; Brisbane, Australia; Toronto, ON, Canada, 1984.
16. Roman, S.M. *The Umbral Calculus*; Academic Press: New York, NY, USA, 1984.
17. Boas, R.P.; Buck, R.C. *Polynomial Expansions of Analytic Functions*; Springer: Berlin/Gottingen/Heidelberg, Germany; New York, NY, USA, 1958.
18. Steffensen, J.F. The poweroid, an extension of the mathematical notion of power. *Acta Math.* **1941**, *73*, 333–366. [CrossRef]
19. Dattoli, G.; Ricci, P.E.; Srivastava, H.M. (Eds.) Advanced Special Functions and Related Topics in Probability and in Differential Equations. In *Applied Mathematics and Computation, Proceedings of the Melfi School on Advanced Topics in Mathematics and Physics, Melfi, Italy, 24–29 June 2001*; Aracne Editrice: Rome, Italy, 2003; Volume 141, pp. 1–230.
20. Dattoli, G.; Germano, B.; Martinelli, M.R.; Ricci, P.E. Monomiality and partial differential equations. *Math. Comput. Model.* **2009**, *50*, 1332–1337. [CrossRef]
21. Weisstein, E.W. Mittag–Leffler Polynomial, From *MathWorld—A Wolfram Web Resource*. Available online: http://mathworld.wolfram.com/Mittag-LefflerPolynomial.html (accessed on 4 February 2019).
22. Sloane, N.J.A.; Plouffe, S. *The Encyclopedia of Integer Sequences*; Academic Press: San Diego, CA, USA, 1995.

© 2019 by the authors. Licensee MDPI, Basel, Switzerland. This article is an open access article distributed under the terms and conditions of the Creative Commons Attribution (CC BY) license (http://creativecommons.org/licenses/by/4.0/).

MDPI
St. Alban-Anlage 66
4052 Basel
Switzerland
Tel. +41 61 683 77 34
Fax +41 61 302 89 18
www.mdpi.com

Axioms Editorial Office
E-mail: axioms@mdpi.com
www.mdpi.com/journal/axioms

www.ingramcontent.com/pod-product-compliance
Lightning Source LLC
LaVergne TN
LVHW071944080526
838202LV00064B/6671